IET ENERGY ENGINEERING SERIES 228

Digital Technologies for Solar Photovoltaic Systems

Other volumes in this series:

Digital Technologies for Solar Photovoltaic Systems

From general to rural and remote installations

Edited by
Saad Motahhir

The Institution of Engineering and Technology

Published by The Institution of Engineering and Technology, London, United Kingdom

The Institution of Engineering and Technology is registered as a Charity in England & Wales (no. 211014) and Scotland (no. SC038698).

© The Institution of Engineering and Technology 2022

First published 2022

The Institution of Engineering and Technology
Futures Place
Kings Way, Stevenage
Hertfordshire SG1 2UA, United Kingdom

www.theiet.org

British Library Cataloguing in Publication Data
A catalogue record for this product is available from the British Library

ISBN 978-1-83953-677-9 (hardback)
ISBN 978-1-83953-678-6 (PDF)

Typeset in India by MPS Limited
Printed in the UK by CPI Group (UK) Ltd, Croydon

Cover image: Richard Newstead/Moment via Getty Images

I dedicate this book to my cousin Meryem El Janah, and to my friend Aderdor Zouhair, whose memories are still with us. Dear readers, I kindly ask you to a make Dua and prayer for their souls.

Contents

About the editor

 **Saad Motahhir** (Eng., PhD, IEEE Senior Member) has previous expertise in the industry as Embedded System Engineer at Zodiac Aerospace Morocco from 2014 to 2019. He has recently become a professor at ENSA, SMBA university, Fez, Morocco. He received his engineering degree in embedded systems from (ENSA) Fez in 2014. In addition to the above, he received his PhD degree in Electrical Engineering from SMBA University in 2018. He has published and contributed to numerous publications in different journals, and conferences in the last few years, most of which are related to photovoltaic (PV) solar energy, and embedded systems. In addition to this, he has published several patents in the Moroccan Office of Industrial, and Commercial Property. He edited many books and acted as a guest editor of different special issues, and topical collections. He is a reviewer and also a member in the editorial board of different journals. He was associated with more than 30 international conferences as a Program Committee/Advisory Board/Review Board member, and he is a member of the Arab Youth Center council. He is the (ICDTA) Conference chair.

Google scholar: https://scholar.google.com/citations?user=G9AuvGgAAAAJ& hl=fr&oi=ao

Acknowledgments

This book could not be that successful without the effort of authors and reviewers, especially Dr Aboubakr El Hammoumi. Therefore, I would like to express my sincere appreciation to all of you who generously supported this book.

S. Motahhir

Chapter 1

Introduction: The role of digital technologies in solar PV systems: from general to rural installations

Aboubakr El Hammoumi[1] and Saad Motahhir[2]

It is widely admitted that the use of fossil energy resources such as oil and gas is condemned to fade. To meet future energy demands, sustainable and renewable energy sources provide next-generation solutions. Among others, solar energy is of paramount interest since it is the most abundant and reliable source. Every day, the sun emits a massive amount of energy onto the earth's surface (e.g., in one hour, the earth receives 172,000TWh of energy from the sun), more than enough to supply the world's energy demands if properly collected. Generation of electricity from the sun can be achieved using concentrating solar thermal power systems that drive conventional turbines or simply using photovoltaic (PV) systems [1]. The latter generate electricity by converting solar radiation directly into electric energy using PV cells (or PV modules), which are the system's most important components. Other components are required to form a PV system that stores and distributes the energy to users as shown in Figure 1.1 [2]. Figure 1.2 presents the main types of solar PV systems, while Figure 1.3 shows different types of PV systems according to their location.

Access to modern energy services remains an important issue for social and economic development. Currently, more than 15% of the world's population do not have access to electricity according to recent statistics provided by the IEA-World Energy Outlook, particularly in rural areas far from grid-connected areas. This is much more significant for the Sub-Saharan African region as more than 43% of the total population in the region is without access to energy [3]. Therefore, when the prospects of being connected to the grid are low, alternative solutions need to be assessed. Solar PV systems are one of the most suitable technologies to meet the demand for electricity in rural areas. For example, solar PV home systems have gained attention as stand-alone PV systems able to expand rural electricity services in developing countries and their rural communities. These have generally been commercialized, used to provide paid electricity services and supported by many

[1]ENSA, Abdelmalek Essaadi University, Tétouan, Morocco
[2]ENSA, Sidi Mohamed Ben Abdellah University, Fez, Morocco

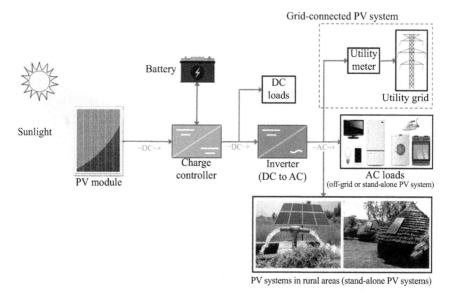

Figure 1.1   Solar PV system components and applications

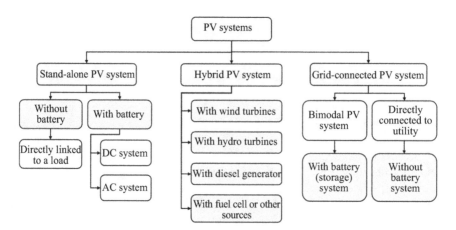

Figure 1.2   PV systems classifications

donor projects as an option for off-grid electrification across the continent [4]. Solar-powered water pumps are also an attractive solution for rural areas that are not connected to the grid. Here, solar panels are being installed for the specific purpose of generating power for the electric pump. As the pump operates during the day, the water is typically stored in a water tank for later use, such as irrigation or drinking water [5,6]. There is no doubt that solar PV systems can contribute to rural development. Proper management and government support can pave the way for

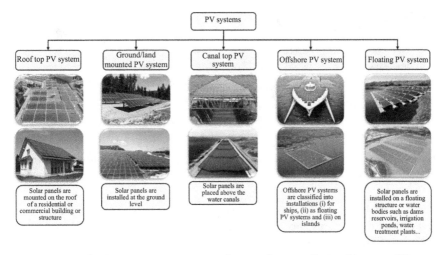

Figure 1.3   PV systems type according to the installation location [2]

economic progress, especially in poorer countries. It is an attractive investment option with a significant advantage: clean energy that does not contribute to environmental degradation and global warming.

Recently, PV systems have been marketed in many countries because of their potential medium- and long-term economic benefits [7], resulting in the rapid growth of the PV industry. By 2024, solar PV systems are expected to generate nearly 57% of all renewable energy sources [8]. However, PV systems still present low efficiencies and high initial costs [2,7]. Therefore, intensive research efforts have been made to improve the performance of PV converters, increase their efficiency and decrease their production cost. These research efforts have been done either at the material level, such as increasing the conversion efficiency of PV panels while minimizing manufacturing costs or at the whole system level, such as maximizing or optimizing the power drawn from the PV panels [9,10]. MPPT (or distributed MPPT) controllers, cooling systems, cleaning systems and solar tracking systems are among the most popular techniques that have been introduced to increase the performance of PV systems and for making the maximum usage possible out of the available solar energy (Figure 1.4) [11–14]. In addition, the performance of PV systems needs to be regularly evaluated over long periods of operation as it is varying according to various electrical and weather parameters as well as the system components functioning [15]. Hence, the monitoring of PV systems especially rural or remote PV installations is important to assess their performance and to ensure the reliable, efficient and stable operation of these systems. Therefore, the use of digital technologies, including emerging ones such as blockchain, cloud computing, artificial intelligence and the IoT, to optimize and control the PV system and decrease the cost is of interest [16–19].

Digital technologies are defined as electronic tools, a set of technologies, systems, devices and resources that generate, transmit, store or process data [20].

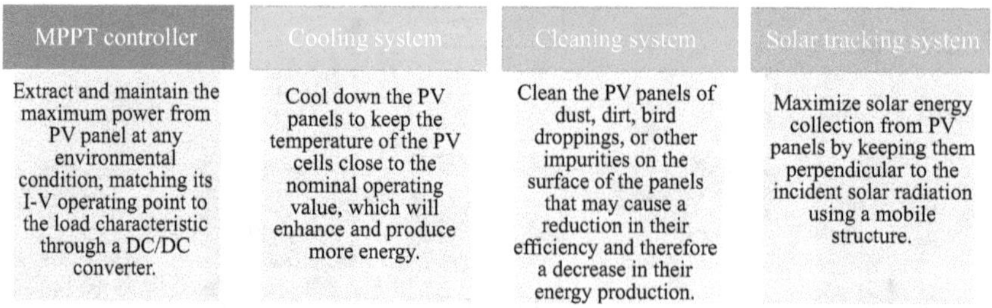

MPPT controller	Cooling system	Cleaning system	Solar tracking system
Extract and maintain the maximum power from PV panel at any environmental condition, matching its I-V operating point to the load characteristic through a DC/DC converter.	Cool down the PV panels to keep the temperature of the PV cells close to the nominal operating value, which will enhance and produce more energy.	Clean the PV panels of dust, dirt, bird droppings, or other impurities on the surface of the panels that may cause a reduction in their efficiency and therefore a decrease in their energy production.	Maximize solar energy collection from PV panels by keeping them perpendicular to the incident solar radiation using a mobile structure.

Figure 1.4   The most commonly used techniques to optimize the power drawn from PV panels

These technologies can be integrated into PV systems to predict, optimize, monitor and control their energy production in the most appropriate and efficient way, for fault detection and diagnosis, as well as for the prediction of PV panel performance, degradation and fatigue failure. Therefore, the main objective of this book is to present relevant works on the integration of digital technologies in PV systems: from general to rural and remote installations. The book focuses on the latest research and developments in remote PV energy system integration. It aims to provide extensive coverage of current research and developmental activities, and new approaches intended to overcome several critical limitations in the use of remote PV systems, such as those connected to grid and water pumps. This using control, monitoring and optimization techniques in the broadest sense, covering the use of new digital technologies such as embedded systems, data connection, Internet of Things, blockchain, cloud computing, artificial intelligence and robotics for PV systems through different applications such as MPPT controller, solar tracker, cleaning system, cooling system, water pumping systems, monitoring systems and solar-powered electric vehicles. Moreover, the applications of PV energy systems in distributed generation, microgrids and smart grid systems are also covered in this book.

In this context, several relevant studies and research solutions have been proposed and presented in this book, the content of which is summarized as follows (Figure 1.5).

In Chapter 2, Demirdelen *et al.* proposed an energy-efficient phase-shifting transformer (PST) as a potential solution for sudden load changes caused by solar PV power plants in rural power systems. The main objective of this work is to maximize the transmission efficiency, control the power flow of the rural power grid by removing line congestions, and foresee power. With this kind of specialized transformer, capacity modification can be made to utilize the power lines as efficiently and with the least amount of loss. Furthermore, a hybrid estimation method based on feedforward artificial neural network (FANN) and particle swarm optimization (PSO) is applied for the apparent power estimation of the proposed PST. For the power estimation study, phase regulation, tap positions, low side voltage and current values were used for the input layer and the apparent power of the PST was used for the output layer. Power estimation allows grid managers to forecast

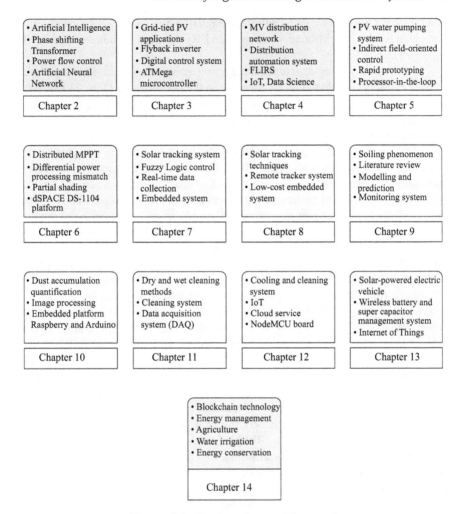

Figure 1.5 Book outline and keywords

and balance energy production and consumption. The method proposed in this work presents accurate power estimation to support safe and reliable operation of the grid.

In Chapter 3, Yaqqob *et al.* presented a low-cost and simple digital control system for a single-stage flyback micro inverter for grid-tied PV applications. This research study aims to design and implement an efficient control circuit based on a low-cost ATMega microcontroller and some analogue operational amplifiers. Moreover, a low-cost and small-size control circuit for injecting an alternating current from the inverter to the grid under different weather conditions is presented. The proposed control circuit is applied to the flyback inverter topology for a 120W prototype in the laboratory. Consequently, the overall components of the proposed control circuit are verified using a Proteus simulation tool. The experimental results are obtained for the different irradiance and temperature values to validate the proposed circuit control.

In Chapter 4, Hoang *et al.* conducted an extensive investigation on the impacts of fault currents contributed by PV systems on the performance of the FLISR (fault location, isolation and restoration service) functionality integrated within the distribution automation system (DAS) being operated in Danang MV distribution network. In addition, this chapter discussed the role of IoT platforms that can be embedded in the DAS to easily control the PV system and find the appropriate operating solution along with support from Big Data. It investigated the impacts of increased PV integration on the existing FLIRS performance when PV systems become more intelligent by integrating inverters with the IoT to efficiently control the system and optimize power generation through maximum power point tracking algorithms. Obviously, the combination of IoT with DAS can provide solutions for remote power system operation, change operation feeder to avoid the fault impact on FLIRS, or recovery of failed feeders can be performed easily and safely, typical with a distribution network with high PV penetration like Danang city.

Many rural areas have suffered from water scarcity, making the use of groundwater a better alternative in the absence of surface water. The use of PV water pumping systems for irrigation installed in rural and remote areas is considered a sustainable and reliable solution to reduce the high rates of lack of access to electricity and offers many advantages including environmental friendliness, operational safety and robustness. In Chapter 5, Errouha *et al.* presented a processor in the loop (PIL) model of a standalone PV water pumping system for rural areas. PIL experimentation is an effective method that can be used to validate the control strategy on the corresponding microcontroller. It is one of the methods utilized in the system development process, which consists of implementing the generated code in the desired processor by ensuring communication between the processor and the computer simulation software to detect early and prevent errors. This technique reduces the cost, accelerates the testing phase and provides a suitable rapid prototyping tool. An indirect field-oriented control (IFOC) is introduced to drive an induction motor powered by solar PV panels. Further, the boost converter is controlled using the step-size incremental conductance technique. Thus, the proposed control strategies are verified on a real microcontroller. The performance of the proposed techniques has been tested through numerical simulations and verified through experimental tests using a low-cost Arduino board.

In the global renewable energy market, PV generation systems (PGS) are gaining popularity due to their clean, low-cost and robust delivery, especially in areas where energy is difficult to use, such as rural and mountain villages. PGS demonstrates high power conversion efficiency under ideal conditions, but it exhibits low output efficiency under a variety of non-ideal conditions, such as mismatches between PV modules or submodules caused by partial shading. The mismatch can cause several problems for PGS, including power losses, multi-peak PV characteristic curve and hot spot effect. In Chapter 6, Guanying *et al.* investigated a novel distributed maximum power point tracking (DMPPT) technique to address the impact of non-ideal conditions on PV generation systems and to improve their performance. This study aims to use a DMPPT control to inject or extract energy through a dedicated submodule-level DC–DC converter to equalize the power of all PV submodules to eliminate the mismatch effect. The advanced control algorithms, time-sharing MPPT, total-minimum-power-point-tracking (TMPPT) and power-balancing-point-tracking (PBPT), have been investigated in this chapter to improve the performance of the proposed DMPPT system, including steady-state characteristics, dynamic convergence speed and system transfer efficiency. Furthermore, the performance of the proposed DMPPT was tested through numerical simulations and then verified by an experimental test using the dSPACE DS-1104 platform.

To maximize the absorption of sunlight and thereby increase energy production, it is necessary to integrate solar tracker systems into conventional solar PV systems, where the solar panels can be fixed on a structure that moves according to the sun's path. In Chapter 7, Bayrak *et al.* proposed an intelligent control method for a dual-axis sensorless solar tracking system (DASTS) using a mathematical model based on fuzzy logic-based decision-making (FDM) approach. The proposed FDM-based DASTS system provides maximum efficiency by adaptively changing the step range without using any sensors according to the Sun's Azimuth/elevation angles, the position of the region and the United States Naval Observatory (USNO) database with soft computing technique. This approach has eliminated sun position detection errors caused by environmental factors in conventional methods. Besides, fabrication cost much less than sensor-based DASTSs. The control of the linear actuators and the entire system is ensured by real-time software developed in the LabVIEW environment with an embedded system. The proposed intelligent sensorless system in this chapter is particularly suitable for rural installations with its cost-effectiveness, low maintenance requirements and high efficiency increase.

In Chapter 8, Rahal and Abbassi investigated the use of a solar tracking system in an off-grid area to maximize the efficiency of solar PV panels for optimal solar energy production. First, this chapter reviews different types of solar tracking techniques. Then, a commercially available two-axis active solar tracking system was placed in a remote area to examine the durability of its mechanism, system efficiency and energy output compared to a traditional fixed system. In addition, this chapter provides the design of a low-cost embedded system-based solar tracking controller suitable for use in rural areas.

The soiling phenomenon is one of the problems that need to be managed and above all to be predicted in PV generation systems. Almost all cleaning systems require information on the soiling effect and on the frequency of cleaning to develop optimal maintenance strategies. This can be obtained through the assessment of the soiling effect and more precisely, its modelling and prediction. In Chapter 9, Laarabi *et al.* presented a review of the different studies conducted to model the soiling phenomenon. A summary of the most important information about each model has been described. This review will help researchers working on this topic gain an overview of the models applied. It will also help to better choose the right direction in modelling and predicting the soiling phenomenon. The aim is to serve as a reference for upcoming works and for better consideration in future monitoring systems.

Solar power plants are often located in desert, rural and arid areas characterized by difficult environmental conditions (e.g. dust scattering, scarcity of rainfall). In fact, the accumulation of dust contamination on PV panels negatively influences their efficiency in such a way that the power output of the panels decreases considerably. Once the concentration of dust contamination exceeds a certain level, the cleaning process becomes a necessity. In Chapter 10, Tribak *et al.* proposed a new cleaning system based on image processing. A processing unit based on an embedded system platform (Raspberry Pi 3 and Arduino), through which the main processing of the proposed system is ensured. The performed experiments have been undertaken in the field under real-like dust accumulation conditions. The proposed system was found to be more advantageous compared to the existing systems, which require complex and expensive equipment.

In Chapter 11, Palpandian *et al.* proposed cost-effective dry and wet cleaning methods for PV panels to restore performance using a vacuum cleaner and pressurized water pump to clean the dust accumulated on the module. In this work, experimental measurements, using a data acquisition system (DAQ), were conducted to analyse the performance of the impact of dry and wet cleaning techniques on mono-crystalline PV modules.

In Chapter 12, Sathwara *et al.* developed an integrated system-based PV module cooling and cleaning device with remote monitoring using IoT and cloud service. The objective of this research study is to improve the conversion efficiency of solar panels by dealing with two important constraints, shading resulting from dust particles and temperature increase of solar panels. A water spray-controlled system based on the NodeMCU board has been developed to be an efficient and cost-effective solution.

In Chapter 13, Ben Said-Romadhane *et al.* investigated the control of solar-powered electric vehicles which are a perfect candidate for mobility in rural areas. In fact, by driving in rural areas, a major constraint for solar cars is eliminated: the urban obstacles (high buildings, tunnels, etc.). In return, the PV generation associated with the car will remedy the problem of the deficiency of electric vehicle charging stations. In this chapter, two control layers of the solar-powered electric vehicle energy storage system are presented. The lower layer concerns power converter control and the higher layer concerns the energy management system. IoT-based wireless battery management system (WBMS) is adopted to establish the battery and the super-capacitor states of charge. In addition, an H-infinity-based controller is proposed for the energy storage system power converters to enhance stability in solar-powered electric vehicles. The suggested controller offers robust stability by ensuring the perfect rejection of disturbances that come from direct current bus fluctuations and parameter variations. Simulations achieved under MATLAB® software are presented and discussed to validate the effectiveness and high performance of both the proposed H-infinity control and energy management strategy.

In Chapter 14, Anand *et al.* summarized recent trends in solar PV systems for agricultural purposes and assessed their significance. The authors discussed emerging digital technologies and algorithms used by blockchain technology for energy conservation. These techniques would maximize productivity, control energy consumption, maximize throughput and provide quality service to farmers. The reported techniques also allow states and central governments to analyse and track the consumption of power used by farmers in rural areas in order to control the unwanted levels of energy that permeate the rural areas. A blockchain-based system enables the sharing of energy at all levels by decentralizing access, which is a successful example of a decentralized mechanism in agriculture, where it is essential to share energy to complete the farming process. The farmers can make use of these techniques in order to conserve energy and reduce the amount of wasted energy during the process of utilizing energy by understanding the potential losses that may occur at each of the stages of the energy utilization process. In this chapter, a number of strategies that will alleviate some of the problems associated with energy in rural and remote areas using data from the literature review are provided. The impact of these problems can be minimized by deploying blockchain and other related technologies in conjunction with it.

References

[1]   Ghirardi E., Brumana G., Franchini G., and Perdichizzi A. The optimal share of PV and CSP for highly renewable power systems in the GCC region. *Renewable Energy*. 2021, vol. 179, pp. 1990–2003.

[2]   El Hammoumi A., Chtita S., Motahhir S., and El Ghzizal A. Solar PV energy: from material to use, and the most commonly used techniques to maximize the power output of PV systems: a focus on solar trackers and floating solar panels. *Energy Reports*. 2022, vol. 8, pp. 11992–12010.

[3]   International Energy Agency (IEA). 'Africa Energy Outlook'. World Energy Outlook Special Report; 2022.

[4]   ACP-EU Energy Facility. Solar PV for Improving Rural Access to Electricity. Available from https://europa.eu/capacity4dev/public-energy/file/10581/download?token=pvMLFRJA [Accessed 01 Oct 2022].

[5]   Allouhi A., Buker M. S., El-Houari H., *et al.* PV water pumping systems for domestic uses in remote areas: sizing process, simulation and economic evaluation. *Renewable Energy*. 2019, vol. 132, pp. 798–812.

[6]   Verma S., Mishra S., Chowdhury S., *et al.* Solar PV powered water pumping system – a review. *Materials Today: Proceedings*. 2021, vol. 46, pp. 5601–5606.

[7]   Choudhary P. and Srivastava R. K. Sustainability perspectives – a review for solar photovoltaic trends and growth opportunities. *Journal of Cleaner Production*. 2019, vol. 227, pp. 589–612.

[8]   International Renewable Energy Agency (IRENA). Renewable energy market analysis: GCC 2019, 2019.

[9]   Kim S., Van Quy H., and Bark C. W. Photovoltaic technologies for flexible solar cells: beyond silicon. *Materials Today Energy*. 2021, vol. 19, p. 100583.

[10] Ahmed R., Sreeram V., Mishra Y., and Arif M. D. A review and evaluation of the state-of-the-art in PV solar power forecasting: techniques and optimization. *Renewable and Sustainable Energy Reviews*. 2020, vol. 124, p. 109792.

[11] Motahhir S., El Hammoumi A., and El Ghzizal A. The most used MPPT algorithms: review and the suitable low-cost embedded board for each algorithm. *Journal of Cleaner Production*. 2020, vol. 246. p. 118983.

[12] Dwivedi P., Sudhakar K., Soni A., *et al.* Advanced cooling techniques of PV modules: a state of art. *Case Studies in Thermal Engineering*. 2020, vol. 21, p. 100674.

[13] Kazem H. A., Chaichan M. T., Al-Waeli A. H., and Sopian K. A review of dust accumulation and cleaning methods for solar photovoltaic systems. *Journal of Cleaner Production*. 2020, vol. 276, p. 123187.

[14] Awasthi A., Shukla A. K., Murali Manohar S.R., *et al.* Review on sun tracking technology in solar PV system. *Energy Reports*. 2020, vol. 6, pp. 392–405.

[15] Gorjian S. and Shukla A. (eds.). Photovoltaic solar energy conversion. In: *Technologies, Applications and Environmental Impacts*, 1st edn. New York, NY: Elsevier Academic Press; 2020. p. 452.

[16] Wang Q., Li R., and Zhan L. Blockchain technology in the energy sector: from basic research to real world applications. *Computer Science Review*. 2021, vol. 39, p. 100362.

[17] Emamian M., Eskandari A., Aghaei M., Nedaei A., Sizkouhi A. M., and Milimonfared J. Cloud computing and IoT based intelligent monitoring system for photovoltaic plants using machine learning techniques. *Energies*. 2022, vol. 15(9), p. 3014.

[18] Zhang S., Wang J., Liu H., Tong J., and Sun Z. Prediction of energy photovoltaic power generation based on artificial intelligence algorithm. *Neural Computing and Applications*. 2021, vol. 33(3), pp. 821–835.

[19] El Hammoumi A., Motahhir S., El Ghzizal A., and Derouich A. Internet of Things-based solar tracker system. In: Motahhir S. and Eltamaly A. M. (eds.), *Advanced Technologies for Solar Photovoltaics Energy Systems. Green Energy and Technology*. New York, NY: Springer, Cham. 2021, pp. 75–95.

[20] IGI Global. What is Digital Technologies. Available from https://www.igi-global.com/dictionary/the-uptake-and-use-of-digital-technologies-and-professional-development/51457 [Accessed: 05 Oct 2022].

Chapter 2

Energy-efficient phase-shifting transformers for rural power systems with solar PV energy sources: the state-of-the-art survey, artificial intelligence-based approach and a case study

Tuğçe Demirdelen[1], Burak Esenboğa[1], İnayet Özge Aksu[2], Abdurrahman Yavuzdeğer[3], Selva Bal[4], Burcu Sakallıoğlu[5], Abdullah Cicibaş[6], Ahmet Kerem Köseoğlu[6], Mahmut Aksoy[6] and Mehmet Tümay[7]

Integration of renewable energy sources (RESs) into electric power grids is increasing day by day. In the world, solar energy potential is at a high level among RESs. Thus, solar PV power plant installations are increasing in many countries. These RESs can bring huge problems to the rural power system grid such as transmission lines congestion. Due to the causative factors nature, congestion can continually happen and maintain for a long commutative time. Thus, transmission efficiency is a key factor when relieving congestion. Phase-shifting transformers (PSTs) can remove these congestions. Compared to other power flow controllers, phase-shifting transformers have advantages in rural power systems such as flexibility and economic characteristic. A phase-shifting transformer controls power flow by changing the phase shift angle between the sending and receiving ends. PST increases the transmission capacity of the power system. A comprehensive study is carried out for the proper selection of the appropriate phase-shifted transformer to improve power quality and energy efficiency, and to ensure grid reliability and power flow control in smart grids

[1]Department of Electrical and Electronics Engineering, Adana Alparslan Türkeş Science and Technology University, Turkey
[2]Department of Computer Engineering, Adana Alparslan Türkeş Science and Technology University, Turkey
[3]Department of Energy System Engineering, Adana Alparslan Türkeş Science and Technology University, Turkey
[4]Department of Electrical and Electronics Engineering, Iskenderun Technical University, Turkey
[5]TEİAŞ 4th Regional Directorate, Turkey
[6]Best Transformer, Turkey
[7]Department of Electrical and Electronics Engineering, Çukurova University, Turkey

emerged with the integration of renewable energy plants into electric grids. This chapter proposes a review of PSTs, consisting of classifications of PST studies and applications according to power rating, types of configurations and winding connection of PST, control and optimization techniques used for PSTs and power systems with PSTs, protection techniques and parallel operation of phase-shifting transformers in literature. A dynamic model of a solar PV power plant integrated with a 10-stage 50 MVA 380-V/154-kV phase shift transformer is presented. Also, artificial intelligence-based apparent power estimation analysis of a PST is realized using manufactured 150 MVA 330/161-kV PST for rural power systems. FANN and PSO-FANN hybrid estimation methods are used to estimate the apparent power of the PST. This study aims to contribute to researchers and application engineers with a comprehensive perspective on the status of PST studies. A list of more than 150 references on the subject will also be added for quick reference.

2.1   Introduction

The whole power system becomes more complicated to operate and less reliable with the development of technology today. Line impedance, voltage magnitude and phase angle offset between the voltages of the source and load side are control variables managing the power flow through an electric grid line. These two control variables can control the power flow for both the active and reactive. Flexible alternative current transmission system (FACTS) technologies are some of the most efficient solutions to achieve these targets. FACTS technologies can assist in damping power system oscillations and developing transient stability. The use of these proper devices can develop the operational effectiveness of an electric transmission network. A PST is a kind of equipment that can be sorted as a series of components of FACTS devices [1].

FACTS-type devices can be applied for fault current suppression with their performance which can regulate power flow functions and with emergency control, serious malfunctions can be prevented. Thyristor-controlled phase-shifting transformer (TCPST) is one of the FACTS devices, and it can adjust the phase angle of the voltage to regulate power flow [2]. Power flow solvers are an essential tool in the operation and planning of power systems. It discussed the voltage, power flows and losses in the rural power systems, and it was detected as unacceptable voltage deviations and specified over-loaded components [3].

To fix the problems in power flow, a PST, which can control active power flow, is proposed in the 1930s. [1]. By phase shifting transformer with adjustment of the angle of phase shift to determine the energy characteristics [4].

In the rural power system, many stakeholders expect versatile performance from their investments. Because of the increment in the cost of the power system, transmission system operators need to consider other technologies. These expectations can be provided with phase-shifting transformers (PST) as solutions that are operational and economic benefits [5].

Some of the functions of PSTs are as follows:

- Reduce the harmonic: displacement of phase between the primary and secondary voltage (line–line).
- Enhance the stability and protection overload of the transmission line: keep the power flow through to the line.
- Provide the secondary voltage regulating.
- Reduce active power losses to minimum values [6,7].

The power flow control diagram of PSTs in the transmission grid is shown in Figure 2.1 [1].

A longitudinal transmission network with PST is shown in Figure 2.2 [8]. The equivalent scheme of power system voltage and currents with PST is shown in Figure 2.3. Detailed explanations can be found in [9,10].

In Figure 2.4, the three-phase circuit of the PST is shown. The PST consists of two windings: an excitation winding BO which is gathered in a triangle, and an

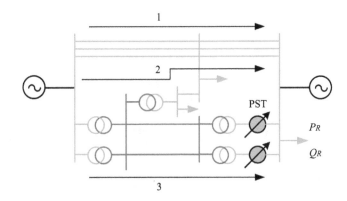

Figure 2.1   Power flow control with PST

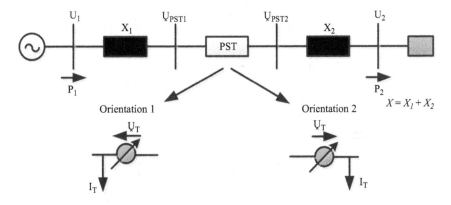

Figure 2.2   PST in a longitudinal transmission network

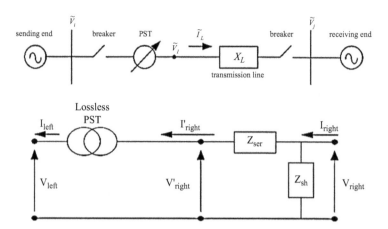

Figure 2.3   The equivalent scheme of a power system with PST

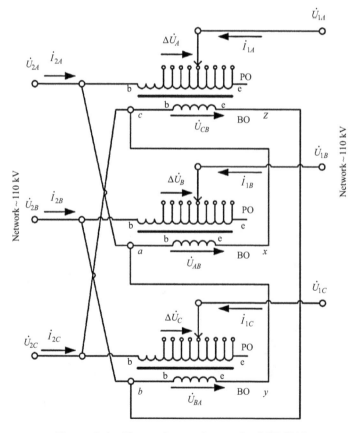

Figure 2.4   Three-phase scheme of a PST [11]

adjusting winding PO which has a general magnetic link with BO and a three-rod magnetic core [11].

The use of FACTS devices in the global power market have attracted more attention in recent years [12]. Using the PSTs benefits in varied applications. One of the applications is the flow control of power [13]. Also, the power transfer can be controlled, and the phase shift can be changed between primary and secondary [14]. PST is connected serially to the circuit and can be counted as a type of FACTs [15,16].

According to structure and design, PSTs are mostly divided into four classes, namely direct, indirect, asymmetrical and symmetrical PSTs.

- Direct PSTs comprise one 3-phase core. This type of transformer presents that a few windings are suitably linked to get the required phase shift.
- Indirect PSTs comprise a different construction with two separate transformers (series and exciting units). The proposed phase shift is achieved by using the proper winding connection of series and variable tap exciting units.
- Asymmetrical PSTs provide an output voltage with the proper phase angle and amplitude compared to the input voltage.
- Symmetrical PSTs are designed.

Concerning the power flow issue, PSTs are competent devices. Some studies on this technique have also been applied.

Furthermore, the aforementioned conventional issues related to differential protection, and PSTs cause new difficulties for differential protection. The novel difficulties contain:

- The off-grade phase shift between the sending and receiving ends;
- Usability of all measurements of winding-current;
- The series winding-core saturation;
- The buried CT replacement/maintenance;
- Differential and limiting currents are dependent on the tap position; turn-to-turn and turn-to-ground fault diagnosis; and turn-to-turn and turn-to-ground fault diagnosis.

PSTs are quite complicated devices. To answer the various requirements of each customer, it is significant to determine the optimum phase-shifting transformer model for different applications in LV, MV and HV power levels by considering some specifications such as power circuit configuration, suitable topology selection, tap changers, control techniques and protection techniques. Therefore, this chapter enables us to meet the highest customer requirements thanks to the comprehensive survey on the phase-shifting transformers and their classification in different categories.

The following is a summary of this chapter's main contributions:

- To decide the selection of the right phase shifter transformer in a grid struggling with undesirable current flows from the electric grid.

- To support the technical success of renewable energy-integrated electric grid operators in a competitive energy market.
- To offer customers a partner and repetitive access to defining features.
- To present the effectivity of the tap change PSTs compared to conventional transformers.

The remainder of this chapter contains seven sections. Begins with an introduction, the following sections include the latest PST technologies, the different configurations used, the control techniques, protection, parallel operation and the concluding remarks. In Section 2.2, phase-shifting transformers are classified as application areas according to the power ratios to get the most out of phase shifters. In Section 2.3, the power circuit configuration and connections of PSTs are examined in detail to achieve improved grid stability and flexibility. Section 2.4 presents the control and optimization techniques because PSTs are an efficient and cost-effective way to control power flow. In Section 2.5, the protection techniques are emphasized to protect the grid system from unbalanced power flows caused by green-energy generation and light loads. In Section 2.6, PSTs are detailed with their parallel operations. Section 2.7 presented the simulation model and analysis of solar PV power plant integrated PST in rural power systems. In Section 2.8, artificial intelligence-based apparent power estimation analysis of a PST used in rural power systems is realized.

2.2   Classification of power rating and applications

Some studies of phase-shifting transformers investigated the structure or operation of PST directly, while the rest is related to power systems, including PST in the literature.

Phase-shifting transformers are divided into classes according to power ratings, power circuit configuration and connection and control techniques as follows.

PST is used in industrial applications such as bi-directional power flow, dynamic power flow controller (DPFC) and unified power flow controller (UPFC). The system contributions have been shown that increase power capacity on the international power transmission grid and protection of international power lines in fault conditions, solution of global grid problems on power transits, enhancement of voltage and loading of feeders, extra-high voltage interconnected transmission lines, multilevel inverter, converter inter-phase transformer, enhancement of voltage control and regulation, dynamic voltage restorer, static frequency converter, voltage supply of variable frequency drives and multi-modular matrix converter, variable frequency transformer, de-icing on high voltage lines, decrease of harmonics in bleach production facility [8,10,19–47,52–54,56–67,70–104,110].

PSTs in literature can be divided into three categories as low power, medium-power and high-power applications. Table 2.1 presents these applications.

Table 2.1   The low-power, medium-power and high-power applications and system contributions of PSTs in literature

Power rating	References	Power system contribution
Low power (<100 kVA)	[43]	• Balancing voltage sags, swells and constant load voltage
	[44]	• Minimizing dimensions and voltage rating by a factor of 1,000
	[45]	• Mitigate the harmonics below the 53rd order
	[46]	• Regulating the output voltage on varying input voltage using the tap changer
	[47]	• Reducing input line current harmonics and output voltage ripple coefficient
	[48]	• Power regulation control
	[49]	• Eliminate asymmetry in the input currents and an increased THDI
	[50,51]	• Creating a simple structure, low cost and high reliability
	[52]	• Decrease mainly the harmonics of the grid voltages, currents and load voltages as well as enhance its functional parameters such as current, voltage, power factor and apparent power
Medium power (>100 kVA, <10 MVA)	[53]	• Regulate the input power factor related to lower switching losses • Ensuring multilevel output voltage waveforms, pure sinusoidal input current
	[54]	• Transforming an autotransformer model to a PST model managed by changing the phase angle between high voltage and medium voltage
	[55]	• Verifying the leakage reactance using FEM • Analysing the core loss, stray loss, placement of magnetic shunt for mitigation of stray loss and steady-state temperature distribution before the manufacturing process of the PST.
	[56]	• Determining the useful tap positions on the frequency response states
	[57]	• Power system harmonic elimination
	[58]	• Providing current with low harmonic, which has better load capacity and can improve the output performance of power conversion device
	[59,60]	• Creating an accurate analytical model by the finite element method (FEM)
High power (>10 MVA)	[10]	• Examining unexpected high-power transits through the Belgian grid and improving grid reliability and efficiency
High power (>10 MVA)	[38]	• Mathematical modelling, phase shift angle control and power control of a dual-core symmetric TCPST
High power (>10 MVA)	[61]	• Technical specifications and design constraints of PST in Meeden and differential protection techniques on PST

(Continues)

Table 2.1   (Continued)

Power rating	References	Power system contribution
	[62]	• Controlling the active power flow with PST
	[63]	• Presenting specifications, design and test procedures (induced test or lightning impulse test) of PST
	[64]	• Examining PST to protect the power lines from overloading substations and to observe power flow control in the Belgian grid
	[65]	• Presenting planning, investment, scheduling and operations of power flow control equipment
	[66]	• Observing power flow control and transient stability studies of asynchronous links using Power System Simulator software
	[67]	• Examining power controller plane method, substation uprating, substation reserve sharing, grid decoupling and power flow control using Assisted PST (APST)
	[68]	• Improving the voltage control, decreasing the electric losses on transmission lines and creating a new connection point for energy reliability in Rio de Janeiro States
	[69]	• Presenting design criteria, specific challenges during dielectric tests, thermal factory tests and selection of the surge arresters for protection of the PST
	[70]	• The power flow optimization and system security enhancement in the Italian transmission grid
	[71]	• Controlling the active power flows by installing a tap change phase shifter between the Czech and Polish power systems
	[72]	• Solving a PST optimization problem in the IEEE 57-Bus System in different European countries
	[73]	• Applying PSTs in the Czech Republic's electrical network, suitable types, main parameters and locations for the PSTs and investigating the impact of PSTs on the neighbouring power networks
	[74]	• Investigating the effects of PSTs on the cross-border transmission capacity of Polish and power networks with power flow calculations and the models of the interconnected power systems in Europe
	[75]	• Designing, manufacturing, testing and commissioning of PST to control the power flow in the Indian power grid
	[76]	• Enhancing harmonic performance as well as the phase switching angle and phase shift control
	[77]	• Calculating the proposed five-stage harmonic filter to depend on IEEE Standard 18 and harmonic suppression methods with PST
	[78]	• Observing power flow, overall system losses and voltages at various nodes by adjusting the phase angle of the PST
	[79]	• Presenting a techno-economic analysis of the Italian network including five PSTs

(Continues)

Table 2.1   (Continued)

Power rating	References	Power system contribution
	[80]	• Enhancing the voltage regulation of the ULCO grid and removing the voltage limit violation (overvoltage/under voltage) in peak and off-peak loading cases
	[81]	• Realizing the design phase of the power controllers (PCs) in the power grid by using a novel method
	[82]	• Presenting the application and test of Standard 87T differential protection for special PST
	[83]	• Discussing usage of 87T differential relays for the PST, 87P and 87S functions during modern numerical differential protection
	[84]	• Proposing the protection scheme, which is applied on two symmetrical, dual-cores PST
	[85]	• Describing the functional properties of a rotary power flow controller using rotary phase-shifting transformers
	[86]	• Presenting the specification and control schemes of PSTs to enhance the interconnection between Northern Ireland and the Republic of Ireland
	[87]	• Comparing a new PST and an upgrade of the present PST at Gronau
	[88]	• Investigating integration of series fixed capacitors and quadrature boosters, which is a kind of PST, in the Great Britain transmission network in the case of sub-synchronous resonance
	[89]	• Considerable improvement of the overloads, beyond other benefits to the operation of the electrical system as load cuts, TTC and losses
	[90]	• Developing a novel IGCT-based cascade HV large capacity converter and mitigating the harmonics in the network
	[91]	• Improving grid stability, loss optimization and system safety with a combined efficient power flow with phase tap change concept
	[92]	• Suppressing harmonic levels with PST and investigating harmonic currents
	[93]	• Performance comparison among PST and examining hybrid flow controller and UPFC for multi-objective optimization on the IEEE 14-bus system
	[94]	• Determination of the optimal place and ratings of 4-TCPST on the Java Bali power system
	[95]	• Demonstrating the positive impact of PSTs' on the maximum power injection capacity of investigated network nodes
	[96]	• Presenting delta-hexagonal PST with a unique design and a new algorithm for the protection
	[97]	• Proposing the series-connected FACTS devices with control methods, which is in literature and a comparison between fixed and static series compensation devices

(Continues)

Table 2.1   (Continued)

Power rating	References	Power system contribution
	[98]	• Demonstrating power flow control with thyristor controlled series capacitor (TCSC), static synchronous series capacitor (SSSC) and PST devices
	[99]	• Examining the fault recognition and discrimination methods for the protection PSTs
	[100]	• Providing better stabilization for external disturbances and more sensitivity to internal faults • Presenting an improved algorithm for protecting phase-shifting transformers
	[101]	• Discriminating the normal and external fault conditions from internal fault and magnetizing inrush conditions by using a novel algorithm
	[102]	• Regulating active power transfer by regulating the voltage at the load side
	[103]	• Understanding of all demands from the tender stage to the finals design stage, testing and operating of the units
	[104]	• Differentiating inrush current and internal fault current in indirect symmetrical PST with a 300 MVA power capacity
	[105]	• Control of power flow impact on the interface and neighbouring system • Limiting short circuit current effect with respect to the impedance of TCPST under varying phase shifting angle as well as system stability performance related issue
	[106]	• Improvement of differential protection operation achieved by harmonic restraint (as in standard protection)
	[107]	• Reduction of the voltage variation in the network during PST energization • Reduction of risk of resonance excitation and corresponding temporary overvoltage
	[108]	• Providing congestion management by deflecting power flows from the congested element
	[109]	• Achieving the regulated power flow control for the concept of line power-flow sensitivity
	[110]	• Achieving phase shift control between voltages and the inductive reactance of a regulating winding
	[111]	• Preventing congestion under different weather load conditions
	[112]	• Realizing active power transfer control of the power system thanks to the flowing strategy realized with the PST tap locked and with the PST automatic control
	[113]	• Providing the reactive power control to reduce the operating cost to satisfy predefined N-1 contingencies
	[114]	• Observing changes in active power transfer and by the step change of the APST

2.3 Classification of the PSTs: power circuit configuration and connections

The phase-shifting transformer provides a phase shift to control power flow in the transmission grid owing to internal impedance. The equivalent circuit of PST is shown in Figure 2.5 [86].

A conventional PST consists of a magnetizing and boosting unit, which is shown in Figure 2.6 [115].

A PST includes series and parallel transformers connected with a mechanical tap changer (Figure 2.7(a)) and an equivalent circuit scheme is also shown in Figure 2.7(b) [8].

PSTs are classified according to the circuit configuration and connection type of windings into categories such as direct/indirect, symmetric/asymmetric, quadrature/non-quadrature, single-core/dual-core, single tank/dual tank and rotary(round) shape.

1. Direct PSTs consist of one 3-phase core. The phase shift operation is obtained by connecting a few windings suitably [62,91,100,106].
2. Indirect PSTs consist of two separate transformers (series transformer and variable tap exciter) which are built into separate tanks (in the dual tank

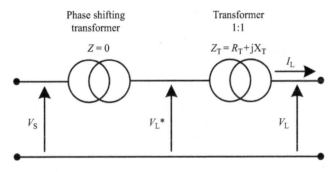

Figure 2.5 The equivalent circuit of PST

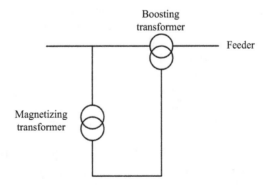

Figure 2.6 The basic layout of PST

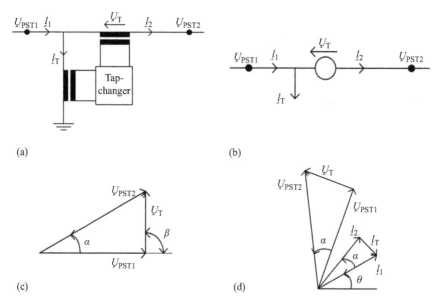

Figure 2.7   *PST model. (a) Series and parallel transformers connected with a mechanical tap changer, (b) equivalent circuit scheme, (c) phasor diagram for the case of a quadrature-boosting transformer application and (d) represents the operation of a phase-angle regulator*

[74,75,115]); the required phase shift is achieved by the proper winding connection of series and exciting units [10,62,65,86,87,91,100,101,104,106]. The winding connections of an indirect PST are shown in Figure 2.8 [86].

3. Symmetrical PSTs generate an output voltage with the phase angle which is desired and with output voltage magnitude being the same as the one of input voltage [10,18,38,62,65,70,74,82,84,86,87,91,96,99–101,104–106,116–119].
4. Asymmetrical PSTs generate an output voltage with the phase angle which is desired and an amplitude which is depending on the input voltage (the output voltage amplitude is not equal to the input voltage amplitude), (winding configuration of an asymmetrical PST is shown in Figure 2.9 [47]) [24,47,62,82,86,91,99,100,106,114,116,117,157].
5. Quadrature PSTs – [8,63,69,116,117]. The phasor diagram and operation of quadrature booster PST are shown in Figure 2.7(c) and Figure 2.7(d) [8].
6. Non-quadrature PSTs – [116].
7. Single-core PSTs – limits of parameters depend on the tap changer of PST [29,63,99,106,116].
8. Dual-core PSTs – [29,38,63,68,70,74,75,82,84,99,100,103,105,116,118–121]. The winding connection of a dual-core symmetrical PST is demonstrated in Figure 2.10(a), Figure 2.10(b) and an example of this type of PST is also shown in Figure 2.10(c) [29,100].

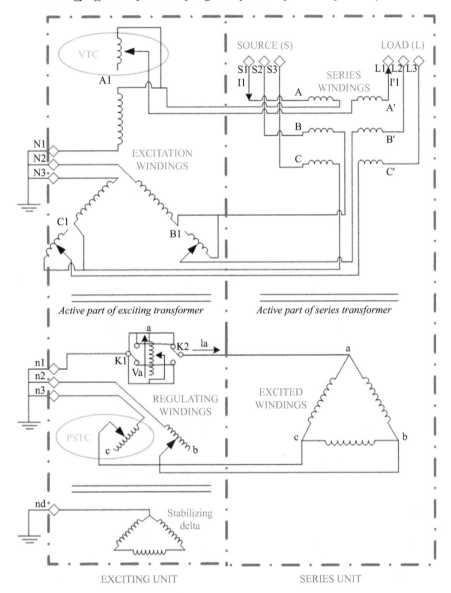

Figure 2.8   Winding connections of indirect PST

9. Rotary (round) shape PSTs – are speedy and economical compared to conventional PSTs [39,85,122,123].

A symmetric PST and an increase of the maximum phase angle can be obtained by inserting an additional regulating winding and a new tap changer to a nonsymmetric quadrature type PST [118].

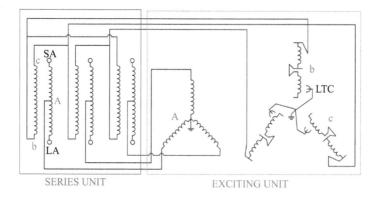

SERIES UNIT   EXCITING UNIT

Figure 2.9   Configuration of an asymmetrical PST

PSTs can have more than two windings and different connection types such as delta, star, zigzag, hexagonal, polygon and Scott connection. Some of these connections are shown in Table 2.2 [38,39,47,54,56,71,76,83,97,101,107,125–131]. Besides, basic connection types of PSTs are presented in [122] in detail.

The most frequently used connection types for the primary, secondary and tertiary windings of the PSTs are seen as delta, star and delta connection, respectively, in literature.

Commonly used types of different PST topologies are indirect, symmetrical, quadrature and dual-core PST devices.

2.4   Classification based on the control technique

PSTs are classified in terms of control techniques as the open-loop controller, closed-loop controller and optimization techniques in this section.

2.4.1   Open-loop controller

Feedback does not occur in open-loop control systems. Therefore, distortions cannot be calculated. Almost there are no open-loop studies of PST; except for the variable frequency transformer (VFT) device. VFT is used as a power flow controller for asynchronous systems. It is based on a rotating machine and the rotor speed of VFT is adjustable. It provides a variable phase shift, similar to PST [66].

2.4.2   Closed-loop controller

Closed-loop control is seen in some PST studies in the literature [8,43,44,126,130]. This type of control is more frequent than open-loop control, while less frequent than optimization methods.

Power flow control on EHV transmission lines by using PSTs is investigated on the Tunisian 400 kV longitudinal transmission network. A reactance and a static phase shifter (SPS) are connected in series in the PST model. Various alternatives

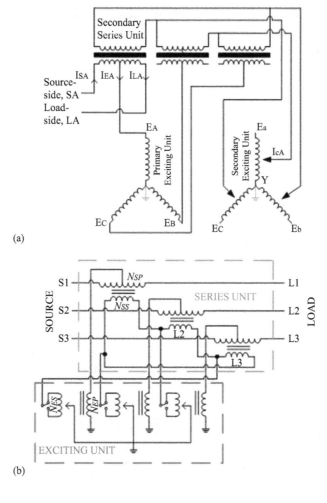

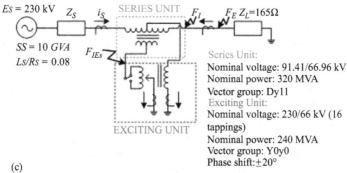

Figure 2.10   *(a) Windings of dual-core symmetrical PST, (b) another representation of windings in dual-core symmetrical PST and (c) an example of a winding connection of dual-core PST: DY11 series unit, Y0Y0 exciting unit*

Table 2.2   The winding connection of phase-shifting transformers

Primary–secondary–tertiary–fourth	Reference
star-star-delta (YNyn0d5)	[106]
star-star-star-delta	[75]
delta-star-delta	[47]
delta-polygon	[54]
star-polygon	[124]
ET: secondary delta, BT: star-star	[38]
zigzag-zigzag-delta-star	[125]
star-star	[39,123,126]
delta-star	[126]
star-delta	[124]
delta-polygon-polygon	[56]
DY11 series unit, Y0Y0 exciting unit	[100]
delta (series unit), star neutral (exciting unit)	[70]
delta-delta-star ($Dd0^{1/4}y1^{1/4}$), ($Dd11^{3/4}y0^{3/4}$)	[82]
delta-hexagonal	[96,127]
extended delta	[124,129]
zigzag	[124,125,127]
Scott	[124]

for the location, operation and control of PSTs are investigated. A closed-loop PI control system obtains phase angle regulation of PST. Although PST can reduce the voltage stability margin, it is inefficient to develop voltage stability [8]. A novel dynamic voltage restorer (DVR) that includes shunt PST, AC-link and vector switching matrix converter is presented with dynamic modelling and controller design. The prototype verifies the proposed model. The closed-loop PI controller is offered to balance voltage sags/swells, by remaining the load voltage constant [43].

A hardware model of a DPFC that includes PST, thyristor switching reactor (TSR)/thyristor switching capacitor (TSC) and mechanically switched capacitor (MSC) is presented. The PST is used to close the operating point of TSC and TSR to a neutral value. A closed-loop PI controller is proposed for DPFC [44]. The 48-pulse multilevel inverter is improved for an interline power flow controller (IPFC) to control the power flow in transmission lines. Eight PSTs (four Y-Y PSTs and four Δ-Y PSTs) are used to couple four 12-pulse voltage source inverters (VSI) with suitable phase shifts to develop a 48-pulse multilevel inverter. PSTs are connected in series to eliminate the lower-order harmonics. A closed-loop PI controller is used for enhancing the voltage profile of the proposed system; low THD level and advanced transfer capability is obtained [126].

The 48-pulse VSC-based SSSC is connected to decrease the harmonics and enhance the power quality in a 400 kV system through four PSTs. The effect of SSSC on the impedance of the transmission line calculated by the distance relay is proposed. The voltage, which is compensated to the line, is controlled by a simple closed-loop PI control system [130].

2.4.3 ANN-based optimization techniques

Optimization of phase-shifting transformers is an essential issue for control to enhance transmission capacity and system reliability, and stability. Because the determination of the best location for PSTs is significant as the design procedure.

Various optimization methods and scenarios are investigated for PSTs: such as particle swarm optimization (PSO), fuzzified PSO (FPSO), analytical optimization, genetic algorithm, fuzzy-genetic algorithm, multi-objective optimization method, differential evolution (DE), multi-objective differential evolution (MODE), gravitational search algorithm (GSA) and flow-based market coupling (FMC) algorithm, selective harmonic elimination (SHE), benders decomposition, improved cross-entropy method are optimization methods for PST. PSS/E, MATLAB®, MATLAB/MOSEK, general algebraic modelling system (GAMS) software and Monte Carlo simulation are used as simulation software; Python, linear programming, nonlinear programming (NLP), mixed-integer nonlinear programming (MINLP) and multi-objective mathematical programming is used as programming languages in PST optimization algorithms [87,89,93–95, 101,132–146].

PST settings are determined to maximize the cross-border transmission capacity with the calculation of the total transfer capacity with PSO and analytical optimization (AO) methods. PSO algorithm is realized through a simulation model, designed in PSS/E and in the Python language. DC load flow approximation-based AO is investigated by using linear programming (LP) [131] and the MOSEK optimization toolbox in MATLAB is used as an LP solver. The transfer capacity increases with a new link, a new PST and the instant PST upgrade at Gronau. Two PSTs are inserted with a 1,400 MVA power capacity. The simulation results show that the variation between the two methods is very little [116].

To overcome the power flow problem, FPSO algorithm is proposed. The proposed algorithm estimates the location and angle of PST to reduce line load. Simulation results on the IEEE 30-bus test system indicate that FPSO is comfortable, useful and reliable [132].

An optimization method that depends on genetic algorithms (GAs) with OPF is proposed to minimize the area for allocation of thyristor-controlled PST (TCPST) with load flow solution and optimum settings of the PST taps. It is tested at a 291-bus system, and it is inferred that the proposed solution is sufficient [89].

An approach based on GA is presented to investigate the optimal location, type, cost and parameters of UPFC, TCSC and TCPST to develop voltage stability and decrease reactive power loss of power systems. The proposed method is tested with case studies on the IEEE 30-bus and IEEE 118-bus systems. The comparative results verify that the performance of power systems is enhanced by the optimal allocation of FACTS equipment [133].

GA is used to identify the optimum location and the effects of some FACTS equipment on the system. Change in the number of TCPSTs does not affect the system while FACTS equipment affects the system [134].

PST controls active power flow by changing the voltage phase angle for variable impedance. The best place and phase angle can be chosen with active power flow sensitivity (APFS) by this improved method [135].

Multi-objective optimization-based fuzzy-genetic algorithms are presented to show the impacts of series capacitors, and PSTs on the reduction of unplanned power flows. The presented technique is realized by using the IEEE 30-bus system in MATLAB. PST gives better results with the proposed method than TCSC for unplanned power flows and power loss grids. Only, it takes a long time due to a genetic algorithm [136].

A simple and multi-objective optimization method based on the parameters of TCPAR devices is proposed for the interaction between PST in the tie-lines and automatic frequency controllers in the interconnected network. The results show that a good suppression of tie-line power and frequency oscillations is provided by the proposed optimization method [137].

A multi-objective optimization technique is given to specify the optimal location and adjustments of PST, hybrid flow controller (HFC) and UPFC. The performance of these three FACTS devices is tested on the 14-bus system IEEE. The proposed method is examined by using MATLAB and the general algebraic modelling system (GAMS). HFC is determined as the best choice in comparison with PST and UPFC in terms of analytical and technical [93].

DE and PSO algorithms are investigated for the optimal place, and control of FACTS equipment consisting of TCSC, SVC and TCPST to develop load-ability in the pool model transmission system. Both approaches are simulated on IEEE 6-bus system and 39-bus New England Test Systems. Load-ability is developed with three SVCs in the IEEE 6-bus system while with one TCPST in New England Test System. DE algorithm is more efficient than the PSO algorithm since it has less computational time and faster convergence properties [138].

The GSA is proposed to determine the optimal place and ratings of 4-TCPST on Java Bali 500 kV power system. The proposed GSA method provides both the reduction of the active/reactive power losses on the Java-Bali 500 kV power system and the enhancement of voltage to confine in 0.95 ± 1.05 pu interval [95].

The impact of PSTs on the FMC algorithm is investigated to obtain optimum cross-border transmission capacity by inserting the phase angle of PST as a variable in this algorithm. The economic efficiency of coupled markets is increased in this way. Congestion can be suppressed by redirecting power flow due to PSTs in the FMC [139].

The operation and harmonic suppression techniques of multi-pulse converters on the FACTS devices at Marcia substation, New York are investigated. SHE is presented and applied to develop the dynamic response of the system. One full-wave modulation and three SHE modulation techniques are presented [140].

A novel formulation for the direct current optimal power flow (DCOPF), which consists of the corrective actions for the PSTs and is based on a corrective power-system rescheduling and load-shedding problem, is proposed. It is expressed that the proposed method is innovative due to the sensibilities of PST in as much as injected powers. It is seen that the proposed approach is proper for the usage of IEEE-RTS 24-buses system as a benchmark network [141].

In the Benelux grid, an optimization algorithm is presented to investigate the switching of PSTs for the day-ahead scheduling of power systems. The proposed method is improved to obtain the optimal phase angles of the PST for the least switching over a day. It can be inferred that the proposed algorithm is appropriate for efficient systems from the simulations of IEEE 9-bus and IEEE 57-bus systems [142].

• An optimization process is proposed to provide more efficient power transfer from wind farms by defining PST tap settings automatically in the West Texas area [143].
• An OPF-based method is proposed to evaluate the maximum power injection capability in the transmission network. The impact of PSTs and HVDC links are presented in this aspect also [95].
• An approach for a techno-economic evaluation and a real options analysis is presented for corrective power flow control with PSTs in the European transmission grid [144].
• A useful system based on linear programming optimization methods is proposed to coordinate tap adjustments of PSTs for preventive and corrective approaches [145].
• A passive filter-based PST is proposed for the design method, and a novel genetic algorithm-based PST method is presented for developing the reliability, effectiveness and accuracy of the filter [147].
• To reduce the power losses and gain voltage stability in a thyristor-controlled phase-shifting transformer-based electrical grid system, the Ageist Spider Monkey Optimization (ASMO) algorithm and a fuzzy logic-based approach are used for standard IEEE 30 and 118-bus systems [148].

The congestion in PST connected power system is removed, thanks to Fuzzy logic [149]. Combining two approaches results in a reduction in unscheduled flows; the discrete PSO and the power flow method [150].

A novel technique based on phase angle shift threshold and optimal radial basis function neural network (ORBFNN) is proposed to prevent differential relay operation of PST depending on varying operation conditions except for internal fault conditions. The developed ORBFNN is compared with the feed-forward back propagation neural network (FFBPNN). It can be referred to as the proposed technique is advantageous with regards to slope classification, generalization ability and without phase shift compensation necessity [101].

Conclusions according to comparisons among PST optimization methods are as follows:

1. FPSO, genetic algorithm, Benders decomposition, ASMO, fuzzy logic-based approach and DCOPF methods are used to overcome the optimal power flow problem in PSTs.
2. The impact of PSO and analytical optimization is similar.

3. FPSO and DE methods are more efficient than the PSO algorithm.
4. MODE gives a better performance compared to the three-algorithm method: nondominated sorting genetic algorithm-II (NSGA-II), strength Pareto evolutionary algorithm 2 (SPEA2) and Pareto differential evolution (PDF).

2.5   Protection techniques

PST is protected with differential, over current, over flux, Buchholz, earth fault, through current backup protection, etc. However, differential protection is the primary method to protect PST; in the literature [74,100,103].

Differential protection is a frequently used technique to protect PST [61,64,75,82,94,96,99–101,104,106,124,125,127,151–154]

A general protection scheme of PST is proposed [62]. 87P (primary differential protection) and 87S (secondary differential protection) are proposed for PSTs [82,145]. 87T differential protection is also used for protecting PST [82,122,152].

The unique design and a novel method for the protection of delta-hexagonal PST are presented to investigate the availability of applying electromagnetic equations. The proposed algorithm has not only the internal/external fault diagnosis and discrimination capacity but also is steady while starting the current of magnetization, a saturation of the current transformer and saturation of the series winding. The non-standard phase shifting can change, and the sensibility of the relay is not affected in case of low fault current and PST at the highest tap; and safety in case of any fault current and PST at the lowest tap. The presented method obtains good sensibility against faults of low turn-to-turn, faults of high resistive and measurement of a significant error. The proposed method needs a reading of the tap position from the PST. The method is tested for varying fault and non-faulted conditions by PSCAD/EMTDC simulation [96].

Fault recognition and discrimination methods are proposed for protection. Not only the proposed method is quick and selective such as conventional differential protection, but also it offers high reliability during cases, like any mismatch between current transformer saturation and zero-sequence current, etc. Besides, it provides a solution to the non-standard phase shift problem independent of the tap changer position. The proposed method obtains the optimal protection for all common types of single/two core symmetrical/asymmetrical PSTs and turn-to-turn fault diagnosis undetectable by the traditional differential method. However, some disadvantages are present such as that the relay data collection system is doubled by extra voltage, and it increases the cost largely. It is preferred that the Process Bus concept in modern protection relays, described by the IEC 61850-9-2, can suppress those disadvantages [99].

The protection problems of PSTs and a dual-core symmetrical PST model are proposed to enhance the protection. The dual and single protection methods are investigated and compared for dual-core symmetrical PSTs with Electro-Magnetic

Transient Program (EMTP). It is concluded that both of them are good for compensation by PST. But they have some challenges, so an enhanced algorithm for the protection of PSTs, which is more stable for external failures and more sensible to an internal fault, is essential [100]:

- A differential protection method of single-core symmetrical PST is presented [106].
- The short circuit fault current mitigation technique, which uses controlling the phase shift angle of TCPST, is presented [153].
- Distance relay protection is proposed for PST [127].
- A novel protection method is proposed because of the disadvantages of traditional differential protection for PSTs [125].
- Variable Speed Drive (VSD) is proposed to eliminate source current harmonics with PST in the textile industry [155].

2.6   Parallel operation

Phase-shifting transformers operate in parallel connected [1,10,65,85,92,103,146,156]. A general scheme of parallel-connected PSTs is shown in Figure 2.11 [156].

The rotor and stator windings of two rotary PSTs are linked in parallel and series, respectively, to control both stator voltage and phase shift [85].

The selection criteria of the On-Load Tap Changer (OLTC) are presented when designing a PST. Some necessities will affect the selection of OLTC: the recovery voltage at the changeover selector and the short circuit impedance of PST. All possible switching conditions should be considered to determine a maximum of the step voltage because overloading of a PST affects the transformer ratings. In parallel operation of more than one transformer including OLTCs, an "out of step" condition takes place for a while owing to the asynchronous operation of the different OLTCs. Thus, the momentary voltage difference between transformers causes a forced circulating current on the circuit. So, the amplitude of the

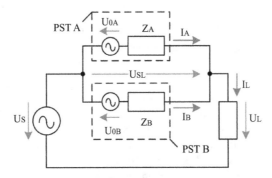

Figure 2.11   Parallel connection of PSTs

Table 2.3   An overview of PST

		Classification of PST
Topology	Symmetric	[10,18,38,62,65,70,74,82,84,86,87,91,96,99–101, 104–106,116–118]
	Non-symmetric (asymmetric)	[47,62,82,86,91,99,100,106,116,117]
	Quadrature	[8,63,69,116,117]
	Non-quadrature	[116]
Power	Single-core	[29,63,99,106,116]
	Dual-core	[29,38,63,68,70,74,75,82,84,99,100,103,105,116, 118–120]
	Single-tank	[116]
	Dual-tank	[74,75,116]
	Direct	[62,91,100,106]
	Indirect	[10,62,65,86,87,91,100,101,104,106]
	Rotary, round shape	[39,85,122,123]
	Low power	[43–47,52–55]
	Medium power	[53,56–60]
	High power	[10,17,38,61–70,73–80,86–108,110–113]

circulating current should also be taken into account concerning the switching capacity of the OLTC [156].

Two PSTs operate with a parallel connection in Meden [65]. The parallel action of PSTs obtains power flow through Belgium by tap positions [10].

It is presented that PSTs can develop the transfer capability of a Cape corridor with parallel lines and at different voltages [1].

It is mentioned that PSTs provide a harmonic cancellation advantage in parallel connection [92].

It is inferred that two single-core PSTs should operate in parallel with additional impedances because of large circulating currents near the neutral tap position [103].

Tables 2.3 and 2.4 summarize the overview of PST articles in the literature.

2.7   Dynamic model and analysis of a solar PV power plant integrated PST for rural power systems

Unreliable electricity supply has an especially negative impact on rural areas. As an alternative, solar power systems may be the solution for rural people such as India blessed with a huge potential for solar energy. The substantial solar photovoltaic penetration inside the inadequate distribution network may result in voltage problems including voltage increase and decrease. Today, solar energy has significantly penetrated the distribution system. The voltage instability in the rural grid is becoming a serious threat unless a detailed study for its mitigation is taken up in all seriousness.

Table 2.4   An overview of PST (Cont'd)

Classification of PST		
Connection	star-star-delta (YNyn0d5)	[106]
Controller	star-star-star-delta	[75]
Protection	delta-star-delta	[47]
	delta-polygon	[54]
	star-polygon	[124]
	ET: secondary delta, BT: star-star	[38]
	zigzag-zigzag-delta-star	[125]
	star-star	[39,123,126]
	delta-star	[126]
	star-delta	[124]
	delta-polygon-polygon	[56]
	DY11 series unit, Y0Y0 exciting unit	[100]
	delta (series unit), star neutral(exciting unit)	[70]
	delta-delta-star $(Dd0^{1/4}y1^{1/4})$, $(Dd11^{3/4}y0^{3/4})$	[82]
	delta-hexagonal	[96,127]
	extended delta	[124,129]
	zigzag	[124,125,128]
	Scott	[124]
	[8,43,44,66,87,89,93–95,101,126,130,132–148]	
	[61,64,75,82,84,96,99–101,104,106,119,124,125,127,149,151–155]	
Parallel operation	[1,10,65,85,92,103,146,156]	
Application	[8,10,39,40,42–47,53–55,58,61–70,73–82,84,122,152,154,155]	

The significant issue that occurs with the increase in the use of renewable energy is the prevention of power imbalances on the grid. In particular, design, modelling and simulation studies to examine the electrical characteristics of PV arrays in detail are performed by using MATLAB®/Simulink® software [158–161]. The instantaneous change of weather conditions creates instabilities in the power produced in solar PV power plants. This causes uncontrolled power flows in rural power systems. Transmission and distribution lines can be overloaded due to uncontrolled power flow in power systems. This power flow is directly related to the impedance of the transmission lines. Because of these uncontrolled power flow and overload problems, power must be controlled. PSTs are suitable elements for solving these problems. Therefore, a solar PV power plant integrated with a PST is modelled and analysed in this study. Figure 2.12 presents the simulation model of a solar PV power plant integrated PST.

In the simulation study, Figure 2.13 shows the solar PV power plant simulation model. A 15-kW solar PV power plant is connected to the rural electric grid with a voltage of 380-V on the LV side. An OLTC transformer is placed between LV and MV. Thus, the positive effect of the OLTC transformer on the grid is observed in case of sudden load change and power instability. Variable temperature and solar radiation are applied to solar PV power modules. The solar PV power plant consists of five parallel strings and each string has six series solar PV panels. The voltage source

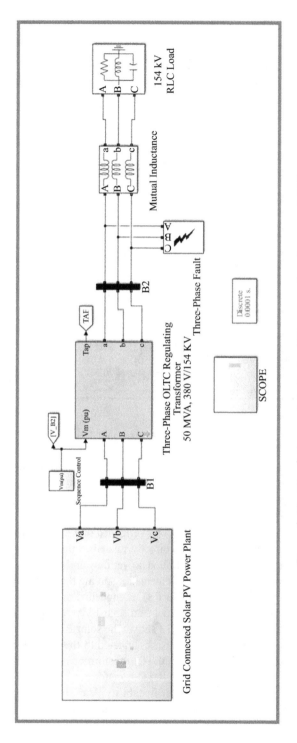

Figure 2.12   Simulation model of solar PV power plant integrated PST

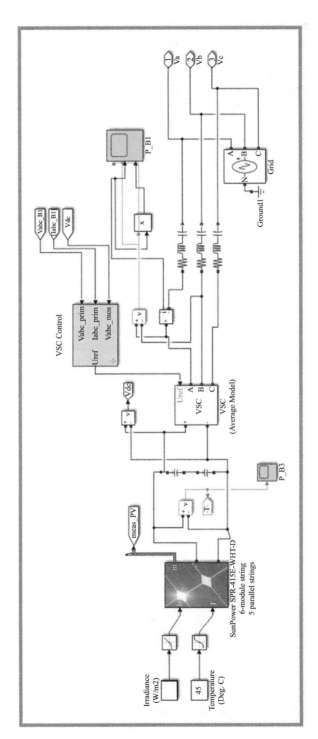

Figure 2.13   15-kW solar PV power plant

inverter is activated by applying the 400-450 V open circuit voltage operating range, The 380-V voltage LV voltage obtained from the inverter output is integrated into the grid. On-grid solar PV power plant LV voltage level is increased to 154-kV with OLTC transformer. The effect of the OLTC transformer is examined to prevent power instability on the rural grid under the sudden changing load condition.

The LV side power changes are observed in Figure 2.14. The solar PV power plant's produced voltage ranges from 400-V to 480-V. The variable temperature and solar radiation level are applied to the solar PV panel to create unbalanced power during 0.5 sec.

On-load tap changers (OLTC) allow the voltage regulation of the loaded transformer to be changed uninterruptedly and safely, without disturbing the voltage stability. The on-load tap changer must provide uninterrupted current flow during the transition from one tap to the next. The current flow must be kept uninterrupted without allowing partial short-circuit in the tap winding. In this simulation study, a 10-stage 50 MVA 380-V/154kV OLTC transformer is used as shown in Figure 2.14. OLTC transformer regulates positive-sequence voltage thanks to the sequence control block at B2 154-kV bus. The reference voltage is set at 1.0 pu. According to the tap change situation, the effects of the voltages at B1 380-V and B2 154-kV are observed. 0.3 second single-phase fault is applied at the secondary side of the OLTC transformer. As shown in Figure 2.14, the transformer does not react by single phase fault. With the tap change feature of the transformer, uninterrupted and safe power flow is ensured without disturbing the voltage stability during the transmission of the electricity.

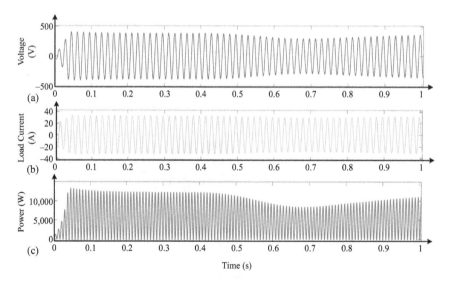

Figure 2.14   15-kW solar PV power plant (a) produced voltage, (b) drawing current from RLC load and (c) produced power

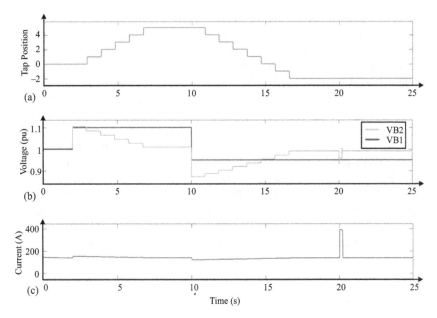

Figure 2.15 *Three-phase OLTC transformer (a) tap positions, (b) primary and secondary voltages and (c) current during the fault*

2.8 Artificial intelligence-based apparent power estimation analysis of a PST used in rural power systems

Artificial neural networks (ANNs) are methods developed for non-linear and difficult-to-solve problems. These structures are preferred in solving complex problems by modelling the functions of the human brain, such as learning, understanding and revealing new information through experience. It is clear from a review of the studies in the literature that the feedforward artificial neural network (FANN) is frequently preferred.

By drawing inspiration from the motions of fish and bird flocks, Dr Kennedy and Eberhart developed the intuitive optimization technique known as particle swarm optimization (PSO) in 1995 [162]. PSO is a well-known meta-heuristic method for resolving optimization problems among the different meta-heuristic algorithms. The approach can be simply applied to numerous issues because it is a straightforward model of each particle's behaviour.

This study aimed to estimate the apparent power obtained from 150 MVA 330/ 161-kV PST. The proposed PST is suitable for use in rural power systems. Phase shifter transformers are for to control the power flow as a response to load flow requirement changes. These transformers are potential solutions for sudden load changes caused by solar PV power plants in rural power systems. To use power lines with minimum loss and maximum efficiency; capacity adjustment could be made with this type of special transformer. For this purpose, a FANN structure is used. In addition, the hybrid estimation method (PSO-FANN) obtained by training the coefficients on the FANN network structure with the meta-heuristic PSO method is used in PST apparent power estimation. At the last stage of the study, the results obtained from these two methods are compared.

The network structure used in the study consists of three layers: input, hidden and output. As it is known, in ANN methods, the number of neurons in the input and output layers is decided according to the structure of the problem. In this power estimation study, the input layer consists of four neurons and the output layer consists of one neuron. In both estimation methods, eight neurons are used in the hidden layer. LV current, LV voltage, phase regulation and tap position data are given to the input layer of the neural network. At the output of the neural network, the apparent power values (S) of the PST are requested. In the PSO-FANN method, which is a metaheuristic-based artificial neural network structure, the weights and biases belonging to the neural network are trained with the PSO algorithm. The Normalized Mean Squared Error (NMSE) criterion is used as the fitness value required during the training:

$$NMSE = \frac{\frac{1}{n}\sum_{i=1}^{n}(x_i - \bar{x}_i)^2}{S^2} \tag{2.1}$$

Here, x_i and x'_i are represents the measured and predicted output values, respectively. S is the variance of the target rows and n is the number of samples.

In both methods, 70% of the data set is used during the training phase. After the training phase is completed, the test phase is started and at this stage and 30% of the dataset is used. Root mean square error (RMSE) and mean absolute percentage error (MAPE) error criteria are used to compare the results obtained from the methods at the end of the study:

$$RMSE = \sqrt{\frac{1}{n}\sum_{i=1}^{n}\left(x_i - x'_i\right)^2} \tag{2.2}$$

$$MAPE = \frac{1}{n}\sum_{i=1}^{n}\left|\frac{x_i - x'_i}{x'_i}\right| \tag{2.3}$$

Here, x_i and x'_i are represents the measured and predicted output values, respectively. n is the number of samples.

In the study, each method is run 20 times and the best results are compared. 150 MVA 330/161-kV PST is designed and manufactured by Balıkesir Electromechanical Industry Facilities (BEST) in Turkey. These PST tests were done in real-time. BEST manufactured two units of 150 MVA 330/161 kV $\pm 8 \times 1.25\%$ tap with $\pm 6°$ phase shifting winding. To control a range of 188–137 kV voltage at the LV side and $\pm 6°$ angle, with an additional phase angle of $+120°$, $-60°$ to make that much of a position and power flow change in a transformer; MR tap changers are used. To change the voltage from 177 kV to 145 kV; three single-column VRC1001 OLTCs are used. VRF1000 type 3 phase 1 head OLTC is used to change $\pm 6°$ change and lastly, COMTAP ARS1000 is used to change phase angle 180°;

Table 2.5 Error-values of the testing phase

	FANN		PSO-FANN	
	RMSE	**MAPE**	**RMSE**	**MAPE**
Training phase	0.0051	0.0026	0.0029	0.0017
Testing phase	0.0095	0.0048	0.0038	0.0021

from $\pm 120°$ to $\pm 60°$. The error results obtained after the completion of the training phase of the network and the error results obtained at the end of the testing phase are given in Table 2.5.

Figure 2.16 shows the results obtained from the methods at the end of the testing phase. Figure 2.16(a) and (b) shows the results obtained by the FANN method and the PSO-FANN hybrid estimation method, respectively.

When the results are analysed, it is concluded that the results obtained from the hybrid model are better than the other approach. Table 2.5 includes the results obtained at the end of the training phase and the results obtained when the test data is passed through the neural network, according to the MAPE and RMSE error criteria. According to both error criteria, it is seen that the results obtained with the PSO-FANN method are more successful. When the results in Figure 2.16 are examined, it is seen that the PSO-FANN hybrid estimation method used is more effective.

2.9 Conclusions

PSTs are frequently used in power flow control to increase transmission systems capacity, system reliability and stability. This chapter proposes a comprehensive review of phase-shifting transformers. However, harmonics in PST is an issue that is used rarely in the literature. The voltage regulation of the PST is insufficient.

High-power PST studies and applications are more common. The most frequently used connection types for the primary, secondary and tertiary windings of the PSTs are seen as delta, star and delta connection, respectively in literature. Commonly used types of different PST topologies are indirect, symmetrical, quadrature and dual-core PST devices. Among optimization methods; FPSO, genetic algorithm, Benders decomposition and DCOPF methods are used to overcome the optimal power flow (OPF) issue in PSTs. The impact of PSO and analytical optimization is similar. FPSO and DE methods are more efficient than the PSO algorithm.

MODE gives a better performance compared to the three algorithms: (NSGA-II), (SPEA2) and PDF. PSS/E, MATLAB, MATLAB/MOSEK, general algebraic modelling system (GAMS) software and Monte Carlo simulation are used as simulation software; Python, linear programming, nonlinear programming (NLP),

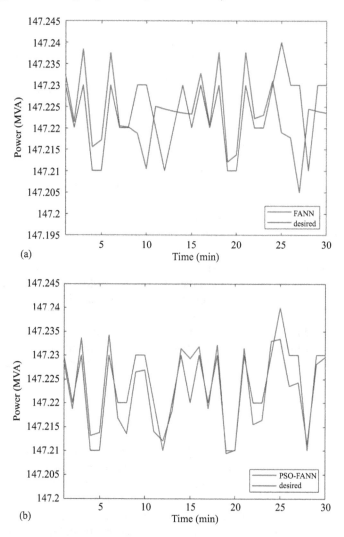

Figure 2.16   Testing phase results. (a) FANN and (b) PSO-FANN

MINLP and multi-objective mathematical programming are used as programming languages in PST optimization algorithms.

A 10-stage 50 MVA 380-V/154-kV OLTC transformer is used for the dynamic simulation. The effect of the OLTC transformer is examined to prevent power instability on the grid under the suddenly changing load condition. With the tap change feature of the transformer, uninterrupted and safe power flow is ensured without disturbing the voltage stability during the transmission of the electricity. Moreover, 150 MVA 330/161-kV PST is designed and manufactured by Balıkesir Electromechanical Industry Facilities (BEST) in Turkey. It is aimed to estimate the apparent power of this PST by using FANN and Particle Swarm Optimization-based FANN (PSO-FANN) hybrid algorithm. The results present that the PSO-FANN hybrid estimation method presents success to forecast the apparent power of the PST.

With grid problems posed by the increasing use of renewable energy sources, the ability to control power flow is rapidly becoming vital. Therefore, smart grids emerge with renewable energy integration into electric grids. It seems that smart transformers with advantages such as improving power quality and energy efficiency, ensuring network reliability and power flow control, voltage regulation, power flow control, bidirectional power flow, fault current limiting, harmonic blocking and galvanic isolation will replace phase-shifting transformers. In future work, R&D studies will be started to design and manufacture OLTC transformers for smart grid-integrated renewable power plants. The efficiency of the OLTC transformer will be examined to prevent power instability on the grid under the sudden changing load condition.

Acknowledgements

The authors would like to acknowledge the Scientific Project Unit of Çukurova University (Project no. FYL-2016-7616) for full financial support. The authors would also like to acknowledge Best Transformer company due to its contribution.

References

[1] Molapo R., Mbuli N., and Ijumba N. 'Enhancement of the voltage stability and steady state performance of the cape corridor using phase shifting transformers'. *IEEE Africon'11*; 2011. pp. 1–6.

[2] Liu J., Hao X., Wang X., Chen Y., Fang W., and Niu S. 'Application of thyristor controlled phase shifting transformer excitation impedance switching control to suppress short-circuit fault current level'. *Journal of Modern Power Systems and Clean Energy*. 2018;6(4):821–832.

[3] Cano J.M., Mojumdar M.R.R., Norniella J.G., and Orcajo G.A. 'Phase shifting transformer model for direct approach power flow studies'. *International Journal of Electrical Power & Energy Systems*. 2017;91:71–79.

[4] Golub I.V., Tirshu M.S., Zaitcev D.A., and Kalinin L.P. 'Characteristics of the phase-shifting transformer realized according to the "Polygon" connection'. *Problemele Energeticii Regionale*. 2017;2017(3):1–8.

[5] Morrell T.J. and Eggebraaten J.G. 'Applications for phase-shifting transformers in rural power systems'. In *IEEE Rural Electric Power Conference (REPC)*; 2019. pp. 70–74.

[6] Yeo J.H., Dehghanian P., and Overbye T. 'Power flow consideration of impedance correction for phase shifting transformers'. In *IEEE Texas Power and Energy Conference (TPEC)*; 2019. pp. 1–6.

[7]   Kochergin A.V., Loktionov S.V., Sharov A.N., Bulatov R.V., and Loktionov G.S. 'The study of phase shifting transformer control angle error due to the change of its reactance'. In *International Youth Conference on Radio Electronics, Electrical and Power Engineering (REEPE)*; 2020. pp.1–4.

[8]   El Hraïech A., Ben-Kilani K., and Elleuch M. 'Control of parallel EHV inter-connection lines using phase shifting transformers'. In *IEEE 11th International Multi-Conference on Systems, Signals & Devices (SSD14)*; 2014. pp. 1–7.

[9]   Yi C., Xiaohui F., Ke Z., and Jian S. 'Transient stability model and pre-ventive control based on phase shifting transformer in power system'. In *34th Chinese Control Conference (CCC)*; 2015. pp. 8926–8930.

[10] Warichet J., Leonard J.L., Rimez J., Bronckart O., and Van Hecke J. 'Grid implementation and operational use of large phase shifting transformers'. *CIGRE Session*, Location: Paris. 2010.

[11] Brilinskii A.S., Evdokunin G.A., Mingazov R.I., Petrov N.N., and Chudnyi V.S. 'Joint regulation of power flow and short circuit current limitations with the aid of a phase-shifting transformer'. *Power Technology and Engineering*. 2018;51(5):584–592.

[12] Brilinskii A.S., Badura M.A., Evdokunin G.A., Chudny V.S., and Mingazov R.I. 'Phase-shifting transformer application for dynamic stability enhance-ment of electric power stations generators'. In *Conference of Russian Young Researchers in Electrical and Electronic Engineering (EIConRus)*; 2020. pp. 1176–1178.

[13] Soliman I.A., El-Ghany H.A.A, and Azmy M.A. 'A robust differential pro-tection technique for single core delta-hexagonal phase-shifting transfor-mers'. *International Journal of Electrical Power & Energy Systems*. 2019;109:207–216.

[14] Zeng J., He Y., Lan Z., Yi Z., and Liu J. 'Optimal control of DAB converter backflow power based on phase-shifting strategy'. *Soft Computing*. 2020; 24(8):6031–6038.

[15] Soliman I.A, H.A. El-Ghany H.A.A., and Azmy M.A. 'A proposed algo-rithm for current differential protection of delta hexagonal phase shifting transformer'. In *Twentieth International Middle East Power Systems Conference (MEPCON)*; 2018. pp. 785–790.

[16] Mohamed S.A. and Abdel-Rahim, A.M.M. 'Phase-shifting transformers' model for optimal load flow formulation'. In *Twentieth International Middle East Power Systems Conference (MEPCON)*; 2018. pp. 589–594.

[17] Ding T., Bo R., Bie Z., and Wang X. 'Optimal selection of phase shifting transformer adjustment in optimal power flow'. *IEEE Transactions on Power Systems*. 2016;32(3):2464–2465.

[18] Elamari K. and Lopes L.A.C. 'Comparison of phase shifting transformer and unified power flow control based interphase power controllers'. In *IEEE Electrical Power and Energy Conference (EPEC)*; 2016. pp. 1–6.

[19] Nishida K., Ahmed T., and Nakaoka M. 'A cost-effective high-efficiency power conditioner with simple MPPT control algorithm for wind-power grid integra-tion'. *IEEE Transactions on Industry Applications*. 2010;47(2):893–900.

[20] Jimenez J.C. and Nwankpa C.O. 'Circuit model of a phase-shifting trans-former for analog power flow emulation'. In *IEEE International Symposium of Circuits and Systems (ISCAS)*; 2011. pp. 1864–1867.

[21] Ara L.A., Kazemi A., and Niaki S.N. 'Modelling of Optimal Unified Power Flow Controller (OUPFC) for optimal steady-state performance of power systems'. *Energy Conversion and Management*. 2011;52(2):1325–1333.

[22] Basu M. 'Multi-objective optimal power flow with FACTS devices'. *Energy Conversion and Management*. 2012;52(2):903–910.

[23] Jiang X., Wang T., Wu X., and Gao W. 'A 12-pulse converter featuring a rotating magnetic field transformer'. In *Proceedings of International Conference on Modelling, Identification and Control*; 2012. pp. 1237–1241.

[24] Rahul R., Jain A.K., and Bhide R. 'Analysis of variable frequency trans-former used in power transfer between asynchronous grids'. In *IEEE International Conference on Power Electronics, Drives and Energy Systems (PEDES)*; 2012. pp. 1–5.

[25] Ghorbani A., Mozaffari B., and Ranjbar A.M. 'Application of sub-synchronous damping controller (SSDC) to STATCOM'. *International Journal of Electrical Power & Energy Systems*. 2012;43(1):418–426.

[26] Vanajaa V.R. and Vasanthi N.A. 'Conceptual study and operational over-view on variable frequency transformer used for grid interconnections'. In *Third International Conference on Computing, Communication and Networking Technologies (ICCCNT'12)*; 2012. pp. 1–7.

[27] Tiejun W., Fang F., Xiaoyi J., and Xiang R. 'Phase-shifting transformers of round shape used for multi-module inverters'. In *IEEE International Symposium on Industrial Electronics*; 2012. pp. 458–463.

[28] Gu C., Zheng Z., Li Y., Ma H., and Gao Z. 'Power balancing control of a multilevel converter using high-frequency multi-winding transformer'. In *Proceedings of the 7th International Power Electronics and Motion Control Conference*; 2012. pp. 1866–1870.

[29] Khan U. and Sidhu T.S. 'A comparative performance analysis of three dif-ferential current measuring principles associated with phase shifting trans-former protection'. In *12th International Conference on Environment and Electrical Engineering*; 2013. pp. 297–302.

[30] Ahmed T., Nishida K., and Nakaoka M. 'A new scheme of full-power con-verter used for grid integration of variable-speed wind turbines'. In *IEEE Energy Conversion Congress and Exposition*; 2013. pp. 3926–3933.

[31] Farmad M., Farhangi S., Gharehpetian G.B., and Afsharnia S. 'Nonlinear controller design for IPC using feedback linearization method'. *International Journal of Electrical Power & Energy Systems*. 2013;44(1):778–785.

[32] Nittala R., Parimi M.A., and Rao K.U. 'A new approach for replenishing DC link energy for interline dynamic voltage restorer to mitigate voltage sag'. In *5th International Conference on Intelligent and Advanced Systems (ICIAS)*; 2014. pp. 1–6.

[33] Vincent P. and Subathra M.S.P. 'A novel approach to wind power grid system with simple MPPT control using MATLAB/Simulink'. In

International Conference on Green Computing Communication and Electrical Engineering (ICGCCEE); 2014. pp. 1–5.

[34]   Ogahara R., Kawaura Y., Ando S., Iwamoto S., Namba M., and Kamikawa N. 'A smart power flow adjustment method with phase shifting transformers for long term generation outages'. In *IEEE PES Asia-Pacific Power and Energy Engineering Conference (APPEEC)*; 2014. pp. 1–6.

[35]   Carvajal G.M., Plata G.O., Picon W.G., and Velasco J.C.C. 'Investigation of phase shifting transformers in distribution systems for harmonics mitigation'. In *Clemson University Power Systems Conference*; 2014. pp. 1–5.

[36]   Sebaa K., Bouhedda M., Tlemcani A., and Henini N. 'Location and tuning of TCPSTs and SVCs based on optimal power flow and an improved cross-entropy approach'. *International Journal of Electrical Power & Energy Systems*. 2014;54:536–545.

[37]   Rao R.S. and Rao V.S. 'A generalized approach for determination of optimal location and performance analysis of FACTs devices'. *International Journal of Electrical Power & Energy Systems*. 2015;73:711–724.

[38]   Xiaonan Y., Hongkun C., Xiaochun Z., *et al.* 'Modeling and control of thyristor controlled phase shifting transformer'. In *IEEE Innovative Smart Grid Technologies-Asia (ISGT ASIA)*; 2015. pp. 1–6.

[39]   Wang T., Wu X., Rao X., and Fang F. 'A novel strategy for composing staircase waveform in multi-module inverters'. In *International Conference on Electrical and Control Engineering*, 2011. pp. 963–966.

[40]   Yao J., Liu F., and Gong J., Li. 'A novel partial units energy feedback cascaded multilevel inverter with bypass control'. In *1st International Future Energy Electronics Conference (IFEEC)*; 2013. pp. 494–499.

[41]   Du L. and He J. 'A simple autonomous phase-shifting PWM approach for series-connected multi-converter harmonic mitigation'. *IEEE Transactions on Power Electronics*. 2019;34(12):11516–11520.

[42]   Shih D.C., Hung L.C., and Young C.M. 'The harmonic elimination strategy for a 24-pulse converter with unequal-impedance phase-shift transformers'. In *1st International Future Energy Electronics Conference (IFEEC)*. 2013. pp. 783–788.

[43]   Garcia-Vite P.M, Mancilla-David F, and Ramirez J.M. 'Dynamic modeling and control of an AC-link dynamic voltage restorer'. In *IEEE International Symposium on Industrial Electronics*; 2011. pp. 1615–1620.

[44]   Häger U., Görner K., and Rehtanz C. 'Hardware model of a dynamic power flow controller'. In *IEEE Trondheim PowerTech*; 2011. pp. 1–6.

[45]   Wang F. and Jiang J. 'A novel static frequency converter based on multilevel cascaded H-bridge used for the startup of synchronous motor in pumped-storage power station'. *Energy Conversion and Management*. 2011; 52(5):2085–2091.

[46]   Jeyaraj S.G., Milne R., and Mitchell G. 'On-site testing of special transformers'. In *10th IET International Conference on AC and DC Power Transmission*; 2012. pp. 1–6.

[47]   Page S., Johnson B., Smith M., *et al.* 'Tap changer for an asymmetrical phase shifting transformer'. *In North American Power Symposium (NAPS)*; 2012. pp. 1–5.

[48]   Meng F., Guo Y., and Gao L. 'Double-star uncontrolled rectifier based on power electronic transformer'. In *22nd International Conference on Electrical Machines and Systems (ICEMS)*; 2019. pp. 1–5.

[49]   Sokol E., Zamaruiev V., Ivakhno V., Voitovych Y., Butova O., and Makarov V. '18-Pulse rectifier with electronic phase shifting and pulse width modulation'. In *IEEE 3rd International Conference on Intelligent Energy and Power Systems (IEPS)*; 2018. pp. 290–294.

[50]   Meng F., Man Z., and Gao Z. 'A 12-pulse rectifier based on power electronic phase-shifting transformer'. In *IEEE International Power Electronics and Application Conference and Exposition (PEAC)*; 2018. pp. 1–4.

[51]   Meng F., Jiang T., and Gao L. 'A series-connected 12-pulse rectifier based on power electronic phase-shifting transformer'. In *22nd International Conference on Electrical Machines and Systems (ICEMS)*; 2019. pp. 1–5.

[52]   Tran T., Luo L., Xu J., Nguyen M., Zhang Z., and Dong S. 'A new phase shifting transformer for multipulse rectifiers'. In *Asia-Pacific Power and Energy Engineering Conference*; 2012. pp. 1–6.

[53]   Wang J., Wu B., Xu D., and Zargari N.R. 'Phase-shifting-transformer-fed multimodular matrix converter operated by a new modulation strategy'. *IEEE Transactions on Industrial Electronics*. 2012;60(10):4329–4338.

[54]   Prakash P.S., Kalpana R., Singh B., and Bhuvaneswari G. 'High-efficiency improved 12kW switched mode telecom rectifier'. In *IEEE International WIE Conference on Electrical and Computer Engineering (WIECON-ECE)*; 2015. pp. 382–386.

[55]   Ugale R.T., Mejari K.D., and Chaudhari B.N. 'Analytical and FEM design of autotransformer with phase shifting capability by intermediate voltage variation'. In *XXII International Conference on Electrical Machines (ICEM)*, vol. 201. pp. 1345–1351.

[56]   Liang X., El-Kadri A., Stevens J., and Adedun R. 'Frequency response analysis for phase-shifting transformers in oil field facilities'. *IEEE Transactions on Industry Applications*. 2013;50(4):2861–2870.

[57]   Abdelsalam I., Adam G.P., Holliday D., and Williams B.W. 'Assessment of a wind energy conversion system based on a six-phase permanent magnet synchronous generator with a twelve-pulse PWM current source converter'. In *IEEE ECCE Asia Downunder*; 2013. pp. 849–854.

[58]   Xu H., Zhao J., and Ma Y. 'Design and application of linear phase-shifting transformer'. *Journal of Physics: Conference Series*. 2020;1549:052101. doi: 10.1088/1742-6596/1549/5/052101

[59]   Guo G., Zhao J., Xiong Y., and Wu M. 'Analytical calculation of open-circuit magnetic field in linear phase-shifting transformer based on exact subdomain model'. *IEEE Transactions on Electrical and Electronic Engineering*. 2021;17(1):72–81.

[60] Guo G., Zhao J., Wu M., and Xiong Y. 'Analytical calculation of magnetic field in fractional-slot windings linear phase-shifting transformer based on exact subdomain model'. *IEEE Access.* 2021;(9):122351–122361.

[61] Kling W.L. 'Phase shifting transformers installed in the Netherlands in order to increase available international transmission capacity'. *CIGRE Paris Symposium*; 2004.

[62] Verboomen J., Van Hertem D., Schavemaker P.H., Kling W.L., and Belmans R. 'Phase shifting transformers: principles and applications'. In *International Conference on Future Power Systems*; 2005. p. 6.

[63] Hurlet P., Riboud J.C., Margoloff J., and Tanguy A. 'French experience in phase-shifting transformers'. *CIGRÉ SC A2-204, Session.* 2006.

[64] Rimez J., Van Der Planken R., Wiot D., Claessens G., Jottrand E., and Declercq J. 'Grid implementation of a 400 MVA 220/150 kV −15°/+3° phase shifting transformer for power flow control in the Belgian network: specification and operational considerations'. *Cigré Sessions* 2006. Paris: Cigré. 2006.

[65] Van Hertem D., Rimez J., and Belmans R. 'Power flow controlling devices as a smart and independent grid investment for flexible grid operations: Belgian case study'. *IEEE Transactions on Smart Grid.* 2013;4(3):1656–1664.

[66] Contreras-Aguilar L., García N., Islas-Martínez M.A., and Adame-Ortiz R. 'Implementation of a VFT model in PSS/E suitable for power flow and transient stability simulations. In *IEEE Power and Energy Society General Meeting*; 2012. pp. 1–8.

[67] Brochu J., Sirois F., Beauregard F., Cloutier R., Bergeron A., and Henderson M.I. 'Innovative applications of phase-shifting transformers supplemented with series reactive elements'. CIGRÉ SC A2-203, Session. 2006.

[68] Vita A.L.N., Bastos G.M., Neto F.C., and Mendes J.C. 'Phase shifting transformer on power control – a brazilian experience'. *CIGRE, A2_202, Session.* 2014.

[69] Carlini E.M., Manduzio G., and Bonmann D. 'Power flow control on the Italian network by means of phase-shifting transformers'. *CIGRE Session*, A2–206. 2006.

[70] Iuliani V., Di Giulio A., Palone F., *et al.* 'New phase shifting transformers in the Italian transmission network'. Design, manufacturing, testing and electromagnetic transients modelling. *CIGRE, A2_207.* 2014.

[71] Sidea D., Eremia M., Toma L., and Bulac C. 'Optimal placement of phase-shifting transformer for active power flow control using genetic algorithms'. *University Polithenic of Buchareste Scientific Bulletin, Series C.* 2018;80 (1):205–216.

[72] Wolfram M., Marten A.K., and Westermann D. 'A comparative study of evolutionary algorithms for phase shifting transformer setting optimization'. In *IEEE International Energy Conference (ENERGYCON)*; 2016. pp. 1–6.

[73] Ptacek J., Modlitba P., Vnoucek S., and Cermak J. 'Possibilities of applying phase shifting transformers in the electric power system of the Czech Republic'. *CIGRE Session*, C2–203. 2006.

[74] Lubıckı W., Kocot H., Korab R., Przygrodzkı M., Tomasık G., and Żmuda K. 'Improving the cross-border transmission capacity of Polish power system by using phase shifting transformers'. *CIGRE, C1_108, Session*. 2014.

[75] Sundar S.V.N., Yuvaraju A., Radhakrishna C., Kumar M., and Sachdeva S. 'Design, testing, commissioning and operational experience of first phase shifting transformer in Indian network'. *CIGRE, A2_205, Session*. 2014.

[76] Zhang L., Kokkinakis M., and Chong B.V.P. 'Generalised predictive control for a 12-bus network using a neutral-point clamped voltage-source-converter UPFC'. In *7th IET International Conference on Power Electronics, Machines and Drives*; 2014. pp. 1–6.

[77] Pragale R., Dionise T.J., and Shipp D.D. 'Harmonic analysis and multistage filter design for a large bleach production facility'. *IEEE Transactions on Industry Applications*. 2011;47(3):1201–1209.

[78] Makhathini D., Mbuli N., Sithole S., and Pretorius J.H.C. 'Enhancing the utilization of the matlala and glencowie 22kV radial feeders by inter-connecting them using a phase shifting transformer'. In *11th International Conference on Environment and Electrical Engineering*; 2012. pp. 844–848.

[79] Martínez-Anido C.B., L'Abbate A., Migliavacca G., *et al*. 'Effects of North-African electricity import on the European and the Italian power systems: a techno-economic analysis'. *Electric Power Systems Research*. 2013; (96):119–132.

[80] Sithole S., Mbuli N., and Pretorius J.H.C. 'Improvement of the Ulco network voltage regulation using a phase shifting transformer'. In *IEEE Africon'11*;2011. pp. 1–6.

[81] Johansson N., Angquist L., and Nee H.P. 'Preliminary design of power controller devices using the power-flow control and the ideal phase-shifter methods'. *IEEE Transactions on Power Delivery*. 2012;27(3):1268–1275.

[82] Gajic Z. 'Use of standard 87T differential protection for special three-phase power transformers—Part I: theory'. *IEEE Transactions on Power Delivery*. 2012;27(3):1035–1040.

[83] IEEE Special Publication, 'Protection of phase angle regulating transformers (PAR)'. A report to the substation subcommittee of the IEEE power system relaying committee prepared by working group K1. 1999.

[84] Gajić Z., Podboj M., Traven B., and Krašovec A. 'When existing recom-mendations for PST protection can let you down'. In *11th IET International Conference on Developments in Power Systems Protection*; 2012. pp.1–6.

[85] Fujita H., Baker D.H., Ihara S., Larsen E.V., and Price W.W. 'Power flow controller using rotary phase-shifting transformers'. In *CIGRE Session*; 2000. pp. 37–102.

[86] Sweeney R., O'Donoghue P., Gaffney P., and Stewart G. 'The specification and control of the phase shifting transformers for the enhanced inter-connection between Northern Ireland and the Republic of Ireland'. In *CIGRE*; 2002. pp. 14–118.

[87]   Verboomen J., Spaan F.J., Schavemaker P.H., and Kling W.L. 'Method for calculating total transfer capacity by optimising phase shifting transformer settings'. In *CIGRE*; 2008. pp. 1.

[88]   Ugalde-Loo C.E., Ekanayake J.B., and Jenkins N. 'Subsynchronous resonance on series compensated transmission lines with quadrature boosters'. In *IEEE Trondheim PowerTech*; 2011. pp. 1–7.

[89]   Wirmond V.E., Fernandes T.S., and Tortelli O.L. 'TCPST allocation using optimal power flow and genetic algorithms'. *International Journal of Electrical Power & Energy Systems*. 2011;33(4):880–886.

[90]   Lan Z., Li C., Li Y., Zhu C., Wang C., and Yang Q. 'The investigation on a novel IGCT-based cascade high voltage large capacity converter'. In *International Conference on Electrical Machines and Systems*; 2011. pp. 1–4.

[91]   Siddiqui A.S., Khan S., Ahsan S., and Khan M.I. 'Application of phase shifting transformer in Indian Network'. In *International Conference on Green Technologies (ICGT)*; 2012. pp. 186–191.

[92]   Pragale R. and Shipp D.D. 'Investigation of premature ESP failures and oil field harmonic analysis'. In *Petroleum and Chemical Industry Conference (PCIC)*; 2012. pp. 1–8.

[93]   Ara A.L., Kazemi A., and Niaki S.N. 'Multiobjective optimal location of FACTS shunt-series controllers for power system operation planning.' *IEEE Transactions on Power Delivery*. 2011;27(2):481–490.

[94]   Purwoharjono P., Penangsang O., Abdillah M., and Soeprijanto A. 'Optimal design of TCPST using gravitational search algorithm'. In *Sixth UKSim/AMSS European Symposium on Computer Modeling and Simulation*; 2012. pp. 323–328.

[95]   Ergun H., Van Hertem D., and Belmans R. 'Identification of power injection capabilities for transmission system investment optimization.' In *IEEE Grenoble Conference*; 2013. pp. 1–6.

[96]   Khan U. and Sidhu T.S. 'New algorithm for the protection of delta-hexagonal phase shifting transformer'. *IET Generation, Transmission & Distribution*. 2014;8(1):178–186.

[97]   Bocovich M., Iyer K., Terhaar R.M., and Mohan N. 'Overview of series connected flexible AC transmission systems (FACTS)'. In *North American Power Symposium (NAPS)*; 2013. pp. 1–6.

[98]   Constantin C., Eremia M., and Toma L. 'Power flow control solutions in the Romanian power system under high wind generation conditions'. In *IEEE Grenoble Conference*; 2013. pp. 1–6.

[99]   Khan U.N. and Sidhu T.S. 'A phase-shifting transformer protection technique based on directional comparison approach'. *IEEE Transactions on Power Delivery*. 2014;29(5):2315–2323.

[100]   Solak K., Rebizant W., and Schiel L. 'EMTP testing of selected PST protection schemes'. In *Proceedings of the 2014 15th International Scientific Conference on Electric Power Engineering (EPE)*; 2014. pp. 85–90.

[101]   Bhasker S.K., Tripathy M., and Kumar V. 'Indirect symmetrical PST protection based on phase angle shift and optimal radial basis function neural

network'. In *Eighteenth National Power Systems Conference (NPSC)*; 2014. pp. 1–6.

[102] Mohsin Q.K., Lin X., Wang Z., Sunday O., Khalid M.S., and Zheng P. 'Iraq network 400kV, 50Hz interconnect with Iran, Turkey and Syria using phase-shifting transformers in control and limit power flow of countries'. In *IEEE PES Asia-Pacific Power and Energy Engineering Conference (APPEEC)*; 2014. pp. 1–6.

[103] Linortner G. 'Phase shifting transformers, specific issues with regard to design and testing'. In *ICHVE International Conference on High Voltage Engineering and Application*; 2014. pp. 1–4.

[104] Bhasker S.K., Tripathy M., and Kumar V. 'Wavelet transform based discrimination between inrush and internal fault of indirect symmetrical phase shift transformer'. In *IEEE PES General Meeting/ Conference & Exposition*; 2014. pp. 1–5.

[105] Cui Y., Yu Y., and Yang Z. 'Application effect evaluation of demonstration project of thyristor-controlled phase shifting transformer in 500kV grid'. In *5th International Conference on Electric Utility Deregulation and Restructuring and Power Technologies (DRPT)*, 2015. pp. 43–1647.

[106] Solak K., Rebizant W., and Schiel L. 'Differential protection of single-core symmetrical phase shifting transformers'. In *16th International Scientific Conference on Electric Power Engineering (EPE)*; 2015. pp. 221–226.

[107] Colla L., Iuliani V., Palone F., Rebolini M., and Zunino S. 'Modeling and electromagnetic transients study of two 1800MVA phase shifting transformers in the Italian transmission network'. In *Proceedings of International Conference on Power System Transients (IPST), Delft*; 2011.

[108] Mănescu L.G., Ruşinaru G., Ciontu M., Buzatu C., Dinu R., and Stroică P. 'Congestion management using dispatch or phase shifting transformers'. In *International Conference on Applied and Theoretical Electricity (ICATE)*; 2016. pp. 1–6.

[109] Kawaura Y., Yamanouchi S., Ichihara M., Iwamoto S., Suetsugu Y., and Higashitani T. 'Phase-shifting transformer application to power-flow adjustment for large-scale PV penetration'. In *IEEE Region 10 Conference (TENCON)*; 2016. pp. 3328–3331.

[110] Smolovik S.V., Brilinskiy A.S., Chudny V.S., Mingazov, R.I., and Petrov N.N. 'Phase-shifting transformer as short-circuit current-limiting device'. In *IEEE Conference of Russian Young Researchers in Electrical and Electronic Engineering (EIConRus)*; 2017. pp. 1585–1589.

[111] Sidea D.O., Toma L., and Eremia M. 'Sizing a phase shifting transformer for congestion management in high wind generation areas'. In *IEEE Manchester PowerTech*; 2017. pp. 1–6.

[112] Pordanjani I.R., Qureshi F., Jafari A., and Cui R. 'Utilizing phase shifting transformer for effective integration of wind generation'. In *IEEE Electrical Power and Energy Conference (EPEC)*; 2017. pp. 1–7.

[113] Cai N., Khatib A.R., Saenz A., and Bottyan J. 'Real-time automation control of a phase-shifting transformer based on mission priorities'. In

IEEE/PES Transmission and Distribution Conference and Exposition (T&D); 2018. pp. 1–9.

[114]   Sakallıoğlu B., Esenboğa B., Demirdelen T., and Tümay M. 'Performance evaluation of phase-shifting transformer for integration of renewable energy sources'. *Electrical Engineering*. 2020;102(4):2025–2039.

[115]   Nittala R., Parimi A.M., and Rao K.U. 'Phase shifting transformer based interline dynamic voltage restorer to mitigate voltage sag.' In *Annual IEEE India Conference (INDICON)*; 2013. pp. 1–6.

[116]   Seitlinger W. 'Phase shifting transformers. "Discussion of specific characteristics"'. *CIGRE, 12_306, Session*. 1998.

[117]   Bauer R., Weidner J., and Salehinajafabadi S. 'Control strategies of phase-shifting transformers in long-term network development'. In *International ETG Congress 2015; Die Energiewende-Blueprints for the New Energy Age*; 2015. pp. 1–7.

[118]   Liang G., Wang L., Gao F., and Liu X. 'A new maximum step voltage calculation method of on-load tap-changer for symmetrical two-core phase-shifting transformer'. *IEEE Transactions on Power Delivery*. 2018;33 (6):2718–2725.

[119]   Bednarczyk T., Szablicki M., Halinka A., Rzepka P., and Sowa P. 'Phase shifting transformer electromagnetic model dedicated for power system protection testing in a transient condition'. *Energies*. 2021;14:627.

[120]   Zhang N., Zhu X., and Liu J. 'Improving the consumption capacity of wind power in distributed network using a thyristor controlled phase shifting transformer'. In *3rd Asia Energy and Electrical Engineering Symposium (AEEES)*; 2021. pp. 80–84.

[121]   Cook B., Thompson M.J., Garg K., and Malichkar M. 'Phase-shifting transformer control and protection settings verification'. In *71st Annual Conference for Protective Relay Engineers (CPRE)*; 2018. pp. 1–15.

[122]   Wang T., Fang F., Jiang X., and Yang L. 'Research on twelve-phase round-shaped transformers applied in rectifier systems'. In *IEEE International Symposium on Electromagnetic Compatibility (EMC)*; 2015. pp. 1392–1395.

[123]   Ba A.O., Peng T., and Lefebvre S. 'Rotary power-flow controller for dynamic performance evaluation—Part I: RPFC modeling'. *IEEE Transactions on Power Delivery*. 2009;24(3):1406–1416.

[124]   Wen J., Qin H., Wang S., and Zhou B. 'Basic connections and strategies of isolated phase-shifting transformers for multipulse rectifiers: a review'. In *Asia-Pacific Symposium on Electromagnetic Compatibility*, 2012. pp. 105–108.

[125]   Dong L., Hao Z., Liu Z., and Zou H. 'Research on differential protection algorithm about special phase-shifting angle rectifier transformer'. In *IEEE 15th International Conference on Environment and Electrical Engineering (EEEIC)*; 2015. pp. 1765–1769.

[126]   Bharathi R. and Rajan C.C.A. 'An advanced FACTS controller for power flow management in transmission system using IPFC'. In *International*

Conference on Process Automation, Control and Computing; 2011. pp. 1–6.

[127] Ghorbani A. and Arablu M. 'Ground distance relay compensation in the presence of delta-hexagonal phase shifting transformer'. *IET Generation, Transmission & Distribution*. 2015;9(15):2091–2098.

[128] Singh P.K. and Chauhan Y.K. 'Performance analysis of multi-pulse electronic load controllers for self-excited induction generator'. In *International Conference on Energy Efficient Technologies for Sustainability*; 2013. pp. 1299–1307.

[129] Gaonkar A.D., Rane R.G., Wagh S.R., and Singh N.M. 'Multi-winding phase-shifting transformer for 36-pulse rectifier: winding turns design and analysis'. In *North American Power Symposium (NAPS)*; 2016. pp. 1–6.

[130] Ara A.L., Kazemi A., and Niaki S.N. 'Multiobjective optimal location of FACTS shunt-series controllers for power system operation planning'. *IEEE Transactions on Power Delivery*. 2011;27(2):481–490.

[131] Ghorbani A., Mozafari B., and Ranjbar A.M. 'Digital distance protection of transmission lines in the presence of SSSC'. *International Journal of Electrical Power & Energy Systems*. 2012;43(1):712–719.

[132] Gill P.E., Murray W., and Wright M.H. *Practical Optimization*. Philadelphia, PA: Society for Industrial and Applied Mathematic. 2019.

[133] Swaroopan N.J. and Somasundaram P. 'Optimal power flow for security enhancement using fuzzified particle swarm optimization'. In *International Conference on Sustainable Energy and Intelligent Systems (SEISCON 2011)*; 2011. pp. 474–479.

[134] Tiwari R., Niazi K.R., and Gupta V. 'Optimal location of FACTS devices for improving performance of the power systems.' In *IEEE Power and Energy Society General Meeting*; 2012. pp.1–8.

[135] Ghahremani E. and Kamwa I. 'Analysing the effects of different types of FACTS devices on the steady-state performance of the Hydro-Québec network'. *IET Generation, Transmission & Distribution*. 2014;8(2):233–249.

[136] Ogahara R., Kawaura Y., and Iwamoto S. 'Using phase shifters for power flow adjustment following large-scale generation loss.' In *IEEE Power & Energy Society General Meeting*; 2013. pp. 1–5.

[137] Dag G.O. and Bagriyanik M. 'A fuzzy-genetic algorithm based approach for controlling unscheduled flows using SC and PST.' In *North American Power Symposium*; 2011. pp.1–5.

[138] Rasolomampionona D. and Anwar S. 'Interaction between phase shifting transformers installed in the tie-lines of interconnected power systems and automatic frequency controllers'. *International Journal of Electrical Power & Energy Systems*. 2011;33(8):1351–1360.

[139] Nagalakshmi S. and Kamaraj N. 'Loadability enhancement for pool model with FACTS devices in transmission system using differential evolution and particle swarm optimization'. In *India International Conference on Power Electronics 2010 (IICPE2010)*; 2011. pp. 1–8.

[140] Mekonnen M.T. and Belmans R. 'The influence of phase shifting transformers on the results of flow-based market coupling'. In *9th International Conference on the European Energy Market*; 2012. pp. 1–7.
[141] Laka A., Barrena J.A., Chivite-Zabalza J., and Rodríguez M.A. 'Analysis and improved operation of a PEBB-based voltage-source converter for FACTS applications'. *IEEE Transactions on Power Delivery*. 2013;28 (3):1330–1338.
[142] Martínez-Lacañina P.J., Marcolini A.M., and Martínez-Ramos J.L. 'DCOPF contingency analysis including phase shifting transformers'. In *Power Systems Computation Conference*; 2014. pp. 1–7.
[143] Thakurta P.G., Van Hertem D., and Belmans R. 'An approach for managing switchings of controllable devices in the Benelux to integrate more renewable sources'. In *IEEE Trondheim PowerTech*; 2011. pp. 1–7.
[144] Hui H., Yu C.N., Surendran R., Gao F., and Moorty S. 'Wind generation scheduling and coordination in ERCOT Nodal market'. In *IEEE Power and Energy Society General Meeting*; 2012. pp. 1–8.
[145] Müller S.C., Osthues M., Rekowski C., Häger U., and Rehtanz C. 'Techno-economic evaluation of corrective actions for efficient attainment of (N- 1)-security in operation and planning'. In *IEEE Power & Energy Society General Meeting*; 2013. pp. 1–5.
[146] Belivanis M. and Bell K. 'Coordination of the settings of phase-shifting transformers to minimize the cost of generation re-dispatch'. In *CIGRE Session*; 2014.
[147] Tümay M., Demirdelen T., Bal S., Doğru B., Cicibaş A., and Köseoğlu A. K. 'Phase shifting transformers: a survey on control and optimization strategies'. In *4th International Conference on Advanced Technology & Sciences (ICAT'Rome)*; 2016. pp. 189–195.
[148] Singh P., Tiwari R., Sangwan V., and Gupta A.K. 'Optimal allocation of thyristor-controlled series capacitor (TCSC) and thyristor-controlled phase-shifting transformer (TCPST)'. In *International Conference on Power Electronics & IoT Applications in Renewable Energy and its Control (PARC)*, IEEE, Mathura, Uttar Pradesh, India; 2020. pp. 491–496.
[149] Dakhare R. and Chandrakar V.K. 'Congestion management by phase shifting transformer using fuzzy logic control'. In *6th International Conference for Convergence in Technology (I2CT)*; 2021. pp. 1–4.
[150] Korab R., Połomski M., and Owczarek R. 'Application of particle swarm optimization for optimal setting of phase shifting transformers to minimize unscheduled active power flows'. *Applied Soft Computing*. 2021;105:107243.
[151] Chitsazan M.A. and Trzynadlowski A.M. 'Harmonic mitigation in interphase power controller using passive filter-based phase shifting transformer'. In *IEEE Energy Conversion Congress and Exposition (ECCE)*; 2016. pp. 1–5.
[152] Sevov L., Zhang Z., Voloh I., and Cardenas J. 'Differential protection for power transformers with non-standard phase shifts'. In *64th Annual Conference for Protective Relay Engineers*; 2011. pp. 301–309.

[153]   Gajic Z. Use of standard 87T 'differential protection for special three-phase power transformers—Part I: theory'. *IEEE Transactions on Power Delivery*. 2012;27(3):1035–1040.

[154]   Abd el-Ghany H.A., Soliman I.A., and Azmy A.M. 'A reliable differential protection algorithm for delta hexagonal phase-shifting transformers'. *International Journal of Electrical Power & Energy Systems*. 2021;127:106671.

[155]   Kastawan I.M.W., Yusuf E., and Fadhilah A. Simulation of source current harmonic elimination technique using phase shifting transformer. In *IOP Conference Series: Materials Science and Engineering*; 2020;830:032028.

[156]   Liu J., Fang W., Duan C., Wei Z., Yang Z., and Cui Y. Fault current limiting by phase shifting angle control of TCPST. In *IEEE Power & Energy Society General Meeting*; 2015. pp.1–6.

[157]   Mojumdar M.R.R., Cano J.M., Assadi M., and Orcajo G.A. Consensus phase shifting transformer model. In *Power & Energy Society General Meeting (PESGM)*, IEEE, Montreal, QC; 2020. pp. 1–5.

[158]   Banik A., Shrivastava A., Potdar R.M., Jain S.K., Nagpure S.G., and Soni M. Design, modelling, and analysis of novel solar PV system using MATLAB. *Materials Today: Proceedings*. 2022;51:756–763.

[159]   Motahhir, S., Chtita, S., Chouder, A., and El Hammoumi, A. Enhanced energy output from a PV system under partial shaded conditions through Grey wolf optimizer. *Cleaner Engineering and Technology*. 2022;9:100533.

[160]   Chalh, A., El Hammoumi, A., Motahhir, S., *et al.* Investigation of partial shading scenarios on a photovoltaic array's characteristics. *Electronics*. 2021;11(1):96.

[161]   Kanagasakthivel B. and Devaraj D. 'Simulation and performance analysis of solar PV-wind hybrid energy system using MATLAB/Simulink'. In *International Conference on Computing and Communications Technologies (ICCCT)*; 2015. pp. 99–104.

[162]   Kennedy J. and Eberhart R. 'Particle swarm optimization'. In *Proceedings of IEEE International Conference on Neural Network, Perth, Australia, IEEE Service Center*, Piscataway, NJ; 1995. pp. 1942–1948.

Chapter 3

Design and practical implementation of a grid-connected single-stage flyback photovoltaic micro-inverter

Salam J Yaqoob[1], Adel A Obed[2], Anand Nayyar[3], Saad Motahhir[4] and Naseer T Alwan[5,6]

Grid-connected photovoltaic (PV) micro-inverters that do not require batteries are becoming increasingly popular in the market. The main issues with this type of inverter are its cost of components, such as the DC/DC converter unit, the number of sensors, and the number of switches. Accordingly, a single-stage flyback PV micro-inverter (FBPVMI) topology is considered an attractive solution for on-grid inverters. This topology requires an expensive digital control scheme to implement the synchronization stage with the grid. This chapter presents a low-cost and small-size control circuit for injecting an AC current from the inverter to the grid under various weather conditions. The proposed control strategy is based on some low-cost operational amplifiers integrated with an Arduino UNO board. The proposed control circuit is applied to the FBPVMI topology for a 120 W prototype in the laboratory. Consequently, the overall components of the suggested control circuit are verified using a Proteus simulation tool. The experimental results are obtained for the different irradiance and temperature values to validate the proposed control. Finally, the proposed topology with this control shows a good practical efficiency of 91% with excellent total harmonic distortion (THD) content of 3.7% in the output injected current. Finally, this chapter is suitable for digital technologies applications such as embedded system technology.

[1]Department of Research and Education, Authority of the Popular Crowd, Iraq
[2]Department of Electrical Engineering, Electrical Engineering Technical College, Middle Technical University, Iraq
[3]Graduate School, Faculty of Information Technology, Duy Tan University, Vietnam
[4]Ecole National des sciences appliquées, University SMBA, Morocco
[5]Ural Federal University, Russia
[6]Technical Engineering College of Kirkuk, Northern Technical University, Iraq

3.1   Introduction

Photovoltaic (PV) power systems have received great attention in recent decades due to the growing concerns in many countries across the world. These systems significantly contribute to the reduction of polluting emissions because they represent one of the renewable energy sources [1–4]. The grid-connected inverters that supply power from a PV system to an AC utility grid are becoming more practical due to the absence of battery costs. Furthermore, grid-connected inverters may be widely accepted by residential and commercial applications due to their high reliability and simple monitor system [5,6]. To ensure that the grid-connected inverter receives enough Harvested energy, the solar PV modules are arranged in a variety of ways that will affect the installation costs and inverter performance. The micro-inverter technology is a combination of a single PV module integrated with a single-phase interactive inverter [7–12]. Due to its many benefits over traditional inverters, this architecture is recognized as a common choice for grid-connected inverters for PV systems [13].

Several studies were presented to design and implement a grid-connected PV inverter. Kim *et al.* [14] presented an on-grid micro-inverter based on a flyback converter. The proposed inverter was used as a voltage sensor for implementing a maximum power point tracking method (MPPT). Only the PV module current has been found to perform an MPPT type with the perturbation and observation (P&O) method. Furthermore, small size and low-cost micro-inverter with an output power of 200 W was proposed.

Kim *et al.* [15] proposed a new robust control circuit of a grid-connected PV flyback inverter. This method was used for an active snubber circuit switch to minimize the voltage stress on an inverter's primary switch. Accordingly, the turn-off losses of the flyback inverter were reduced. The micro-inverter was verified by mathematical analysis and simulation results. Reference [16] presented a PV grid-connected micro-inverter with a high-efficiency digital control technique for an interleaved flyback inverter. Soft switching was used to operate this method, which was based on the synchronous rectifier that served as the main switch; consequently, switching losses can be decreased. The simulation results for the proposed inverter have been obtained by PSIM software.

Özturk and Çadirci [17] designed and implemented a flyback inverter. The proposed inverter was controlled by using a direct digital synthesis technique based on the dsPIC board to implement the MPPT method. The experimental results were obtained on a 120 W prototype system. This inverter was a suitable choice for AC PV module applications. Zhang *et al.* [18] presented a grid-connected inverter based on two flyback converters connected in parallel to enhance the transferred power from the inverter to the grid. Two open-loop phase synchronization methods were used for synchronizing this inverter with the grid. The simulation results with an experimental 200 W prototype were obtained to validate the performance of the proposed inverter.

Reddy *et al.* [19] presented a two-stage single-phase inverter that consisted of a DC/DC conversion stage and a push-pull converter operating at the grid frequency. Three semiconductor switches were used to build this inverter. The first switch was switched ON/OFF at a high switching frequency. Meanwhile, the other switches were triggered at a grid frequency. Analog devices were used for building the gate sequences of the switches. The proposed inverter was tested in the stand-alone and grid-connected modes.

Sher *et al.* [20] presented a single-stage flyback with a hybrid MPPT method. The flyback converter of this inverter was designed to work with a hybrid MPPT method that combines the P&O method and fractional short circuit current (FSCC). This MPPT method was simulated using co-simulation between Simulink and power simulation (PSIM) software. The results indicated enhanced power harvesting compared with the classical P&O technique. Reference [21] presented a single-stage flyback micro-inverter. An auxiliary snubber circuit was suggested to decrease the turn OFF losses across the primary switch. The proposed inverter depends on a soft switching condition for the primary flyback switch. Furthermore, the theoretical analysis, operating principle, and design circuit for the proposed inverter were discussed. The proposed inverter was verified through simulation.

Liao *et al.* [22] proposed a single-phase inverter that depended on a new decoupling power configuration that consisted of active and passive circuits to minimize the steady-state oscillation in the injected power. The required decoupling capacitance and switching losses were reduced to acceptable levels of 50 µF and 24%, respectively. Lodh and Agarwal [23] presented a single-phase inverter with a multi-port converter. This inverter could be used for grid-connected or stand-alone PV power applications. Based on a single-stage DC/AC conversion, the proposed inverter topology provided galvanic isolation between the PV module, the battery, and the load. This mechanism resulted in higher efficiency and increased battery lifetime with a small ripple in the output power. The simulation results of the proposed inverter were achieved through MATLAB®/Simulink®.

Kalilian and Guglilmi [24] proposed a single-stage grid-connected inverter using a zero-current switching (ZCS) capability for the primary device of the flyback. The main switch was protected from the spike voltage during the turn OFF time. The simulation results of the proposed inverter were achieved through PSIM.

Trujillo *et al.* [25] presented a two-stage grid-tied inverter. However, a push-pull DC–DC converter was used in the first stage of the inverter. Meanwhile, the second stage corresponds to the H-bridge inverter with a sinusoidal pulse width modulation (SPWM) was investigated as a current control scheme. Moreover, the inverter circuit was synchronized with the utility grid and delivers MPPT by the PV module. The micro-inverter was simulated with different solar irradiance levels and ambient temperatures to show the stability and transient response of the micro-inverter under various weather conditions. Sukatjasakul and Ngam [26] presented a two-stage PV micro-inverter with a simple MPPT controller. The MPPT method was simple to implement under changing solar irradiance by only adjusting the PV module voltage. The proposed inverter was built to make a sinusoidal current with minimal THD by using the phase-locked loop (PLL) synchronization method.

In summary, Table 3.1 shows and summarizes the commented literature works in terms of the inverter type or the main circuit structure, the main controller unit used to control the inverter's circuit, and the MPPT method used to obtain the MPP of the PV system.

Table 3.1   Comparison between the literature works

Reference	Inverter type	Main controller unit	MPPT technique
Kim *et al.* [14]	Single-stage flyback, 200 W	DSP TMS320-F28035	P&O
Kim *et al.* [15]	Interleaved flyback, 100 W	DSP TMS320-F28035	P&O
J. Jan *et al.* [16]	Interleaved flyback, 250 W	N.A	Modified P&O
Özturk and Çadirci [17]	Single-stage flyback, 120 W	dsPIC board	Constant voltage tracking
Zhang *et al.* [18]	Interleaved flyback, 200 W	FPGA EP3C10E	NA
Reddy *et al.* [19]	Front-end with Push-pull	Simple analog circuit	NA
Sher *et al.* [20]	Single-stage flyback, 200 W	NA	FCSS with P&O
Mojtaba *et al.* [21]	Soft-switching flyback, 100 W	NA	P&O
Liao *et al.* [22]	Single-phase active decoupling, 200 W	NA	NA
Lodh and Agarwal [23]	Multi-port flyback, 100 W	Set of logical gates	NA
Kalilian and Guglilmi [24]	Two-stage flyback, 100 W	NA	P&O
Trujillo *et al.* [25]	Two-stage interleaved flyback, 200 W	NA	NA
Sukatjasakul and Ngam [26]	Two-stage with H-bridge, 1,200 W	NA	Simple P&O

NA: not specified.

In this chapter, a single-stage flyback PV micro-inverter (FBPVMI) has been implemented to inject the pure sine wave current into the utility grid. Proteus software provides many electronic boards and microcontrollers in its library. Accordingly, the suggested inverter control was tested using Proteus software, which provides more accurate and close to real-time implementation software. Therefore, the overall components of the control circuit are simulated and tested in the Proteus software tool before implementation to verify the implementation parts.

The following is a list of the chapter's major contributions:

- Due to the implementation of effective micro-inverter architecture, various advantages have been realized, including the direct conversion of adequate DC power to AC power and the amplification of low DC input voltage levels to high voltage levels without the use of an extra DC/DC converter.
- A single-stage inverter with excellent reliability and minimal cost was suggested to provide the grid with a pure sinusoidal current.
- An inexpensive control circuit based on an embedded Arduino board was suggested.

The rest of this chapter was organized as follows: Section 3.2 reports the grid synchronization methods. Section 3.3 introduces the practical implementation procedure. Section 3.4 practical design of control scheme. Section 3.5 introduces the simulation and experimental results while Section 3.6 presents the conclusion and future works.

3.2 Grid synchronization methods

The major important function of the control scheme is to synchronize the inverter current with the voltage of the main grid. The output current and grid voltage should have the same phase angle and frequency value [14–16]. If any deviation in these parameters occurs, then the proposed PV inverter should be removed from the grid. Many methods for grid synchronizing are presented in references [17,18]. With these methods, the micro-inverter should be able to:

- Deliver power at PF within the limits of the standards;
- Reduce the harmonics current content (THD < 5%);
- Minimize the grid connection transients.

The two methods are discussed in the following sections.

3.2.1 Zero cross detection

This technique finds the grid voltage's zero crossing. Next, the phase angle is determined. The technique is simple to use and does not require many calculations [19,20]. Because the approach is dependent on zero-crossings detection, the phase angle is only updated twice each cycle of the grid voltage frequency; as a result, the dynamic performance of this method is poor. By immediately using a unique high-order predictive filter, this scenario may be avoided; however, the complexity increases [21].

Using an Arduino board and comparator, which will be detailed later in this chapter, the ZCD will be realized in this work.

3.2.2 Phase locked loop

The phase and frequency between the input and the output are synchronized using a PLL, which is a negative feedback control mechanism. To achieve a power factor that is close to unity, a PLL in a PV inverter aims to synchronize the phase angle of the inverter output current with the grid voltage angle by the least amount of error [22,23]. Figure 3.1 displays the PLL's fundamental block diagram. A phase

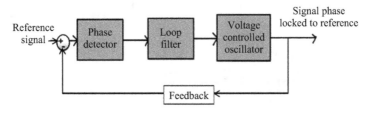

Figure 3.1   PLL block diagram

detector, a loop filter, and a voltage-controlled oscillator make up the fundamental PLL circuit.

By employing PLL and the arctangent function, it is possible to determine the phase difference between the reference signal and the phase-locked signal. The precise phase difference is provided by this method. The loop filter may function as a regulator to eliminate phase inaccuracy. PI regulators are frequently used as this filter [24–26]. After the loop filter, a voltage-controlled oscillator is used, with frequency as the output. The phase-locked angle is outputted by this device, which is often a simple integrator [25].

3.3   Practical implementation procedure

Figure 3.2 shows the block diagram of the proposed micro-inverter. This diagram consists of the PV module, P&O MPPT controller, synchronization method, single-stage flyback DC/AC converter, and an output filter. The following primary requirements determine how the recommended inverter will be implemented in hardware:

- PV output voltage: 33–38 V.
- RMS grid voltage: 220 V.
- Fundamental frequency: 50 Hz.
- peak output power: 120 W.

The implementation components are designed based on the following sections to simplify the implementation procedure.

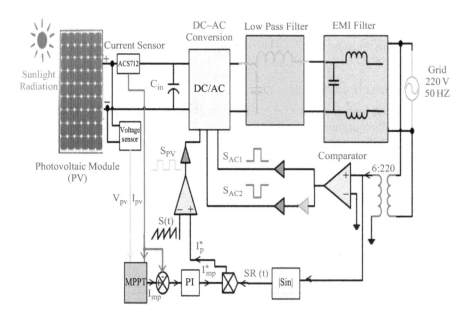

Figure 3.2   Block diagram of the proposed FBPVMI

3.3.1   Switching devices and snubber circuit

As previously mentioned, insulated gate bipolar transistors (IGBTs) and metal oxide semiconductor field effect transistors (MOSFET) are the most common power semiconductors used as switches for FBPVMI [27–29]. Each switch has its advantages, and a high degree of overlap is observed in the specifications of the two. IGBTs are typically used in high-voltage applications (approximately up to 600 V), and they are utilized at lower frequencies because they lack the high-switching frequency capability of MOSFETs [27].

MOSFETs can operate at frequencies up to 200 kHz, which is a greater switching frequency than IGBTs. MOSFETs may be employed at voltages lower than 500 V and less than 500 W and have comparable capabilities for high voltage and high current applications [29]. In this design, two IGBTs are used in the power circuit at the secondary side of the transformer. The first IGBT S_{AC1} is used for the positive half of the output current and the second IGBT, while S_{AC2} is used for the negative half. The IGBTs used in this work (Table 3.1) can handle voltage stress up to 1,500 V and 40 A.

On the primary side of the transformer, the main switch S_{PV} that is used to operate with a high switching frequency (30 kHz) is a MOSFET. Hence, the MOSFET is used as the main switch for FBPVMI, which handles 500 V and 32 A. The power diodes connected across the IGBTs at the secondary side must withstand voltage stress $\left(V_{D1} = V_{D2} = \widehat{V}_g + \frac{V_{pv}}{N} = 311 + \frac{33}{0.1} = 641\text{V}\right)$.

For this reason, the selected type of power diode, which can handle voltage 1,200 V and 75 A, is shown in Table 3.2.

If the spike voltage across S_{PV} is over the MOSFET rating voltage when the main MOSFET S_{PV} is turned off, then the device has failed. Accordingly, the resistor-capacitor-diode (RCD) snubber is adapted to clamp the voltage spike for decreasing the voltage spike on the main MOSFET [30]. The energy stored in the leakage inductance is dissipated by the RC network during each cycle. The capacitance of the snubber circuit capacitor can be calculated based on the maximum ripple of the snubber circuit voltage. The resistance of the snubber circuit resistor can be adjusted based on the power loss and acceptable voltage spike in practical applications [28].

3.3.2   Input capacitor design

The ripple frequency is twice the grid frequency because the output of the flyback is a rectifier sine wave [31,32]. The input capacitor balances the different instantaneous powers in the system. The size of the decoupling capacitor can be

Table 3.2   Selected switching devices used in this study

Device name	Model number
IGBT (module with an anti-parallel diode)	FGL40N120D
MOSFET	FDH50N50
Power diode	RHRG75120

calculated, as presented in the equation below [31]:

$$C_{in} = \frac{P_{pv}}{2\,\pi\,f\,V_{pv}\,\Delta_V} \tag{3.1}$$

$$C_{in} = \frac{P_{pv}}{2\pi f V_{pv}\Delta_V}\,C_{in} = \frac{163}{2\times\pi\times 50\times 33\times 1.5} = 10{,}487\ \mu F.$$

In this work, a set of parallel capacitors ($5\times 2{,}200\ \mu F$, 50 V) is used to form an input capacitor.

3.3.3   Output LC filter design

An output low-pass filter (LPF) is required in grid-connected micro-inverter applications to link with the utility grid. The two main functions of LPF are to reduce the harmonic current content and protect the micro-inverter from transients [33].

Many types of output filters are used in grid-connected micro-inverter applications [33–35]. These topologies include L, LC, and LCL filters, as shown in Figure 3.3. According to [33–35], the most crucial characteristics for the design and operation of filters are used to emphasize the benefits and drawbacks. Among the performance characteristics of various filter topologies that are explored in this research include harmonic attenuation, improved decoupling between the filter and the grid impedance, and system dynamics.

When L filters are utilized, the attenuation of the harmonics is not especially noticeable, a voltage drop is seen, and a large inductor is needed for the implementation [33]. The LC filters have a good response in terms of voltage-to-current conversion with acceptable damping of the high-frequency (HF) value. Meanwhile, the damping of HF is reduced by using an L–C–L filter, but it requires extra inductance. In this design, a simple and low-cost LC filter is used. The LC filter is designed to avoid the harmonics of current around the switching frequency. The output filter inductance can be calculated as follows [34]:

$$L_f = \frac{V_{in}(1 - m_a)m_a}{4i_{out}\,\Delta_{ripple}\,f_s}, \tag{3.2}$$

where V_{in} is the inverter input voltage, m_a is the modulation index, i_{out} is the rated output current, and Δ_{ripple} is the maximum ripple magnitude of the output current (5%–25%). The value of the output filter inductance L_f is:

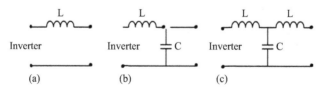

Figure 3.3   Output filter topologies: (a) L filter, (b) LC filter, and (c) L–C–L filter

$$L_f = \frac{342 \times (1 - 0.5) \times 0.5}{4 \times 0.68 \times 0.25 \times 30,000} = 4.2 \text{ mH.}$$

The value of the output filter inductance is 4 mH.

The ferrite material is the optimal material to implement an output filter in grid-connected micro-inverter applications. The output filter capacitance C_f can be calculated as follows [34]:

$$C_f = \frac{1}{(2\pi f_r)^2 L_f}, \tag{3.3}$$

where f_r is the resonant frequency of the LC filter. The resonant frequency should be selected in the range:

$$f_r > \frac{1}{k} f_s \text{ and } k > 15 \tag{3.4}$$

The value of the output filter capacitance C_f can be calculated by taking $f_r = 2,000$ Hz and substituting the value of L_f in Eq. (3.3), which yields:

$$C_f = \frac{1}{(2 \times \pi \times 2,000)^2 \, 4.2 \times 10^{-3}} = 1.5 \text{ μF} \tag{3.5}$$

Meanwhile, the output voltage of the micro-inverter is AC; hence, a non-polarized capacitor with low equivalent series resistance is used. Figure 3.4 shows the practical circuit of the output LC filter.

Figure 3.4   Practical circuit of an output LC filter

3.3.4   Electromagnetic interference filter

Printed circuit board traces, inductors, capacitors, and resistors are some of the electronic components that EMI noise may flow over [36,37]. The two main categories of conducted EMI are conducted and radiated. Due to large currents and low voltages, magnetic fields are the most typical kind of EMI that is emitted.

A common mode noise (CMN) and a differential mode noise make up the conducted EMI (DMN). Power electrical equipment generates EMI waves with frequency ranges up to 1 GHz. High-frequency noise that is in phase with each other and has circuit routes to the ground is referred to as line-to-ground noise (CMN) [37]. The parasitic capacitances in the heat sink and transformer interwinding capacitance charge and discharge, cause the production of the CMN. Common-mode EMI noise is produced by parasitic capacitances between those points of the system and the ground. Additionally, the CMN current returns through the ground conductor and travels in the same direction on both power conductors. A typical EMI filter (Figure 3.5) shows CMN and DMN of EMI noise. The CMN is suppressed by using dual-wound toroid-type inductors. These inductors are wound on a single ferrite toroid core (L_{CM1} and L_{CM2}). This mechanism provides a high impedance to the in-phase CMN on each AC conductor. Moreover, this mechanism absorbs the noise and dumps it to the ground through shunt capacitors (C_{Y1} and C_{Y2}) with low impedance. Figure 3.6 shows the practical circuit of an EMI filter used in this work. The selected values of the EMI filter components are presented in Table 3.3.

3.3.5   Flyback transformer

The flyback transformer that was designed in this chapter is implemented with a ferrite core type EE/55/28/21-C90. The primary winding turns equals $N_p = 21$, and

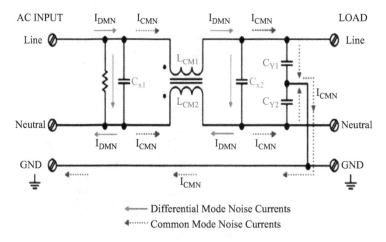

Figure 3.5   Schematic of a typical EMI used to filter the EMI noise

the secondary winding is $N_{s1} = N_{s2} = 196$. The wire size of the primary winding is number #19, while the secondary winding is winded with wire size number #25. Figure 3.7 shows the practical flyback transformer that is used in the flyback converter circuit.

Figure 3.6   Practical circuit of the EMI filter

Table 3.3.   Selected values of the EMI filter components

Parameter	Symbol	Value
Common mode choke	$L_{CM1} = L_{CM2}$	3 mH
Common mode capacitor	$C_{Y1} = C_{Y2}$	4.7 μF
Differential mode capacitor	$C_{X1} = C_{X2}$	10 pF

Figure 3.7   Practical flyback transformer

3.3.6   Hardware power stage circuit

The hardware power stage circuit for the proposed FBPVMI is implemented using the design specification, as presented in Figure 3.8. The voltage of the PV module has a range between 33 V and 38 V that is connected to the input terminals (J1) of the DC/AC power circuit. In addition, a voltage divider and ACS712 current sensor are used to sense the PV module voltage and current, respectively. The signals of these sensors are fed to the Arduino board A/D pins (A1 and A0) to implement the MPPT algorithm. A bank of the input electrolytic capacitors consisting of five 2,200 µF were connected across the PV input side. The main MOSFET (S_{PV}) of the FBPVMI with RCD snubber circuits are connected across the primary side of the flyback center-tap transformer. Furthermore, the secondary side of the transformer is coupled with two super-fast diodes and two IGBTs. The two RCD snubber circuits are used to protect the IGBTs against the stress voltage during the turn-off state. The design specifications of the switching devices used in this work to implement the hardware part are presented in Table 3.1.

The output terminals of FBPVMI are connected across the LC filter to reduce the harmonic content in the output current that is injected into the main grid. Thereafter, the EMI filter is used to prevent the EMI noise. Relays 1 and 2 are connected across the grid line and grid-neutral terminals, respectively, to avoid false grid synchronization. The phase detection of the grid voltage is implemented using an Arduino board. Then, the latter controls the relays based on the grid voltage phase; if an error phase shift occurs between the output current and the grid voltage, the relays trip off the terminals of the grid and remove the synchronization process. Furthermore, the RMS grid voltage is sensed using a small transformer (220/6 V, 50 Hz, 0.5 A). The output voltage of this transformer feeds the precision diode rectifier circuit. Finally, the terminals of the grid voltage are connected to the junction (J2) after the overload protection devices (F1 and F2) are connected to protect the circuit from the short circuit current problem.

3.4   Practical design of control scheme

3.4.1   Maximum power point tracking and PI controllers

The Arduino UNO board is used to implement the MPPT algorithm. Moreover, the microcontroller board known as Arduino UNO is built on the ATmega328. As seen in Figure 3.9, this board features 14 digital input/output pins, including a 16 MHz crystal oscillator, 6 PWM output pins, 6 analog inputs, a USB connection, a power jack, an In-circuit Serial Programming (ICSP) header, and a reset button. Furthermore, the Arduino integrated development environment or Arduino IDE contains a text editor for writing code, a message area, a text console, a toolbar with buttons for common functions, and a series of menus. This environment makes it easy to write code and upload it to the board. As a result, the voltage of the PV module is sensed with a simple voltage divider while the current of the PV module is sensed with the Hall effect current sensor (ACS712 (20A)). Due to the lack of the FTDI USB-to-serial drive chip, Arduino UNO has an advantage over current

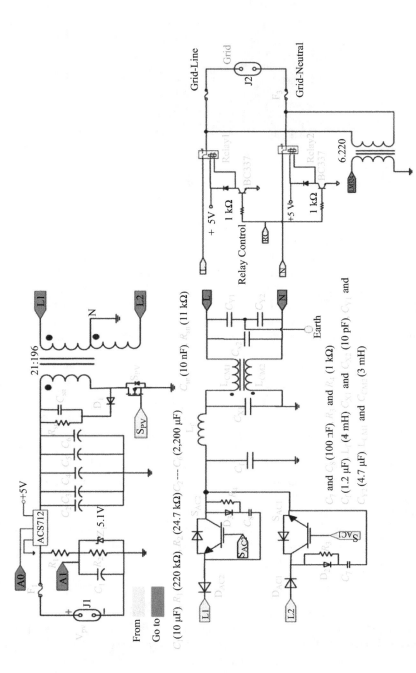

Figure 3.8   Power stage circuit for the proposed FBPVMI

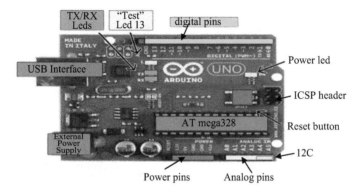

Figure 3.9   Schematic view of an Arduino UNO [38]

boards [38]. For this reason, the Arduino UNO serves as the primary brain of the control circuit in this work.

These sensed voltage and current are fed to the analog to digital (A/D) pins of the board. The digital output signal of the board is converted to an analog value that is used to control the level value of the sinusoidal reference signal by using a digital-to-analog (DAC) device. In this work, the digital output of Arduino UNO can be converted to analog output by writing suitable code on the Arduino board. This code is based on the conversion of the PWM signal to the analog sinewave signal without a negative axis.

The Sallen-key circuit is used to filter the analog signal that is fed to the multiplier circuit. The circuit parameters can be calculated based on the cutoff frequency of the circuit. The higher the cutoff frequency, the faster the change in the analog output. The PWM fundamental frequency is set as high as possible to ensure flexibility when choosing a cutoff frequency. For this reason, the PWM frequency of Arduino UNO in this work is 31.5 kHz. Circuit parameters $R_1 = R_2 = 4.7$ kΩ and $C_1 = C_2 = 100$ nF are used in this work.

The operational amplifier of the circuit used in this work is LM358. The complete circuit of MPPT with a DAC converter is shown in Figure 3.10. This circuit will be tested later in Proteus software.

The PI controller aims to ensure that the actual output current of the PV module $I_{PV}(t)$ follows the reference value $I_{ref}(t)$ by reducing the error signal $e(t)$, where:

$$e(t) = I_{ref}(t) - I_{pv}(t) \tag{3.7}$$

In this work, a digital PI controller is implemented using the Arduino board. The program control code of the PI controller is written using Arduino IDE and directly loaded into the microcontroller. After implementing the P&O algorithm in the Arduino board, the error signal between the desired current and the actual current of the PV module enters the PI controller, which adjusts the duty cycle of the output signal by using the analogWrite function in the Arduino board.

3.4.2   Precision diode rectifier circuit

The main MOSFET S_{PV} should be turned ON and OFF with a complete sinusoidal rectified signal to synchronize the proposed FBPVMI with a grid voltage. The sample voltage is extracted from the grid voltage with the small signal transformer and fed to the precision diode rectifier circuit, which produces a sinusoidal rectified signal feedback to the multiplier circuit to generate sequences of gate pulses to the main MOSFET. The PWM signal of the main MOSFET is generated by comparing the sinusoidal rectified signal $SR(t)$ with a high-frequency sawtooth signal $S(t)$ (30 kHz).

The sinusoidal rectified signal is applied to utilize a precision diode rectifier circuit, as presented in Figure 3.11, which is implemented using an LM324N (ON-Semiconductor®).

3.4.3   PWM circuit

The PWM-integrated circuit combines from the comparator with a fixed switching frequency and a fixed amplitude saw-tooth-wave oscillator, which are implemented using an SG3525A PWM control circuit from (ON-Semiconductor®), to provide a

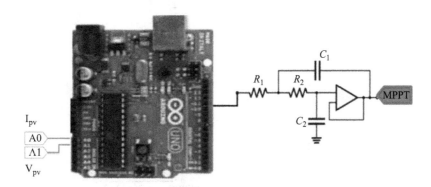

Figure 3.10   Practical circuit of the P&O MPPT algorithm

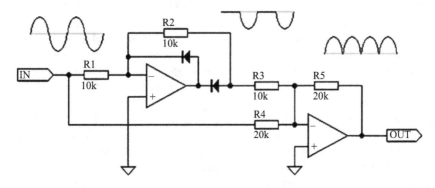

Figure 3.11   Precision diode rectifier circuit

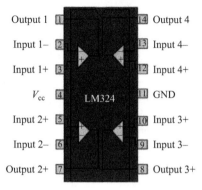

Figure 3.12   Top view schematic of LM324N

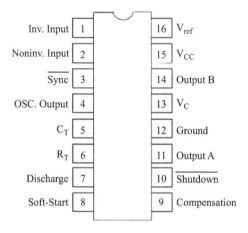

Figure 3.13   Top view schematic of SG3525A

PWM signal to the main MOSFET, as shown in Figure 3.13. The SG3525A offers improved performance and lower external part count when utilized to control all types of switching. In addition, the output of the multiplier, which is a rectified sin wave with a variable amplitude-based MPPT output is compared with a high-frequency saw-tooth signal S(t) through the SG3525A. As a result, the output is a PWM signal with a variable duty cycle ratio that is applied to the main MOSFET through the optocoupler and gate driver circuit. The oscillator frequency of the saw-tooth signal can be set by connecting a capacitor and a resistor (C_T and R_T). This PWM IC has dual outputs equal in magnitude and different in phase. Accordingly, the switching frequency of the output is equal to the half oscillator frequency. For this reason, the dual output pins are followed by an OR-gate to combine the two outputs and make the switching frequency equal to the oscillator frequency, as will be presented later in this chapter. So, the OR-gate is implemented using CD4075BE from Texas Instruments[®].

3.4.4   Optocoupler and gate driver circuits

Three gate driver optocouplers are used in this chapter to isolate the control circuit from the power circuit in the proposed micro-inverter. The two-gate driver opto-couplers are used for the two IGBTs S_{AC1} and S_{AC2}, and the other gate driver optocoupler is utilized for the main MOSFET S_{PV} on the PV module. The gate driver optocoupler is implemented using an HCPL-3120 integrated circuit from Avago Technologies. These optocouplers are ideally suited for driving power MOSFETs and IGBTs used in inverter applications. The voltage and current sup-plied by these optocouplers make them ideally suitable for directly driving IGBTs with ratings up to 1,200 V/100 A. The gate driver optocouplers turn ON the MOSFET and IGBTs with (+12 V).

3.4.5   Dead time circuit

A dead time is necessary to ensure that the IGBTs in the grid side of FBPVMI never simultaneously turn on. The dead time circuit topology is shown in Figure 3.14. The PWM input signal from the comparator (LM 339) has a low-frequency 50 Hz square wave signal. These signals are applied to the Schmitt trigger (CD40106B IC) to generate two proper signals (pin 4 and pin 8) that feed the RC circuit network, which is used to delay the edges of the two signals. The diodes (D1 and D2) are used to ensure that only one of the two edges is delayed. The Schmitt trigger is needed to clearly define the point where the output changes due to the slow alteration of the output of the RC circuit. The digital buffer circuit is used to provide a suitable voltage value for the output signals that feed the IGBTs. The values of the dead time circuit parameters that are used in this work are $C_1 = C_2 = 100$ nF and $R_1 = R_2 = 220$ kΩ. The diode (D1 and D2) types used in this work is 1N4007.

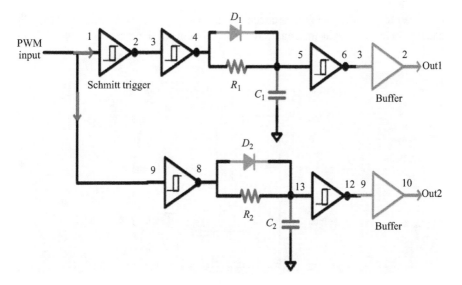

Figure 3.14   Dead time circuit

3.4.6   Implementation of the ZCD method

As previously mentioned, the ZCD method is a common method that is used to synchronize the output current of an inverter with a grid voltage due to its simplicity and easy implementation. The zero-crossing detector detects the rising edge of a grid voltage. The ZCD method generates a periodic pulse signal that represents the frequency and the positive signal of the grid voltage, as will be illustrated later in this chapter. The two IGBTs of the inverter should be turned on and off according to the grid state to synchronize the proposed inverter with the grid.

3.4.7   Comparator circuit

As suggested by the name, comparators are employed to contrast two voltages. As detectors, these circuits can be used in a variety of situations. These circuits could, for instance, contain a reference voltage on one input and a detected voltage on another. The comparator's output will turn on even though the observed voltage is higher than the reference. The comparator's state will change if the measured voltage is lower than the reference, and this might be utilized to signal the problem. A large number of integrated circuits (ICs) have been created expressly for use as comparators. In this work, LM339N (from ON-Semiconductor®) is a quad comparator that is designed to operate from a single power supply over a wide range of voltages used, as shown in Figure 3.15. The 220 V RMS grid voltage is attenuated to be a 6 V RMS voltage by a small 50 Hz voltage transformer. The RMS grid voltage signal is converted to a square wave signal with a comparator; thus, a zero-crossing voltage is detected. These square waves are used to turn on the two IGBTs with the gate driver.

3.4.8   Schmitt trigger and digital buffer circuits

The Schmitt trigger circuit is used to generate square waves from the comparator (LM339) to produce the gate sequence pulses for grid side switches (IGBTs). The square waves are equal in magnitude and different in phase by 180°. An IGBT is

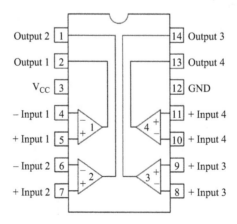

Figure 3.15   Top view schematic of LM339N

turned ON during the positive half of the grid voltage, which is responsible for the positive half of the output current signal that is injected into the grid. The other IGBT is turned ON during the negative half of the grid voltage, which is responsible for the negative half of the output current signal that is injected into the grid. In this chapter, the Schmitt trigger is implemented using CD40106B (Texas Instruments®), as shown in Figure 3.16. Furthermore, a digital buffer is a special electronic circuit that is used to isolate the input side from the output side, forbidding the impedance of one circuit from changing the impedance of another. Moreover, this type of circuit is used to provide power amplification to the digital signal extracted from the Schmitt trigger to provide a square wave with a suitable voltage level that is used to drive the IGBTs. In this chapter, the digital buffer is implemented using CD4050B, as shown in Figure 3.17.

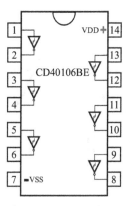

Figure 3.16   Top view schematic of CD40106B

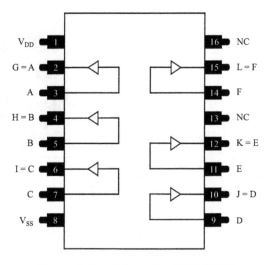

Figure 3.17   Schematic of CD4050B

3.4.9 *Phase detector*

In FBPVMI, phase detection of a grid voltage is an important part to implement the ZCD to avoid an error phase shift between the output current of the inverter and the grid voltage. This phase detector can be implemented with many topologies. In this work, an Arduino board is used to implement the phase detector of the grid voltage. Figure 3.18 shows the zero-crossing pulses at 0, 180 two pulses per cycle for a grid frequency of 50 Hz. These generated pulses feed the digital pins of the board (pin 2). The grid voltage state is detected according to this state; the output of the comparator is changed. When the voltage grid condition changes from negative to positive, it alters the state of the D/A pins of the Arduino board from low to high, and vice versa, as shown in Figure 3.19. The output of the digital pin (pin 5) of the Arduino board feeds relay control to flag the phase difference between the output

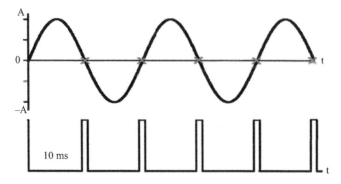

Figure 3.18   Zero-crossing pulses of the grid voltage

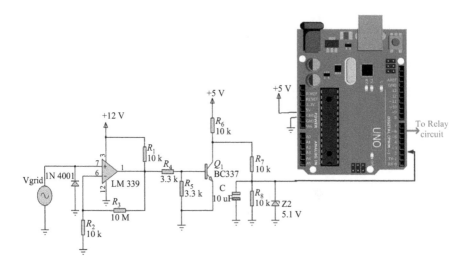

Figure 3.19   Zero cross-detection using Arduino

current and the grid voltage. When Arduino detects a fault, the relay circuit will cut off and isolate the synchronization process of the inverter with the grid.

3.4.10 Proposed controller schematic

Figure 3.20 shows the proposed control block diagram of FBPVMI under a DCM operation. In this work, many digital and analog devices are used to implement the

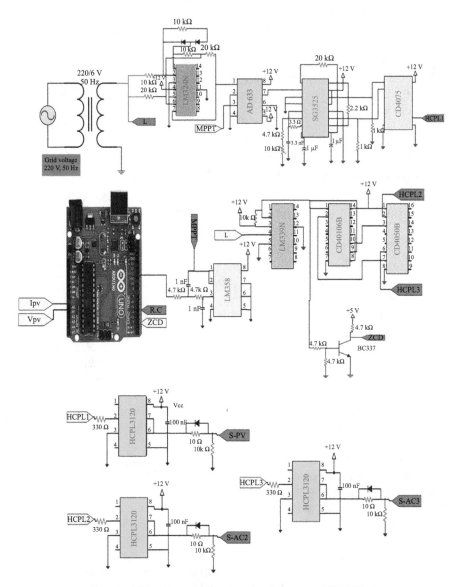

Figure 3.20 Proposed control schematic of FBPVMI

proposed control scheme. As shown below, a small signal transformer is used to lower the grid voltage (220/6 V, 50 Hz). This voltage converts the reference rectified signal by LM324N to generate a PWM to the main MOSFET that depends on the grid voltage state using SG3525 with an isolated gate driver optocoupler.

The sinusoidal voltage signal is converted to a square wave signal using LM339N. This signal is converted to dual square waves using Schmitt trigger CD40106B with digital buffer CD4050B; both are used to turn on IGBTs with a gate driver optocoupler. The DAC provides an analog signal of MPPT to the multiplier circuit, which multiples this signal with an output of LM324N. The output of LM324N, which is a sinusoidal reference SR(t) signal, is used to produce a primary reference current I_p^* signal. The primary reference signal that has a rectified sine wave shape is fed to the PWM IC (SG3525) and compared with the sawtooth signal S(t) to produce a gate sequence pulse to the main MOSFET via the driver optocoupler circuit.

Finally, the phase detection of the grid voltage that is implemented in the Arduino board (pin 5) manages the relay state that connects across the grid terminals. This signal cutoffs the relays and isolates the synchronization process when a fault occurs in the phase detection of the grid voltage.

3.5   Simulation and experimental results

3.5.1   *Proteus simulation results*

Every part in the proposed control circuit is simulated by Proteus software to explain the work of these circuits in practical implementation. This step is also carried out to simplify and understand the proposed control scheme with the same experimental components that are used in the practical implementation of this chapter.

Figure 3.21 shows the proposed control scheme under Proteus software, which includes all parts of the proposed control scheme. The simulations results for every part will be shown as follows: Figure 3.22 shows the input and output voltage waveforms for the LM324 IC; Figure 3.22(a) shows the sample grid voltage waveform, while Figure 3.22(b) presents the sinusoidal reference signal SR(t) that has a shape-like rectified sine wave. This signal feeds the multiplier circuit AD633.

Figure 3.23 shows the high-frequency saw-tooth signal, S(t) (30 kHz), from SG3525 IC, which is compared with the primary current reference signal (I_p^*) to generate the gate pulses to the main MOSFET (S_{PV}), which is shown in Figure 3.24. Figure 3.25 shows the PWM output circuit; Figure 3.25(a) shows the dual PWM outputs from the SG3525, while Figure 3.25(b) illustrates the combination of these outputs by the OR gate. This signal feeds to the gate driver of the main MOSFET, S_{PV}, to turn on this switch.

Figure 3.26 shows the DAC convert circuit signal from the Arduino board. The analog signal represents the MPPT value of the PV module, which is used to adjust the peak value of the sinusoidal reference signal, SR(t). The PWM signals of the

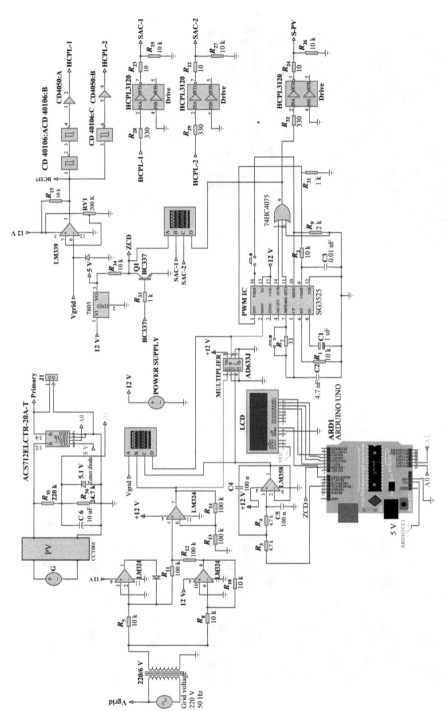

Figure 3.21   Proposed control scheme under Proteus software

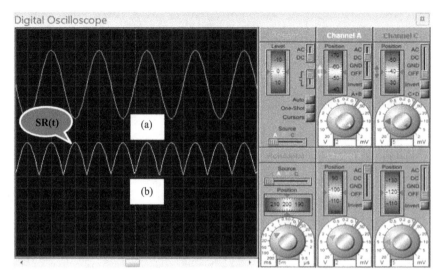

Figure 3.22   Input and output voltages for the diode rectifier circuit: (a) sample from the grid voltage waveform; (b) sinusoidal reference signal SR(t)

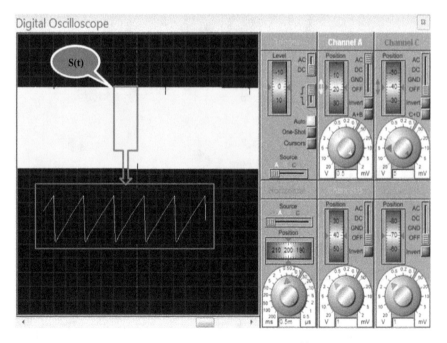

Figure 3.23   High-frequency sawtooth signal, S(t) using SG3525IC

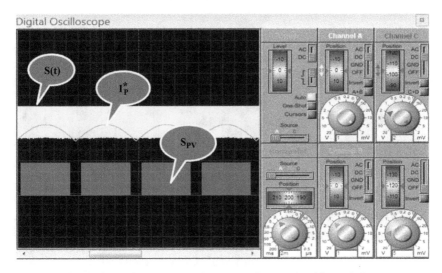

Figure 3.24   Comparator between the sawtooth signal, $S(t)$, and the primary current reference signal, I_p^* using SG3525IC

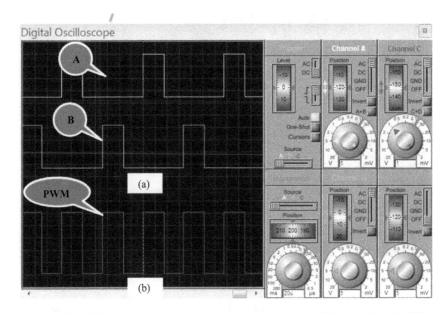

Figure 3.25   PWM output circuit: (a) dual outputs of the SG3525 IC; (b) PWM combined with the dual outputs by the OR gate

two IGBTs are shown in Figure 3.27. These signals are generated by Schmitt trigger CD40106B and digital buffer CD4050B, synchronized with the grid voltage. Figure 3.28 shows the zero-crossing pulses from the ZCD circuit with an Arduino board.

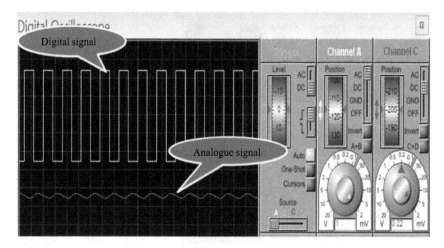

Figure 3.26   DAC converter circuit signal with an Arduino board digital output

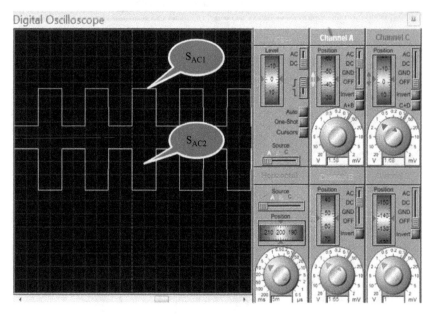

Figure 3.27   Gate pulses for the two IGBTs, S_{AC1} and S_{AC2}, from the Schmitt trigger and digital buffer circuits

3.5.2   Experimental results

The prototype setups of the FBPVMI circuits are presented in Figures (3.8) and (3.20) and are tested under different weather conditions. The FBPVMI is experimentally examined on a 240 W PV module with a voltage range of 33–38 V. The

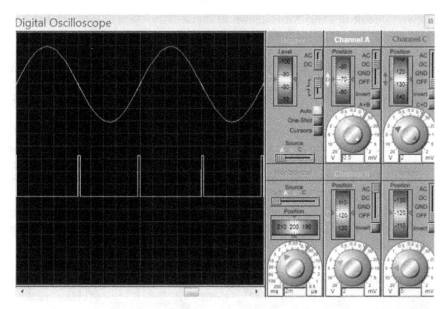

Figure 3.28 Zero-crossing pulses of the grid voltage by the Arduino board

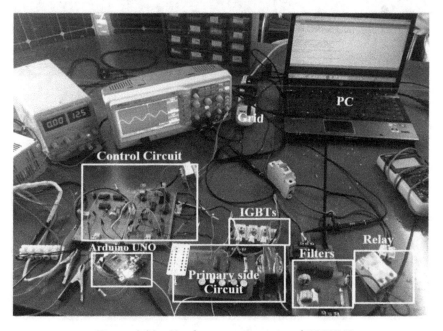

Figure 3.29 Hardware components of FBPVMI

hardware circuit of FBPVMI is shown in Figure 3.29. An automatic digital lux-
meter (model LX1010B) is used to monitor sun irradiation, and a thermometer is
used to gauge the temperature. A digital storage oscilloscope (TSO 1022, 25 MHz,
500 MSa/s) is used to capture the measurements. Figure 3.30 shows the output
voltage from the PV module, $V_{pv} = 34.8$ V for solar irradiance ($G = 650$ W/m^2)
and ambient temperature ($T = 20\,^\circ$C). A little value of voltage ripple $\Delta_v = 2$ V is
observed.

Figure 3.31 shows the output current from the PV module, $I_{pv} = 3.7$A. A little
value of the current ripple $\Delta_i = 0.16$A can be observed. Figure 3.32 shows the
sinusoidal reference signal SR(t) that is generated by the precision diode rectifier
circuit from IC LM324N. Figure 3.33 shows the high-frequency saw-tooth signal
(30 kHz), S(t), from IC SG3525.

Figure 3.34 shows the comparator process between the saw-tooth signal, $S(t)$,
and the primary current reference signal, I_p^*, in IC SG3525, which produces the
required gate pulses to the main MOSFET, S_{PV}, as shown in Figure 3.35.

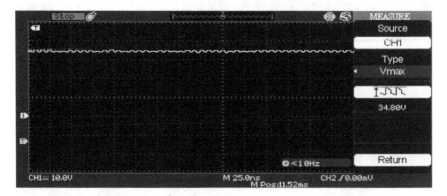

Figure 3.30   Output voltage from the PV module (10 V/div.)

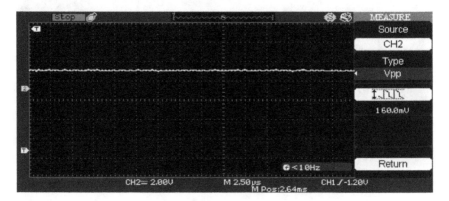

Figure 3.31   Output current from the PV module (2 A/div.)

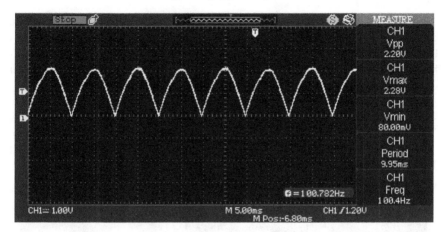

Figure 3.32   Sinusoidal reference signal SR(t) from the LM324 (1 V/div.)

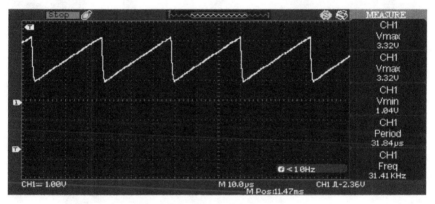

Figure 3.33   Sawtooth signal, S(t), from SG3525 (1 V/div.)

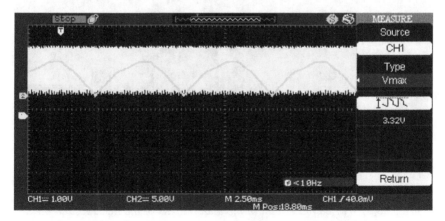

Figure 3.34   Comparator between the saw-tooth signal, S(t), and the primary current reference signal, I_p^*, in SG3525 (5 V/div.)

Figure 3.36 shows the dual outputs of the IC SG3525 and the combined outputs by the OR gate (IC CD4075BE). Figure 3.37 shows the gate pulses for the two IGBTs (S_{AC1} and S_{AC2}) from the Schmitt trigger and digital buffer circuits with grid frequency (50 Hz). These pulses have a dead time of approximately 680 µs between them, as shown in Figure 3.38, to avoid the short circuit problem across the IGBTs.

Figure 3.39 shows the primary current of the FBPVMI that has a rectified sine waveform due to the sinusoidal modulation. The peak value of this current is $\widehat{I}_p = 26.4A$.

Figure 3.40 shows the current stress across the main MOSFET of the FBPVMI. The peak value of this current is $\widehat{I}_1 = 13.6A$.

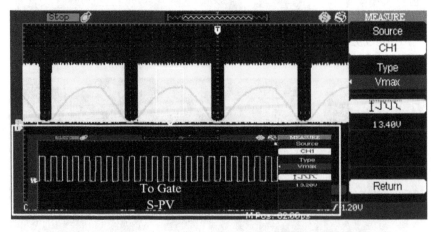

Figure 3.35   Gate pulses fed to the main MOSFET, S_{PV} (5 V/div.)

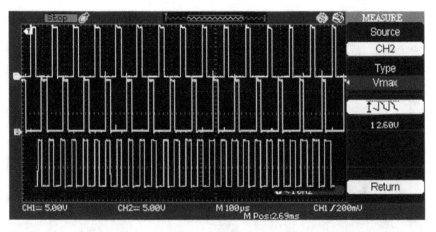

Figure 3.36   PWM output circuit of the dual outputs of the SG3525 and the combined dual outputs by the OR gate (5 V/div.)

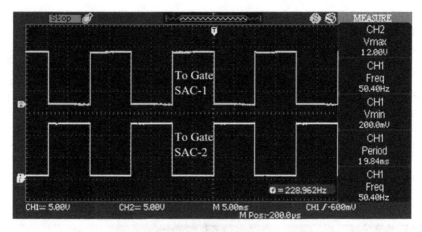

Figure 3.37   Gate pulses for the two IGBTs, S_{AC1} and S_{AC2} (5 A/div.)

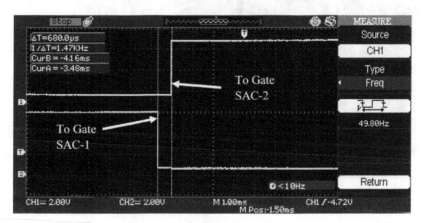

Figure 3.38   Dead time waveform between the IGBTs

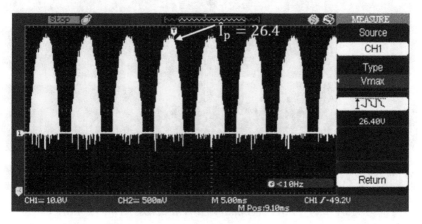

Figure 3.39   Primary current in the FBPVMI (5 A/div.)

Figure 3.41 shows the secondary current across the IGBT across the positive half cycle, S_{AC1}. The peak of this current is $\hat{I}_2 = 3.6$ A. Figure 3.42 shows the drain-source voltage (V_{ds}) of S_{PV} for FBPVMI. The peak value of this voltage is $V_{ds} = 64.4$ V.

Figure 3.43 shows the AC output current fed to the grid in phase with the grid voltage for radiation, $G = 650$ W/m^2, and temperature, $T = 20$ °C. The output current value is $I_{out,rms} = 0.54$ A, power factor PF $= 0.98$, and THD $= 3.7\%$. The output power transferred to the grid is $P_o = 118.8$ W with respect to the given weather conditions.

Figure 3.44 shows the AC output current fed to the grid within phase with the grid voltage for $G = 400$ W/m^2, and $T = 30$ °C. The output current value is

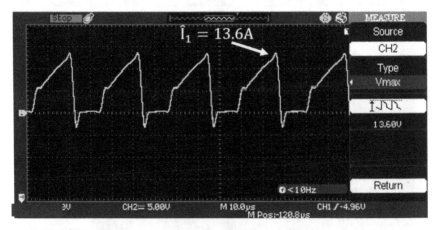

Figure 3.40   Current stress across the main MOSFET, S_{PV} (5 A/div.)

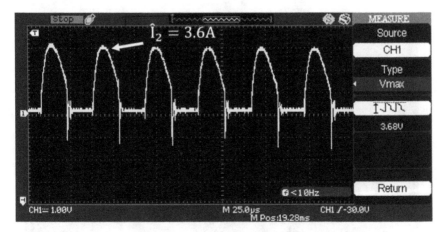

Figure 3.41   First secondary current across, S_{AC1} (1 A/div.)

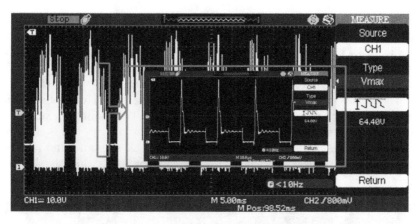

Figure 3.42   Drain-source voltage (V_{ds}) of S_{PV} (10 V/div.)

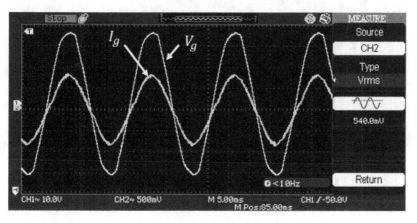

Figure 3.43   AC output current fed to the grid, $I_{out,rms} = 0.54$ A (0.5 A/div.)

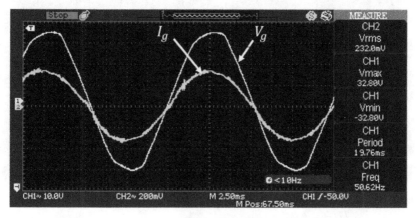

Figure 3.44   AC output current fed to the grid, $I_{out,rms} = 0.23$ A (0.2 A/div.)

$I_{out,rms} = 0.23$ A, power factor PF $= 0.95$, and THD $= 4.2\%$. The output power transferred to the grid is $P_o = 50.6$ W with respect to the given weather conditions.

Figure 3.45 shows the Fast Fourier Transform (FFT) of the AC output current fed to the grid for $G = 650$ W/m^2 and $T = 20$ °C. Figure 3.46 shows the FFT of the AC output current fed to the grid for $G = 400$ W/m^2 and $T = 30$ °C.

The total system's brain was represented by the digital controller device. Compared with the high-cost digital controllers used in flyback inverter studies, the proposed control scheme-based Arduino board shows a low cost and is simpler in structure. As a result, dependent on the type of flyback micro-inverter topology. Additionally, the system efficiency in this study was 91%, which is similar to

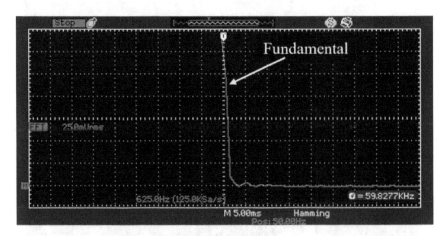

Figure 3.45   FFT for the output current fed to the grid, $I_{out,rms} = 0.54$ A, THD $= 3.7\%$

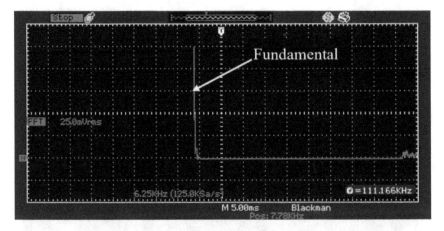

Figure 3.46   FFT for the output current fed to the grid, $I_{out,rms} = 0.23$ A, THD $= 4.2\%$

existing works. Finally, the suggested control scheme circuit is efficient, particularly at high power levels.

3.6 Conclusion and future works

In this chapter, a single-stage flyback micro-inverter with a simple and low-cost control strategy was presented. The overall cost of the implementation can be reduced by using the low-cost Arduino UNO board in the control circuit instead of an expensive FPGA or TMS320F28035 microcontroller. Accordingly, the flyback micro-inverter control circuit was implemented using different operational amplifiers and ATMega 328 microcontroller embedded in Arduino UNO. To validate the suggested work, a prototype 120 W PV module is used to supply the proposed inverter in the experiment. The Proteus simulation and experimental results are achieved under different weather conditions to prove the proposed control strategy. The results demonstrate that the proposed control is robust and has a small size and low cost. Finally, a pure sinusoidal current was injected into the utility grid with good power factor $PF = 0.98$ and lower THD $= 3.7\%$.

Future work related to this study includes:

- Hardware implementation of a new MPPT technique to validate and improve the inverter performance.
- Design and implement a robust control scheme using a DSP control card.
- Increase the performance of the inverter using a hybrid PV system with a fuel cell or lithium battery.

References

[1] A. Khaligh and O. Oner, *Energy Harvesting Solar, Wind, and Ocean Energy Conversion Systems*, New York, NY: Talyer and Franceis Group, 2010, ISBN: 978-1-4398-1508-3.

[2] H. G. Lee, *"Optimal design of solar photovoltaic systems,"* Ph.D. thesis, University of Miami, Miami, FL, August 2015.

[3] A. N. Celik and N. Acikgoz, "Modeling and experimental verification of the operating current of mono-crystalline photovoltaic modules using four and five parameter models," *Applied Energy*, vol. 84, pp. 1–15, 2007.

[4] M. Villalva, J. Gazoli, and E. Ruppert, "Modeling and circuit-based simulation of photovoltaic arrays," *Brazilian Journal of Power Electronics*, vol. 14, No. 1, pp. 35–45, 2009.

[5] M. G. Villalva, J. R. Gazoli, and E. R. Filho, "Comprehensive approach to modeling and simulation of photovoltaic arrays," *IEEE Transactions on Power Electronics*, vol. 42, No. 5, p. 1198, 2009.

[6] S. Motahhir, A. Chalh, A. El Ghzizal, S. Sebti, and A. Derouich, "Modeling of photovoltaic panel by using proteus," *Journal of Engineering Science and Technology Review*, vol. 10, No. 2, pp. 8–13, 2017.

[7]   H. Hu, S. Harb, N. H. Kutkut, Z. J. Shen, and I. Batarseh, "A single stage micro-inverter without using electrolytic capacitors," *IEEE Transactions on Power Electronics*, vol. 28, No. 6, pp. 2677–2687, 2013.

[8]   M. Dong, X. Tian, L. Li, D. Song, L. Wang, and M. Zhao, "Model-based current sharing approach for DCM interleaved flyback micro-inverter," *Energies*, vol. 11, No. 7, pp. 1685, 2018.

[9]   M. A. Rezaei, *"Design, implementation and control of a high efficiency interleaved flyback micro-inverter for photovoltaic applications,"* Ph.D. thesis, North Carolina State University, Raleigh, NC, 2015.

[10] M. E. Basoglu and B. Cakir, "Comparisons of MPPT performances of isolated and non-isolated DC–DC converters by using a new approach," *Renewable and Sustainable Energy Reviews*, vol. 60, pp. 1100–1113, 2016.

[11] C. M. Lai, M. J. Yang, and W. C. Liu, "Parallel-operated single-stage flyback-type single-phase solar micro-inverter," In *IEEE, International Conference in Intelligent Green Building and Smart Grid (IGBSG)*, 2014, pp. 1–5.

[12] S. J. Yaqoob, S. Motahhir, and E. B. Agyekum, "A new model for a photovoltaic panel using Proteus software tool under arbitrary environmental conditions," *Journal of Cleaner Production*, vol. 333, p. 130074, 2022.

[13] S. Motahhir, A. Chalh, A. El Ghzizal, and A. Derouich, "Development of a low-cost PV system using an improved INC algorithm and a PV panel Proteus model," *Journal of Cleaner Production*, vol. 204, pp. 355–365, 2018.

[14] Y. Kim, J. Kim, Y. Ji, C. Won, and T. Lee, "Flyback inverter using voltage sensor-less MPPT for AC module systems," In *IEEE Power Electronics Conference (IPEC)*, 2010, pp. 948–953.

[15] Y. Kim, J. Kim, Y. Ji, C. Won, and T. Lee, "A new control strategy of active clamped flyback inverter for a photovoltaic AC module system," In *IEEE, 8th International Conference on Power Electronics Conference (ECCE)*, Korea, May 30–June 3, 2011, pp. 1880–1885.

[16] J. Jan, Y. Kim, D. Ryu, C. Won, and Y. Jung, "High efficiency control method for interleaved flyback inverter with synchronous rectifier based on photovoltaic AC modules," In *IEEE, 38th Annual Conference in Industrial Electronics Society (IECON)*, October 2012, pp. 5720–5725.

[17] S. Öztürk and I. Çadirci, "DSPIC microcontroller based implementation of a flyback PV micro-inverter using direct digital synthesis," In *IEEE Energy Conversion Congress and Exposition (ECCE)*, Denver, September 15–19, 2013, pp. 3426–3433.

[18] Z. Zhang, M. Chen, W. Chen, C. Jiang, and Z. Qian, "Analysis and implementation of phase synchronization control strategies for BCM interleaved flyback micro-inverters," *IEEE Transactions on Power Electronics*, vol. 29, No. 11, pp. 5921–5932, 2014.

[19] B. D. Reddy, M. P. Selvan, and S. Moorthi, "Design, operation, and control of S3 inverter for single-phase micro-grid applications," *IEEE Transactions on Industrial Electronics*, vol. 62, No. 9, pp. 5569–5577, 2015.

[20] H. A. Sher, K. E. Addoweesh, and K. Al-Haddad, "Performance enhancement of a flyback photovoltaic inverter using hybrid maximum power point tracking," In *IEEE, 1st Annual Conference in Industrial Electronics Society (IECON)*, Yokohama, November 9–12, 2015, pp. 005369–005373.

[21] M. Khalilian, M. M. Rad, E. Adib, and H. Farzanehfard, "New single-stage soft switching flyback inverter for AC module application with simple circuit," In *IEEE, 6th International Power Electronics Drive Systems and Technologies Conference (PEDSTC)*, February 3–4, Tehran, Iran, 2015, pp. 41–46.

[22] J. Liao, J. Su, L. Chang, and J. Lai, "A mixed decoupling power method for single-phase grid-connected inverters," In *IEEE, 7th International Symposium on Power Electronics for Distributed Generator Systems (PEDG)*, June 2016, pp. 1–5.

[23] T. Lodh and V. Agarwal, "Single stage multi-port flyback type solar PV module integrated micro-inverter with battery backup," In *IEEE, International Conference in Power Electronics, Drives, and Energy Systems (PEDES)*, December 2016, pp. 1–6.

[24] M. Kalilian and P. Guglilmi, "Single stage grid-connected flyback inverter with zero current switching for AC module application," In *IEEE, Annual Conference in Industrial Electronics Society (IECON)*, October 2016, pp. 2390–2395.

[25] C. L. Trujillo, F. Santamaria, and E. E. Gaona, "Modeling and testing of two-stage grid-connected photovoltaic micro-inverter,"*Renewable Energy*, vol. 99, pp. 533–542, July 2016.

[26] S. Sukatjasakul and S. Po-Ngam, "The micro-grid connected single-phase photovoltaic inverter with simple MPPT controller," In *IEEE, 5th International Electrical Engineering Congress*, Pattaya, Thailand, March 8–10, 2017, pp. 1–4.

[27] M. H. Ahmed, M. Wang, M. A. S. Hassan, and Ullah, I., "Power loss model and efficiency analysis of three-phase inverter based on SiC MOSFETs for PV applications," *IEEE Access*, vol. 7, pp. 75768–75781, 2019.

[28] S. J. Yaqoob, A. Obed, R. Zubo, *et al.*, "Flyback photovoltaic micro-inverter with a low cost and simple digital-analog control scheme," *Energies*, vol. 14, no. 14, pp. 4239, 2021.

[29] D. Karimi, H. Behi, J. Jaguemont, M. El Baghdadi, J. Van Mierlo, and O. Hegazy, "Thermal concept design of MOSFET power modules in inverter subsystems for electric vehicles," In *2019 9th International Conference on Power and Energy Systems (ICPES)*. IEEE, 2019, pp. 1–6.

[30] Zhang, F., Yang, X., Chen, W., and Wang, L. (2020). Voltage balancing control of series-connected SiC MOSFETs by using energy recovery snubber circuits. *IEEE Transactions on Power Electronics*, vol. 35, No. 10, pp. 10200–10212.

[31] A. Ch. Kyristis, E. C. Tatakis, and N.P. Papanikolaou, "Optimum design of the current-source flyback inverter for decentralized grid-connected photovoltaic systems," *IEEE Transactions on Energy Conversion*, vol. 23, No. 1, pp. 281–293, 2008.

[32]   Y. H. Ji, D. Y. Jung, J. H. Kim, C. Y. Won, and D. S. Oh, "Dual mode switching strategy of flyback inverter for photovoltaic AC modules," In *IEEE, International Power Electronics Conference (IPEC)*, Sapporo, June 21–24, 2010, pp. 2924–2929.

[33]   A. Reznik, M. G. Simões, A. Al-Durra, and S. M. Muyeen, "LCL filter design and performance analysis for grid interconnected systems," *IEEE Transactions on Industry Applications*, vol. 50, No. 2, pp. 1225–1232, 2014.

[34]   H. Cha and T. K. Vu, "Comparative analysis of low-pass output filter for single-phase grid-connected photovoltaic inverter," In *IEEE, 25th Annual Applied Power Electronics Conference and Exposition (APEC)*, Palm Springs, February 21–25, 2010, pp. 1659–1665.

[35]   A. A. Ahmad, A. Abrishamifar, and M. Farzi, "A new design procedure for output LC filter of single phase inverters," In *IEEE, 3rd International Conference on Power Electronics and Intelligent Transportation System (PEITS)*, Egypt, May 20–24, 2010, pp. 86–91.

[36]   F. Luo, D. Boroyevich, and P. Mattavelli, "Improving EMI filter design within circuit impedance mismatching," In *IEEE, Twenty-Seventh Annual in Applied Power Electronics Conference and Exposition (APEC)*, February 2012, pp. 1652–1658.

[37]   A. Majid, J. Saleem, H. B. Kotte, R. Ambatipudi, and K. Bertilsson, "Design and implementation of EMI filter for high frequency (MHz) power converters," In *IEEE, International Symposium on Electromagnetic Compatibility (EMC EUROPE)*, Rome, September 17–21, 2012, pp. 1–4.

[38]   ATmega328P Data Sheet, Atmel Corporation, 2009. pdf online: http://atmel. com / images/ atmel - 8271- 8 - bit - Avr- Microcontroller -Atmega48a-48pa-88a-88pa-168a-168pa-328-328p_datasheet complete.pdf data sheet Arduino.

Chapter 4

Assessment of influences of high photovoltaic inverter penetration on distribution automation systems: Vietnam distribution network case study

Tran The Hoang[1], Ngoc Thien Nam Tran[2], Thi Minh Chau Le[3], Tuan Le[4] and Minh Quan Duong[5]

Stand-alone solar photovoltaic (PV) systems are a convenient way to provide electricity for people far from the electric grid or for people who want electric power with a slight dependence on the utility grid, to run their usual activities either at home or at businesses. These PV systems, combined with storage components, are considered distributed energy resources (DERs). In fact, DERs connected to distribution grids are increasing, especially PV systems, which pose several challenges to fault location, isolation, and restoration service (FLIRS) integrated into distribution automation systems (DAS). While most studies paid attention to issues focused on background knowledge or impedance-based fault location, others only adopted simplified PV models that cannot ensure accurate presentation of actual PV fault behaviors. Therefore, this chapter aims to extensively investigate the impacts of increased PV integration on the existing FLIRS performance, when PV systems become more intelligent by integrating inverters with the Internet of Things (IoT) to efficiently control the system and optimize power generation through maximum power point tracking algorithms. Since Danang Power Company put a commercial advanced FLIRS function integrated with DAS in operation along with the orientation of transforming into a smart grid, considering the extremely rapid development of PV systems in medium-voltage (MV) distribution networks, it has become a typical model for studying faults occurring in MV systems when PV penetration is high. A total of 2,700 error scenarios were carried out for comprehensive investigation purposes. All the network components relevant to the study are simulated with great detail in an environment of PowerFactory/

[1]The University of Auckland, New Zealand
[2]Delta Electronic Inc, Tainan, Taiwan
[3]Hanoi University of Science and Technology, Vietnam
[4]Roberval Laboratory, University of Technology of Compiegne, France
[5]The University of Danang-University of Science and Technology, Vietnam

DIgSILENT software, which can communicate with DAS through communication methods. From then, can be built IoT solutions that support modern PV system monitoring and control, which can predict and evaluate outputs from available data, possibly from Big Data.

4.1   Introduction

Renewable energy has recently attracted much attention due to increasing energy demand, growing concern about environmental sustainability, and the rising price of fossil fuels. Solar energy stands out among all renewable sources as it is the cleanest and inexhaustible energy source, and its use is also environmentally friendly. The current worldwide energy demand is relatively less than the available potential of solar energy [1].

The nature of renewable in general and solar energy is that the input energy will be stable if the generators are dispersed in the spatial domain [2]. Therefore, technology for DERs has been engaging in recent years for efficiently mining such energy resources. DERs have gained significant attention for their successful attempts to supply electrical power to autonomous off-grid rural areas, and many implementations have been successfully done worldwide. Moreover, owing to zero sound pollution and greenhouse gas emissions, it contributes to the sustainability of the environment [3].

One of the highlight technologies for such renewable distributed energy systems is the smart grid based on the Internet of Things (IoT) [4]. Smart grid-oriented sustainable development of power systems requires control and operation higher and stricter. With this strategy, power companies must develop advanced solutions such as IoT integration for monitoring, forecasting and data evaluation to determine feasible operating scenarios. Moreover, power supply regulations present more stringent requirements regarding system reliability indices such as SAIDI, SAIFI, and CAIDI to operators [5]. In order to meet new operational legislation, the existing DAS should be upgraded, in which the FLISR functionality are of primary interest. Typically, the operation of the FLISR function, after the operation of protective relays and respective circuit breakers (CBs), mainly relies upon the fault information reported by the fault indicators (FIs) installed along the faulted feeders [5]. However, most off-the-shelf FLIRS schemes, even the most advanced ones, have been designed to function in conventional single-source networks.

The increasing penetration of DERs such as PV systems or wind turbines (WTs) has transformed the traditional passive distribution networks into more active multi-sources ones. Under new circumstances, fault currents might be no longer uni-directional, which is from the external transmission system to the fault point but bidirectional, contributed by diverse connected DERs. This current can constantly change in magnitude and direction by fluctuations in the output of PV or WT due to uncertain factors such as weather, wind speed, and temperature. Consequently, FIs located downstream of the fault position might be triggered and report fault detection signals to the FLIRS, leading to false fault location [6–13].

As mentioned, the high and increasing PV penetration is slowly becoming a concern for FLIRS and DAS. Inverter technologies are improved and have become smarter to solve this problem partially. Some studies have integrated IoT into monitoring and controlling PV systems via inverters [14,15]. In addition, Big Data and deep learning methods are also applied to the process of predicting, evaluating the output, and diagnosing errors of PV [16]. Nevertheless, the effects on system failure or availability for the FLIRS function have not been considered.

Therefore, the impacts of high solar inverter integration on the performance of the FLIRS function operated in MV distribution network should be thoroughly investigated. The state-of-the-art review shows that a majority of recently published papers including [6–10] have been focused on the impedance-based or knowledge-based fault location methods which are currently rarely deployed in the field for the distribution network due to its heterogeneity. In the meanwhile, other studies reported in [11–13] did not take into account the difference in dynamic response of inverter-based DER in comparison with rotating generation, instead average models which could not provide accurate fault behavior were adopted.

Located in the middle of Vietnam, Danang city, according World Bank statistics, is a region with high capability of solar energy with, on average, about 2,283 h of sunshine per year and approximately with solar irradiance of about 4.89 kWh/m^2/day [17]. As is obvious, recent years have been observed a rapid increase in the installed solar capability in the network. Until 2021, a total of 2,528 PV systems is under operation with an overall installed power of around 80 MWp. In several 22 kV feeders, the solar generations have reached the feeder ratings and still keep increasing, and thus at some peak instants during a day, dispatch centers even require PV curtailment to reduce the grid congestion and overloading. Performance of the FLIRS, certainly, is at high risk of maloperation if faults occur during this period, resulting in reduced network reliability indices and even personnel hazards.

With all these aforementioned rationales, an extensive investigation on the performance of the FLIRS function currently operated in the Danang MV distribution network is of vital importance. The objective of the chapter, therefore, is to examine whether with such a high integration of inverter-based PV systems, the FLIRS scheme is still able to ensure its accuracy and reliability in locating the faulty section. In order to achieve high accuracy of analysis results, highly detailed models of PV systems are developed and built in the software environment DIgSILENT/PowerFactory. Moreover, MV feeder named Cau Do F474 is considered for the case study since it is one of the MV feeders with the highest PV penetration level. Its parameters including conductors, loads consumption, PV power, protection settings, and system short-circuit power are provided with high accuracy by the Danang Power Company. The results presented in this chapter are mainly captured from those studies reported in [18].

Considering the long-term target, IoT platforms can be embedded in the DAS in combination with FLIRS to easily control the PV system and find the appropriate operating solution along with support from Big Data.

4.2   The MV distribution network under study

Da Nang PC is currently managing and operating 100 MV of 22 kV feeders with 1,205 km length; 4,194 secondary MV/LV distribution substations are operating at voltage levels of 22/0.4 kV with a total capacity of 2,200.6 MVA (including customer facilities). In 2015, the average capacity reached 515 kVA per substation, and reached 541.6 kVA per substation in 2021. Most of MV feeders are overhead lines with some being underground cables. The cross-sectional areas of the feeders are from 50 mm^2 to 300 mm^2 for aluminum conductors and 38 mm^2 to 500 mm^2 for copper conductors.

4.2.1   The studied F474 MV feeder

The F474 MV Cau Do feeder is selected for the case study due to its high penetration of inverter-based PV systems compared with other feeders in the Danang MV network. Overall power production delivered by the connected PV systems reaches around 8 MW at 12 pm which is equal to the feeder power rating. The feeder is 2.2 km length and consists of mainly 240 mm^2 conductors. MV/LV substations and PV systems are connected distributively along the feeder. The feeder single-line diagram modeled in DIgSILENT is provided in Figure 4.1.

4.2.2   Profiles of load and PV systems connected to F474 feeder

Typical daily characteristics of LV aggregated loads as well as of the connected PV systems are illustrated in Figure 4.2. Load and PV generation values were provided by the Danang Power Company.

4.2.3   Settings of MV feeders overcurrent protection

The feeder protection relay setting parameters are provided by Central Dispatch Company and tabulated in Table 4.1.

4.2.4   Settings of PV undervoltage protection

The setting parameters for PV interface undervoltage protection are tabulated in Table 4.2.

4.3   Component modeling in DIgSILENT

4.3.1   Solar photovoltaic systems

The structural diagram of a three-phase inverter-based PV system is illustrated in Figure 4.3. The control scheme adopted for the PV inverters under study consists of two control circuits, including outer voltage loop and inner current loop [19]. The former assumes responsibility for regulating the inverter DC

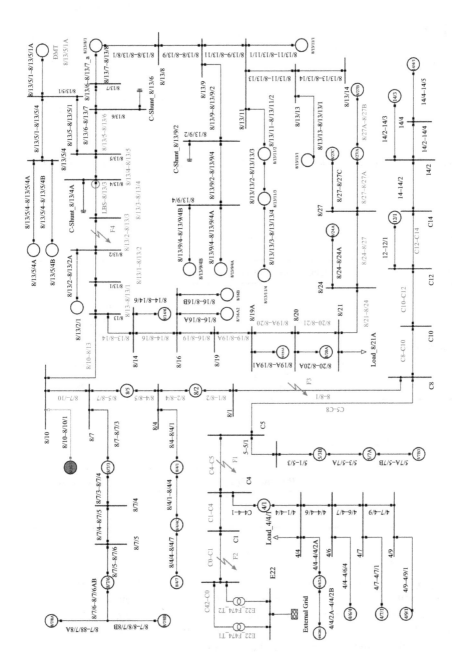

Figure 4.1 Single-line diagram of F474 MV Feeder modeled in DIgSILENT

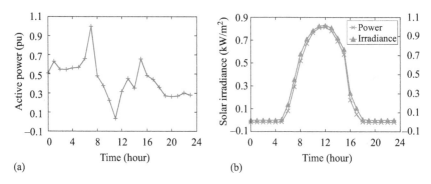

(a)   (b)

Figure 4.2   PV and load profiles on F474 MV feeder. (a) Load profile and (b) PV system profile

Table 4.1   Protection settings of F474 feeder

Level	Current threshold (A)	Operation time (s)
Phase element		
I	$I >= 600$	$t >= 0.15\frac{0.14}{M0.02-1}$
II	$I >> = 4{,}800$	$t>>=0.42$
III	$I >>>=8{,}000$	$t>>>=0.25$
Ground element		
I	$Ie > 72$	$te >= 0.4\frac{0.14}{M0.02-1}$
II	$Ie>>=4{,}800$	$te>>=0.42$
III	$Ie>>>=8{,}000$	$te>>>=0.25$

Table 4.2   PV interface protection settings

Level	Voltage threshold (pu)	Operation time (s)
I	$0.45 < V < 0.8$	3
II	$V < 0.45$	0.3

voltage (V_{DC}) to track the reference DC voltage (V_{DCref}) provided by the MPPT system. In the meantime, the latter regulates the inverter AC currents (i_d and i_q) to their respective reference currents generated by the voltage control loop, i.e. (i_{dref} and i_{qref}). Outputs of this control loop are the modulation indices, m_{dq}, which are then used for computing the switching pulses for IGBT gates of the solar inverters. Apart from V_{DC} control for extracting maximum active power from PV panels, today's smart inverters also provide a wide range of advanced

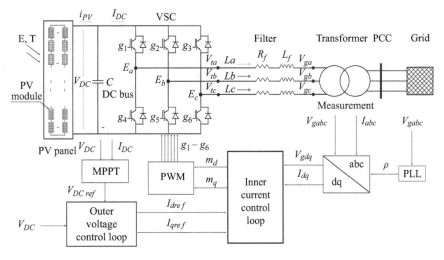

Figure 4.3   Schematic diagram of a PV system

grid support functions such as vol-var control, voltage control and capability of dynamic voltage support, fault ride through.

4.3.1.1   Inner current control loop

By observing Figure 4.3, grid voltage can be related with inverter AC-side current as

$$\frac{di_d}{L_f df_{i_d}} = L_f \omega i_q - R_f + V_{td} - V_{gd}$$

$$\frac{di_q}{L_f df_{i_q}} = L_f \omega i_d - R_f + V_{td} - V_{gq}$$

(4.1)

where V_{td} and V_{tq} are voltages at inverter terminal; V_{gd} and V_{gq} are grid voltages; i_d and i_q are inverter AC-side currents; L_f and R_f are filter inductance and resistance.

On the other hand, the inverter terminal voltages can be represented by its DC voltage by

$$V_{tdq} = \frac{V_{DC}}{2 \cdot m_{dq}}$$

(4.2)

In order to eliminate the dynamic coupling between d- and q-components in (4.1), new control inputs u_d and u_q are introduced as:

$$m_d = \frac{2}{V_{DC}}(u_d - L_f \omega_0 i_q + V_{gd})$$

$$m_q = \frac{2}{V_{DC}}(u_q - L_f \omega_0 i_d + V_{gq})$$

(4.3)

By substituting (4.3) into (4.1), we obtain

$$L_f \frac{di_d}{dt} = -R_f i_d + u_d + V_{gd}$$

$$\hspace{8cm} (4.4)$$

$$L_f \frac{di_q}{dt} = -R_f i_q + u_q + V_{gq}$$

Equation (4.4) shows that dynamics of d- and q-component currents, i.e. i_d and i_q, are linear and independent from each other, and thus can be regulated by two separate control loops. The schematic diagram of the PV inverter control scheme is shown in Figure 4.4. It can be seen that, u_d is the output of the *d*-axis PI controller that processes the deviation between i_d and i_{dref}. Likewise, u_q is the output of the *q*-axis PI controller that processes the deviation between i_q and i_{qref}. Transfer functions of these controllers are designed, as follows:

$$PI_d(s) = K_{pd} + \frac{K_{id}}{s}$$

$$\hspace{8cm} (4.5)$$

$$PI_q(s) = K_{pq} + \frac{K_{iq}}{s}$$

where K_{pd} and K_{pq} are proportional coefficients; K_{id} and K_{id} are integral coefficients.

4.3.1.2   Outer voltage control loop

The voltage loop regulates the voltage across the DC-link in accordance with the reference voltage (V_{DCref}) normally provided by the MPPT system as shown in Figure 4.3.

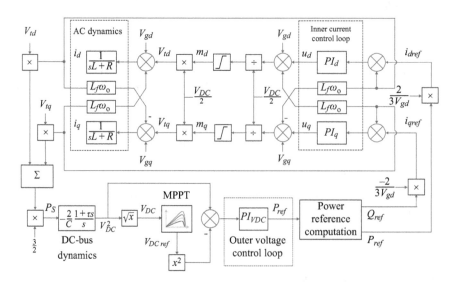

Figure 4.4   Schematic diagram of the PV inverter control scheme

This allows the DC power from the solar battery string to be extracted to the AC grid after DC/AC conversion. During this process, the charge and hence the voltage of the DC-link capacitor continues to decrease due to losses in the IGBT gates of the PV inverter. Therefore, the DC link capacitor needs to charge some amount of power to maintain the DC voltage at the reference value V_{DCref}. During the day, the voltage controller uses a small amount of DC power from the solar panels to compensate for the power loss and deliver the remaining power to the grid.

Equation of power balance at the DC side of the PV inverter can be expressed as

$$\frac{d}{dt}\frac{1}{2}CV_{DC}2 = P_{PV} - P_{INV,in} \tag{4.6}$$

where P_{PV} is DC power output of the PV panel and $P_{INV,in}$ is the DC input power of the PV inverter.

Equation (4.6) proves that with a certain DC output power of the considered PV panel, DC voltage of the DC-link capacitor can be controlled by regulating the input power of the PV inverter which in turn, can be regulated by inverter AC-side current (i_d). The compensation for the error between V_{DC}^2 and V_{DCref}^2 is realized by a PI controller as illustrated in Figure 4.4.

4.3.2 Aggregated loads

In this case study, all household and commercial loads connected to the secondary MV/LV distribution transformers are considered as aggregated loads and modeled by using a dynamic load model. This model improves the accuracy of simulation results compared to the static load model, i.e., the ZIP model, which only includes constant impedance loads, constant current loads, and constant power loads. In addition to static load, the dynamic load model also consists of motor, 22/0.4 kV transformer, total impedance of the equivalent LV feeder, and electronic load. The composition of the modeled aggregated load is provided in Figure 4.5.

4.3.3 External grid

As the main study interest lies in the distribution network, the external grid is modeled as a voltage source with an internal impedance. The model implies that all synchronous

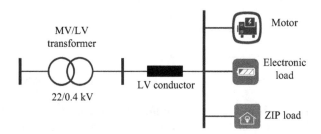

Figure 4.5   Aggregated load model

generating units connected to the grid are in remote areas far from the distribution networks under study. Values of the internal impedance are calculated following the short-circuit calculation results provided by the Central Dispatch Company.

4.4   Case study and discussions

In order to study the influence of the solar PV system on performance of the FLISR function installed in the Danang MV distribution networks, different fault types occurring at different times over a day with various fault impedance values are considered. However, because there is no solar radiation at night and PV systems in Danang network are not required to participate in generating reactive power to control voltage at night (Q-night mode) we only need to take into consideration a few hours of the day with the highest solar radiation (Figure 4.2(b)), typically 10, 12, 14, and 16 h.

4.4.1   Faults at the beginning of the feeder

The fault is located in the middle of section C0–C1, around 120 m from the 22 kV busbar (F2 in Figure 4.1). Fault impedance considered varies from 0 to 2 Ω. For faults of larger impedances, voltage drop is negligible, and thus PV fault currents are not significantly different from the pre-fault values and are therefore not shown.

4.4.1.1   Three-phase faults

In all fault scenarios simulated, the short-circuit currents flowing through the feeder circuit breaker are greater than instantaneous setting threshold of the feeder protection, i.e., 8,000 A in Table 4.1. Consequently, all faults are cleared after 0.25 s as shown in Figure 4.6(a) for the 12 h scenario. However, the voltage drops are clearly different upon different fault impedances, with a deep drop when $R_f \leq 0.5$ Ω and a negligible decrease when $R_f > 0.5$ Ω (Figure 4.6(b)).

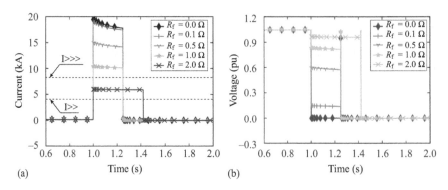

(a)

(b)

Figure 4.6   Three-phase short circuit at the beginning of the F474 feeder (F2 in Figure 4.1) at 12 p.m. (a) Currents flowing through feeder CB, and (b) Voltage at the PV 8/13/5/1A terminal

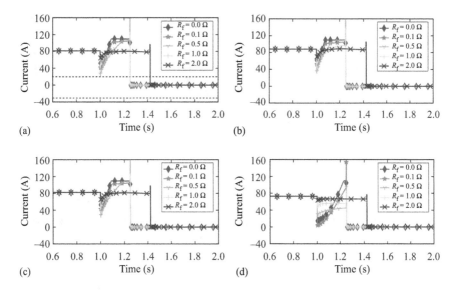

Figure 4.7   *Fault currents through FI 8/13/3 for a three-phase fault at the beginning of the feeder at different hours during a day. (a) Fault at 10 h, (b) Fault at 12 h, (c) Fault at 14 h, and (d) Fault at 16 h*

The fault currents passing through the considered fault indicator FI 8/13/3, placed inside LBS 8/13/3 in Figure 4.1 at 10, 12, 14, and 16 h are shown in Figure 4.7. The fault currents are mainly supplied by PV inverters installed downstream of the LBS in the direction away from the 22 kV busbar. It can be seen that there are different fault current levels depending on the value of the fault impedance, which, in turn, cause different voltage drops at the inverter PCCs. However, in general, all fault currents recorded by FI 8/13/3 are less than 120 A, which is lower than its 600 A setting threshold. Therefore, FI 8/13/3 is not activated for all three-phase faults at the beginning of the feeder.

4.4.1.2   Ground faults

As with the three-phase fault case above, except for the case where $R_f = 2$, phase-to-ground faults at F2 in Figure 4.1 with $R_f = 1$ or less all cause a fault current exceeding 8,000 A, which corresponds to level III of the overcurrent protection function (Table 4.1). All faults are eliminated after 0.25 s as shown in Figure 4.8(a). Only for $R_f = 2 \ \Omega$ that the fault current resembles the level II of the feeder relay and thus is tripped after 0.42 s. It should be noted that even though it is a single-phase ground fault, phase current is large enough for the 50/51P function to operate. Therefore, it is unnecessary to consider the operation of the 50/51N ground function.

The voltage drop of the faulted phase in single-phase ground faults also varies with the fault impedance, as shown in Figure 4.9(b). With such diverse voltage drops, the PV faults current are different, and therefore the levels of fault current

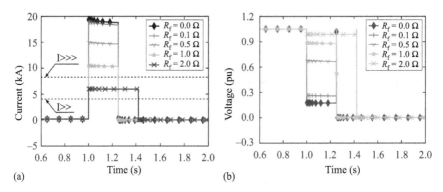

Figure 4.8   Single-phase short circuit at the beginning of the F474 feeder (F2 in Figure 4.1) at 12 h. (a) Currents flowing through feeder CB, and (b) Voltage at the PV 8/13/5/1A terminal

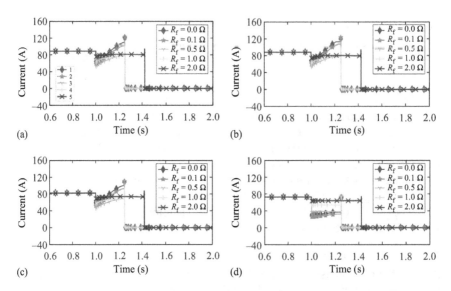

Figure 4.9   Fault currents through LBS 8/13/3 for a ground fault at the beginning of the feeder at different hours during a day. (a) Fault at 10 h, (b) Fault at 12 h, (c) Fault at 14 h, and (d) Fault at 16 h

flowing in FI 8/13/3 (Figure 4.1) will be different. However, the simulation results show that the fault currents are also limited and still below the 600 A fault detection threshold of FI 8/13/3 as evidently shown in Figure 4.9. Therefore, FI 8/13/3 is not be activated for all these fault scenarios.

4.4.2   Faults in between feeder CB and FI 8/13/3

The fault is located in the middle of section 8/8/1, i.e. F3 in Figure 4.1 with fault impedance ranging from 0 to 2 Ω.

4.4.2.1   Three-phase faults

All faults are isolated after 0.25 s, except for R_f = 2 Ω which is eliminated after 0.42 s. There is a significant difference in voltage drop between the different fault impedances, with a considerable drop when $R_f \leq 0.5$ Ω and negligible drops when $R_f > 0.5$ Ω. At R_f = 2 Ω, the voltage at the PV 8/13/5/1A terminal is almost negligible and is in the dead band of its inverter controller. The fault currents flowing through FI 8/13/3 in all scenarios are shown in Figure 4.10. It can be seen that there is a different increase in the fault current depending on fault impedances; however, all the currents are under 120 A threshold, which is not sufficient enough to activate the FI 8/13/3.

4.4.2.2   Ground faults

Similar to the three-phase fault scenario, the voltage drop between fault phases for a one-phase ground short circuit also varies depending on the value of the fault impedance. However, simulation results show that the current values through the LBS 8/13/3 are also limited to less than 120 A as can be seen in Figure 4.11. Therefore, the fault detection function of this FI is not activated.

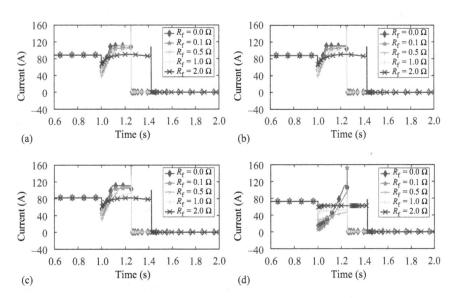

Figure 4.10   Fault currents through FI 8/13/3 for a three-phase fault in between feeder CB and itself at different hours during a day. (a) Fault at 10 h, (b) Fault at 12 h, (c) Fault at 14 h, and (d) Fault at 16 h

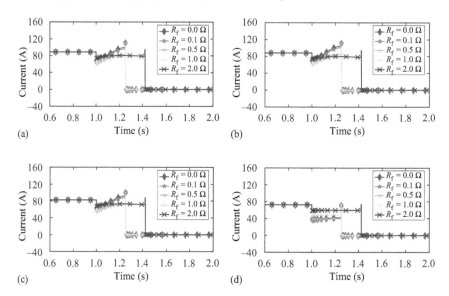

Figure 4.11   Fault currents through FI for a one-phase ground fault in between feeder CB and FI 8/13/3 at different hours during a day. (a) Fault at 10 h, (b) Fault at 12 h, (c) Fault at 14 h, and (d) Fault at 16 h

4.4.3   Faults closed to FI 8/3/13

The fault considered in this case is located at the end of section 8/13/2–8/13/3 next to the considered FI 8/13/3 which is fault point F4 in Figure 4.1.

4.4.3.1   Three-phase faults

For this fault scenario, only fault with $R_f = 2\,\Omega$ is cleared after 0.42 s as the fault current is greater than 4,200 A but less than 8,000 A. Other faults with less R_f values are isolated after 0.25 s. We see that the largest fault current values through FI 8/13/3 are at 12 and 14 hours with $R_f = 0\,\Omega$, corresponding to Figures 4.12(b) and 4.12(c), respectively. However, all these currents are smaller than 120 A, and thus do not trigger the operation of FI 8/13/3.

4.4.3.2   Ground faults

In contrast to the previous ground faults at F2 and F3, this ground fault scenario causes smaller fault currents due to considerable increase in impedance of the faulted circuit. Therefore, faults are removed by 50/51N function at different delay times and different voltage drops are thus observed.

However, as is evident, the fault location is close to those PV systems located downstream of the considered FI 8/13/3 and thus causes a significant voltage drop at their terminals. In certain scenarios, the voltage can reduce to zero, for instance, in the case of a direct or 0, 1 Ω fault. As a result, the fault currents contributed by the PV systems running across FI 8/13/3 are only marginally increased compared to the pre-fault currents, as demonstrated in Figure 4.13. The result again helps confirm that the PV fault currents are unable to activate the FI 8/13/3 operation.

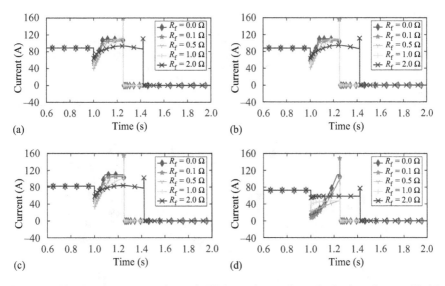

Figure 4.12   Fault currents through FI for a three-phase fault closed to itself. (a) Fault at 10 h, (b) Fault at 12 h, (c) Fault at 14 h, and (d) Fault at 16 h

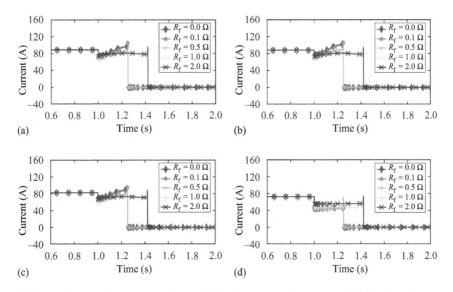

Figure 4.13   Fault currents through FI for a one-phase ground fault closed to itself. (a) Fault at 10 h, (b) Fault at 12 h, (c) Fault at 14 h, and (d) Fault at 16 h

Table 4.3.   Parameters for all analyzed fault scenarios

Fault type	Phase-to-ground, two-phase, two-phase-to-ground, three-phase
Distance to source (m)	100–1,500 with increments of 100 m
Time (h)	8, 9, 10, 11, 12, 13, 14, 15, 16
Fault impedance (Ω)	0, 0.1, 0.2, 0.5, 1

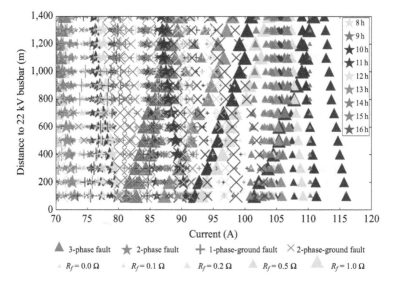

Figure 4.14   Overall presentation of all 2,700 analyzed scenarios

4.4.4   Overall presentation and conclusion

The parameters for all simulated fault scenarios for the F474 Cau Do feeder are provided in Table 4.3. A total of 2,700 fault cases were simulated and analyzed. All the average values of the short-circuit current flowing through the FI 8/13/3 over a full cycle window length are systematically presented in Figure 4.14.

As is obvious from Figure 4.14, the fault current supplied by the PV systems flowing through FI 8/13/3 peaks at 11 h in the case of a three-phase 0.5 Ω fault located close to 22 kV busbar. The same fault conditions, but at 10 h, also cause higher fault currents than the remaining conditions. Two-phase faults are the least dangerous because the fault currents detected are much smaller compared to other fault types.

4.5   Perspectives

In near future, there are other types of distributed non-inverter-based generation, such as type 3 wind generators (DFIG) or waste-to-energy plants, which will be connected to 22 kV MV distribution networks. For generation of this type, especially synchronous waste-to-energy plants, the characteristics of the fault current largely depend on the total impedance of the faulted circuit. Thus, the magnitude of the fault current may be greater than the threshold value of 600 A, causing the FI located downstream of the fault point inaccurately report the fault detection, the FLISR function fails to determine the exact location of the faulted section. Moreover, it might present challenges to relay protection coordination which has been historically carried out based on the single-source philosophy. Therefore, further research involving waste-to-energy plants or DFIG wind turbines needs to be conducted.

On the other hand, it is necessary to have comprehensive control over the operation and development of this type of generator due to its instability. Embedded solutions for IoT communication need to be integrated into specialized software that gives operators real-time data for quickly changing the DAS settings and improving the stability of the whole system. Parallel with that, increasing the accuracy of renewable energy forecasting and power output assessment technologies is also key to enhancing system sustainability, where Big Data and artificial intelligence can help.

When the market share structure of renewable energy sources is rising, gradually take over fossil energy proportion, IoT and Big Data application for system management has been placed in sustainable system development vision of Vietnam Electricity (EVN). The orientation suggests that IoT application to multi-faceted communication between operators, consumer and devices can monitor and control the system strictly, especially transparency in energy transaction when the electricity market is gradually expanding. The combination of IoT with DAS can provide solutions for remote power system operation, change operation feeder to avoid the fault impact on FLIRS, or recovery of failed feeders can be performed easily and safely, typical with distribution network with high PV penetration like Danang city. Big Data will be applied for improving the quality of forecasting output capacity of renewable energy in order to provide safety scenarios for the power system. Along with that is calculating energy storage solutions for electricity market operation and ancillary services.

4.6   Conclusion

This chapter conducted extensive investigation on the impacts of fault currents contributed by PV systems on the performance of the FLISR functionality integrated within the DAS system being operated in Danang MV distribution network. The results obtained indicate that, due to the limitation of the PV inverter maximum current up to 1.2 times their nominal currents, fault currents contributed by the downstream-connected PV systems unlikely present any challenges to the FLISR performance.

Even if all PV systems connected to the F474 Cau Do feeder are assumed to be located behind FI 8/13/3, the maximum fault current that flow through FI 8/13/3 reaches only about 260 A. If we increase the maximum output current limit of the involved PV inverter to 1.6 pu, a maximum value of only about 340 A can be achieved. This confirms that the short-circuit currents generated by the PV systems connected to 22 kV feeders, considering the feeder maximum hosting capability, do not affect the performance of the existing FLIRS function.

Acknowledgment

This research was funded by the Ministry of Education and Training under project number CT 2022.07.DNA.06.

References

[1] Akikur RK, Saidur R, Ping HW, *et al.* Comparative study of stand-alone and hybrid solar energy systems suitable for off-grid rural electrification: a review. *Renewable and Sustainable Energy Reviews*. 2013;27:738–752.

[2] Brown TW, Bischof-Niemz T, Blok K, *et al.* Response to Burden of proof: a comprehensive review of the feasibility of 100% renewable-electricity systems. *Renewable and Sustainable Energy Reviews*. 2018;92:834–847.

[3] Joshi AS, Dincer I, and Reddy BV. Performance analysis of photovoltaic systems: a review. *Renewable and Sustainable Energy Reviews*. 2009; 13(8):1884–1897.

[4] Kumar NM, Chand AA, Malvoni M, *et al.* Distributed energy resources and the application of AI, IoT, and blockchain in smart grids. *Energies*. 2020; 13(21):5739.

[5] Hoang TT, Tran QT, and Besanger Y. A multiagent and IEC 61850-based fault location and isolation system for distribution network with high PV integration – a CHIL implementation. In: *2019 IEEE International Conference on Environment and Electrical Engineering and 2019 IEEE Industrial and Commercial Power Systems Europe (EEEIC/I&CPS Europe)*. IEEE; 2019. p. 1–6.

[6] Menchafou Y, El Markhi H, Zahri M, *et al.* Impact of distributed generation integration in electric power distribution systems on fault location methods. In: *2015 3rd International Renewable and Sustainable Energy Conference (IRSEC)*; 2015. p. 1–5.

[7] Jahanger HK, Sumner M, and Thomas DWP. Influence of DGs on the single-ended impedance based fault location technique. In: *2018 IEEE International Conference on Electrical Systems for Aircraft, Railway, Ship Propulsion and Road Vehicles International Transportation Electrification Conference (ESARS-ITEC)*; 2018. p. 1–5.

[8] Perez R, Vásquez C, and Viloria A. An intelligent strategy for faults location in distribution networks with distributed generation. *Journal of Intelligent & Fuzzy Systems*. 2019;36(2):1627–1637.

[9] Dashti R, Ghasemi M, and Daisy M. Fault location in power distribution network with presence of distributed generation resources using impedance based method and applying π line model. *Energy*. 2018;159:344–360.

[10]   Jiang Y. Data-driven fault location of electric power distribution systems with distributed generation. *IEEE Transactions on Smart Grid*. 2020; 11(1):129–137.

[11]   Seuss J, Reno MJ, Broderick RJ, *et al.* Determining the impact of steady-state PV fault current injections on distribution protection. *Sandia National Lab.(SNL-NM), Albuquerque, NM (United States)*; 2017.

[12]   Bhagavathy S, Pearsall N, Putrus G, *et al.* Performance of UK distribution networks with single-phase PV systems under fault. *International Journal of Electrical Power & Energy Systems*. 2019;113:713–725.

[13]   Hariri A and Faruque MO. A hybrid simulation tool for the study of PV integration impacts on distribution networks. *IEEE Transactions on Sustainable Energy*. 2016;8(2):648–657.

[14]   Gupta J, Singla MK, Nijhawan P, *et al.* An IoT-based controller realization for PV system monitoring and control. In: *Business Intelligence for Enterprise Internet of Things*. New York, NY: Springer; 2020. p. 213–223.

[15]   Mansouri M, Trabelsi M, Nounou H, *et al.* Deep learning based fault diagnosis of photovoltaic systems: a comprehensive review and enhancement prospects. *IEEE Access*. 2021;10:13852–13869.

[16]   Dash SK, Garg P, Mishra S, Chakraborty S and Elangovan D Investigation of adaptive intelligent MPPT algorithm for a low-cost IoT enabled standalone PV system. *Australian Journal of Electrical and Electronics Engineering*. 2022;19(3): 261–269, doi: 10.1080/1448837X.2021.2023251

[17]   Vietnam – Solar Irradiation and PV Power Potential Maps; 2021. [Online; accessed Oct 21, 2021]. https://datacatalog.worldbank.org/search/dataset/0041731.

[18]   Hoang TT, Tran QT, Le HS, *et al.* Impacts of high solar inverter integration on performance of FLIRS function: case study for danang distribution network. In: *2022 11th International Conference on Control, Automation and Information Sciences*. IEEE; 2022. p. 1–6.

[19]   Varma RK. *Smart Solar PV Inverters with Advanced Grid Support Functionalities*. New York, NY: John Wiley & Sons; 2021.

Chapter 5

Processor-in-the-loop implementation for PV water pumping applications

Mustapha Errouha[1], Quentin Combe[2] and Saad Motahhir[3]

In the past few years, photovoltaic (PV)-based water pumping systems for rural areas are receiving huge attention because of economic and environmental considerations compared to diesel-powered water pumping. To achieve a fast code verification before the hardware implementation on real test bench, a processor-in-the-loop (PIL) simulation of a PV water pumping system based on two stages of power conversion is proposed. An indirect field-oriented control (IFOC) is introduced to drive an induction motor powered by solar PV panels. Further, the boost converter is controlled using the step-size incremental conductance technique. The PIL approach consists of combining the microcontroller and the computer simulation. The embedded C code generated from the simulated mathematical formulations is launched into a microcontroller board, Arduino Mega. To validate the proposed design, the obtained MATLAB®/Simulink® results are compared to those using the PIL test. The obtained results show that the model in the loop test's output is in accordance with processor in the loop test's output for the presented control strategies.

5.1 Introduction

The life of inhabitants in many countries would be under real threat due to water scarcity. For rural sites, like urban areas, access to water is important for many energy services such as water pumping, especially for developing countries in which the subsistence of the population depends on agriculture [1]. Many rural areas have suffered from water scarcity, which makes the use of groundwater a better alternative in the absence of surface water.

So, the agricultural sector depends strongly on electrical energy availability. According to WEO-2016, nearly 15% of the global population do not have access

[1]Plasma and Conversion of Energy Laboratory, ENSEEIHT, University of Toulouse, France
[2]LEMTA, University of Lorraine, Vandœuvre-lès-Nancy, France
[3]ENSA, SMBA University, FEZ, Morocco

to electricity, and most of this population lives in rural areas [2]. Moreover, the absence of electricity reflects the lack of sustainable irrigation systems [3]. The extension of a centralized grid can be considered as a possible solution in remote and rural areas, but this option needs huge capital costs. Renewable sources of electricity represent a suitable solution to attenuate the negative effects of non-access to electricity for irrigation.

The use of PV water pumping systems for irrigation installed in remote and rural areas is considered as a sustainable and reliable solution to reduce the high rates of lack of access to electricity and offers numerous advantages, including environmental consciousness, operation safety, and robustness.

The induction motor (IM) is preferred for solar PV water pumping systems due to its rugged construction and maintenance-free. Various control strategies are implemented to drive the induction motor [4]. In [5], V/f technique is used. This method is easy to implement and simple. In [6], direct torque control is introduced. However, it suffers from torque and flux oscillations [7,8]. IFOC provides a faster and more good response. Consequently, the IFOC method is employed to operate the motor that drives the pump. It consists of controlling the engine based on the stator currents, in which the direct stator current serves to control the flux while the quadrature stator current is used for torque control [9].

The proposed power converter and controllers are designed and simulated in MATLAB/Simulink software. Then, validation tests must be effectuated using real hardware.

However, developing a prototype is not easy due to the complexity of the topology and circuit and the high cost of the components, of which parameter mismatches or unexpected conditions can degrade the elements.Moreover, it is a long process. On the other hand, the deviations of the system parameters and time delays influence the performance which can be different from the embedded software operating on the controller. The model-based testing and model-based design are advanced methods utilized in different disciplines. These techniques are elaborated to reduce the time of the testing and development process. The PIL test is one of the methods utilized in the system development process which consists of implementing the generated code in the desired processor by ensuring the communication between the used processor and computer simulation soft-ware to detect early and prevent the error. This technique reduces the cost and accelerates the testing phase. The validation of the method using the PIL approach is employed in different fields. In [10], the PIL test is used in grid-connected PV systems. In [11], the authors used the PIL for wind energy conversion applications. In [12], the validation technique is employed for electric vehicles. Moreover, various embedded boards are used in PIL test such as Raspberry Pi 4B board [13], TMS 320F28379D control board [14,15], embedded system F28M35xx [16], and STM32F4 discovery board [17]. Therefore, a design and experimental verification of the PIL approach using a low-cost Arduino mega board for PV water pumping system applications is presented. The proposed PV system consists of an IFOC method for the induction motor driving the pump combined with a variable step size incremental conductance method that extracts the maximum power from the panels. The validation of the proposed methods presented in this chapter is supported by several simulations under different operating conditions. The PIL test is applied to the two proposed control strategies. Moreover, the obtained MATLAB/Simulink simulation results are compared to those using the PIL platforms.

The chapter is structured as follows: Section 5.2 presents the PV water pumping system description. The proposed control strategies are discussed in Section 5.3. The PIL approach is described in Section 5.4. Simulation and PIL test results are presented in Section 5.5. Finally, Section 5.6 concludes the chapter.

5.2 PV water pumping system description

The proposed system is composed of PV panels associated with DC/DC converter which is controlled by a variable step size incremental conductance method. DC/AC inverter is employed to control the induction motor driving a centrifugal pump (Figure 5.1).

5.2.1 PV panel

To achieve the appropriate power, a series module is realized using PV panels. Therefore, the current–voltage relationship can be expressed by [18,19]:

$$I_{pv} = I_{ph} - \frac{(R_s I_{pv} + V_{pv})}{R_{sh}} - I_0 \left(\exp \frac{q(R_s I_{pv} + V_{pv})}{CKTN_s} - 1 \right) \tag{5.1}$$

where
I_{pv}, V_{pv}: PV panel current and voltage
I_{ph}, I_0: panel photocurrent and diode saturation current
R_s, R_{sh}: series and shunt resistances
a: diode's ideality factor
K: Boltzmann constant
T: Junction temperature

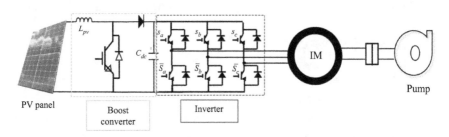

Figure 5.1   PV water pumping system

N_s: number of panels in series
q: electron charge

5.2.2   DC/DC boost converter

The DC/DC converter is inserted to avoid the undesired effects of PV power output and allows it to operate at the maximum power point. In this work, the boost converter is utilized to increase the voltage, thus, the relationship between its outputs and inputs characteristics [20]:

$$V_{pv} = V_{dc}(1 - \alpha) \tag{5.2}$$

$$I_{dc} = I_{pv}(1 - \alpha) \tag{5.3}$$

where
V_{dc}, I_{dc}: the outputs voltage and current of DC/DC converter
α: duty cycle

5.2.3   DC/AC inverter

The two-level inverter is inserted to convert the DC voltage into AC voltage which feeds the induction motor. Thus, it is composed of six transistor switches and the generated voltages can be expressed by [21]:

$$V_a = (2Sa - Sb - Sc)\frac{V_{dc}}{3}$$

$$V_b = (-Sa + 2Sb - Sc)\frac{V_{dc}}{3} \tag{5.4}$$

$$V_c = (-Sa - Sb + 2Sc)\frac{V_{dc}}{3}$$

where
V_{dc}: DC link voltage
Sa, Sb, Sc: inverter switching functions

5.2.4   Induction motor

The IM dynamic equations in the dq coordinate can be reported as follows [22]:

$$\begin{cases} V_{ds} = R_S i_{ds} + \dfrac{d\phi_{ds}}{dt} - w_s\phi_{qs} \\ V_{qs} = R_S i_{qs} + \dfrac{d\phi_{qs}}{dt} + w_s\phi_{ds} \\ V_{dr} = R_r i_{dr} + \dfrac{d\phi_{dr}}{dt} - (w_s - pw_r)\phi_{qr} \\ V_{qr} = R_r i_{qr} + \dfrac{d\phi_{qr}}{dt} + (w_s - pw_r)\phi_{dr} \end{cases} \tag{5.5}$$

where

V_{ds}, V_{qs}: stator voltage components
V_{dr}, V_{qr}: rotor voltage components
i_{ds}, i_{qs}: stator current components
i_{dr}, i_{qr}: rotor current components
R_S, R_r: stator and rotor resistances
ϕ_{ds}, ϕ_{qs}: stator flux components
ϕ_{qr}, ϕ_{qs}: rotor flux components
p: number of pole pairs

The stator and rotor flux can be expressed by:

$$\begin{cases} \phi_{ds} = l_s i_{ds} + l_m i_{dr} \\ \phi_{qs} = l_s i_{qs} + l_m i_{qr} \\ \phi_{dr} = l_r i_{dr} + l_m i_{ds} \\ \phi_{qr} = l_r i_{qr} + l_m i_{qs} \end{cases} \tag{5.6}$$

where

l_s, l_r, l_m: stator, rotor, and mutual stator–rotor inductances
The electromagnetic torque is given by:

$$T_{em} = \frac{P l_m}{l_r} \left(\phi_{ds} i_{qs} - \phi_{qs} i_{ds} \right) \tag{5.7}$$

5.2.5 Centrifugal pump

The proposed water pumping system utilizes a centrifugal pump in which the developed load torque can be calculated as follows [23]:

$$T_r = K \, \Omega^2 \tag{5.8}$$

where
K: pump constant.

5.3 Control strategies

5.3.1 Indirect field-oriented control

To achieve better control of the induction motor, an indirect field-oriented control is used. It consists in decomposing the machine currents into two components. The first one is i_{ds} which controls the flux and the second one is i_{qs} which controls the torque. The control strategy scheme is illustrated in Figure 5.2. In this technique, the rotor flux space is aligned with the direct axis (*d*), this allows

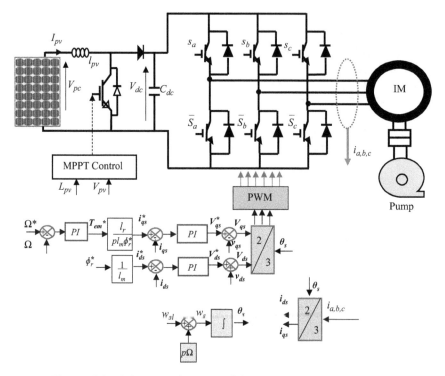

Figure 5.2   Schematic diagram of the PV water pumping system

to keep the quadrature axis (*q*) rotor flux always at zero [24,25]. Hence, the rotor flux components can be expressed as follows:

$$\phi_{dr} = \phi_r; \phi_{qr} = 0 \tag{5.9}$$

Then, the stator voltage components are decoupled and given by:

$$V^*_{ds} = (R_s + s\sigma l_s)i_{ds} = V_{ds} + w_s\sigma l_s i_{qs} = V_{ds} + v_{ds}$$
$$V^*_{qs} = (R_s + s\sigma l_s)i_{qs} = V_{qs} - \left(w_s \frac{l_m}{l_r}\phi_r + w_s\sigma l_s i_{qs}\right) = V_{qs} - v_{qs} \tag{5.10}$$

Consequently, the rotor flux and the torque are written as follows:

$$T^*_{em} = p\frac{l_m}{l_r}\phi_r i_{qs} \tag{5.11}$$

$$\phi^*_{dr} = l_m i_{ds} \tag{5.12}$$

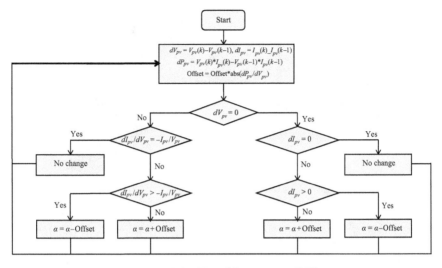

Figure 5.3   Variable step size INC

5.3.2   Maximum power point tracking algorithm

Various maximum power point tracking (MPPT) techniques are proposed in the literature to extract as much power as possible from PV panels. In this chapter, variable step
size incremental conductance (INC) is employed. The instantaneous and incremental
conductance of the PV panel are compared. The maximum power point is located when
the tangent slope $(\frac{i}{v} + \frac{di}{dv})$ of the power–voltage characteristic equals zero. Contrary to the
conventional INC, this method uses a variable step size which is adjusted automatically
to track the maximum power point as illustrated in Figure 5.3 [26].
where:
Offset: Step size

5.4   PIL test

PIL test is an effective method that can be used to validate the control strategy on a
corresponding microcontroller. During the PIL verification, the generated code is
tested in real-time while the plant model runs on a computer (Figure 5.4) which allows
to detect and correct possible errors. The PIL test platform is composed of Arduino
mega board and a computer. The Arduino mega board is connected with the PV water
pumping system modeled and run in the Simulink environment. The control board is
composed of an ATmega2560 microcontroller which is used for the controller's
programming, thanks to the fast-prototyping technique provided by the Simulink
environment. The elaborated model is utilized to create the C code using embedded
coders and Simulink real-time workshop, then it will be downloaded to the processor.
Moreover, the discrete versions are generated from the appropriate controllers.

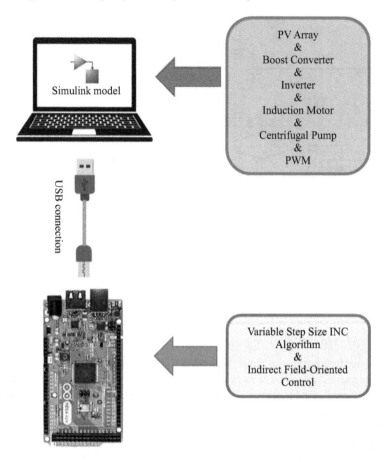

Figure 5.4   Processor-in-the-loop simulation block diagram

5.5   Simulation results

5.5.1   MATLAB®/Simulink® results

The proposed PV water pumping system consists of eight PV panels connected in series and is modeled and simulated using MATLAB/Simulink software. The parameters of the proposed PV system are given in Tables A.1 and A.2. The first step consists in evaluating the performance of the system based on the MPPT technique for various levels of irradiance. The irradiance is reduced from 1,000 W/m^2 to 600 W/m^2 at $t = 2$ s. Figure 5.5(A) and 5.5(B) illustrate the output power of PV panels with and without the proposed MPPT technique. According to Figure 5.5(A), variable step size INC technique forces the PV generator to function at the MPP despite weather changes. The obtained PV power is greater with the MPPT technique than the classical technique without the MPPT method.

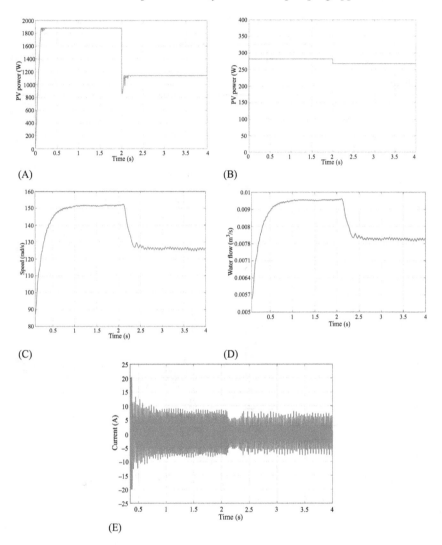

Figure 5.5   Simulation results. (A) PV power using MPPT; (B) PV power without MPPT; (C) motor speed; (D) water flow; (E) motor current

Figure 5.5(C) illustrates the motor speed. As can be observed, the motor speed is varying depending on the irradiance level. It increases from zero to about 153 rad/s when the radiation increases from zero to 1,000 W/m². In addition, it can reach optimal values for each level of irradiance. The water flow and stator current are shown in Figure 5.5(D) and 5.5(E), respectively. As seen, the optimal values are reached after a short transient for different operating conditions.

Table 5.1   Comparison between variable step size INC and P&O methods

Criteria	[27]	[28]	Used technique
Efficiency (%)	98.3	98.6	99.46
Oscillations	High	High	Neglected

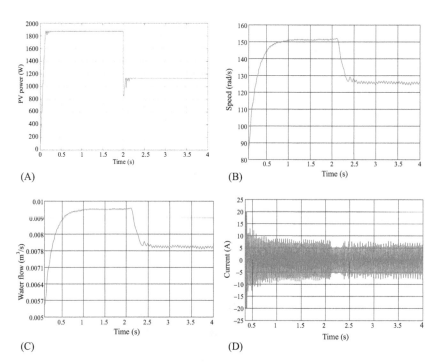

Figure 5.6   PIL results. (A) PV power using MPPT; (B) motor speed; (C) water flow; (D) motor current

Table 5.1 summarizes the performance comparison of the used method and the perturb and observe (P&O) technique utilized in different studies. We note that they do not refer to the same conditions because it is very difficult to find different studies realized under the same conditions.

5.5.2   PIL test results

This step consists to test the proposed control strategies which are variable step size INC and IFOC on a hardware processor. Therefore, Arduino mega is utilized as an embedded controller in this test.

Figure 5.6(A) illustrates that the power produced by the PV generator reaches its maximum for various levels of irradiance. The speed and stator current of the IM are shown in Figure 5.6(B) and 5.6(D), respectively. The pumped water is illustrated in Figure 5.6(C). It can be observed that the obtained simulation results using MATLAB/Simulink are similar to those of the PIL test. Hence, the control algorithm is verified on a real microcontroller.

5.6 Conclusion

A processor in the loop model of a standalone PV water pumping system for rural areas is presented. This technique aims to offer a proper rapid prototyping tool. The proposed PV system consists of two stages of power conversion which are controlled by variable step size INC and IFOC techniques. A microcontroller board based on the ATmega 2560 is utilized as an embedded controller in the test to perform the control process. Therefore, the proposed control strategies are verified on a real microcontroller. The performance of the proposed techniques was tested through numerical simulations and verified through experimental tests using a low-cost Arduino board. In comparing the results obtained from the simulation and PIL test, we conclude that the proposed controllers are practically implementable. A future study can be extended to prepare an experimental PV water pumping system for irrigation purposes. Moreover, robust control strategies will be developed to improve the dynamic performance of the system. The PIL approach will be effectuated before the hardware implementation on a real test bench to detect mismatch between PIL and practical results in the aim to provide a cost-effective and efficient PV water pumping system that can be used in rural areas.

Appendix A

Table A.1   Csun 235-60p PV panel characteristics

Maximum power	235 W
Open circuit voltage	36.8 V
Short circuit current	8.59 A
Maximum power voltage	29.5 V
Maximum power current P_{max}	7.97 A

Table A.2   IM characteristics

R_s, R_r	4.85 (Ω), 3.805 (Ω)
l_s, l_r	0.274 (H), 0.274 (H)
Nominal power	1.5 (KW)
P	2
Inertia moment	0.031 (kg m^2)
Viscous friction	0.00114 (N m s/rad)

References

[1] R. J. Chilundo, G. A. Maúre, and U. S. Mahanjane, "Dynamic mathematical model design of photovoltaic water pumping systems for horticultural crops irrigation: a guide to electrical energy potential assessment for increase access to electrical energy," *J. Clean. Prod.*, vol. 238, p. 105888, 2019, doi: 10.1016/j.jclepro.2019.117878.

[2] T. M. Tawfik, M. A. Badr, and O. E. Abdellatif, "Optimization and energy management of hybrid standalone energy system: a case study," *Reinf. Plast.*, vol. 25, pp. 48–56, 2018, doi: 10.1016/j.ref.2018.03.004.

[3] R. J. Chilundo, D. Neves, and U. S. Mahanjane, "Photovoltaic water pumping systems for horticultural crops irrigation: advancements and opportunities towards a green energy strategy for Mozambique," *Sustain. Energy Technol. Assessments*, vol. 33, pp. 61–68, 2019, doi: 10.1016/j. seta.2019.03.004.

[4] M. Errouha, S. Motahhir, Q. Combe, A. Derouich, and A. El Ghzizal, "Fuzzy-PI controller for photovoltaic water pumping systems," in *7th International Renewable and Sustainable Energy Conference (IRSEC), Nov. 27–30, 2019 – Agadir, Morocco*, 2019, pp. 0–5, doi: 10.1109/ IRSEC48032.2019.9078318.

[5] N. Yussif, O. H. Sabry, A. S. Abdel-Khalik, S. Ahmed, and A. M. Mohamed, "Enhanced quadratic V/f-based induction motor control of solar water pumping system," *Energies*, vol. 14, p. 101, 2021.

[6] S. A. A. Tarusan, A. Jidin, and M. L. M. Jamil, "The optimization of torque ripple reduction by using DTC-multilevel inverter," *ISA Trans.*, vol. 121, pp. 365–379, 2021, doi: 10.1016/j.isatra.2021.04.005.

[7] X. Wang, Z. Wang, and Z. Xu, "A hybrid direct torque control scheme for dual three-phase PMSM drives with improved operation performance," *IEEE Trans. Power Electron.*, vol. 34, no. 2, pp. 1622–1634, 2019, doi: 10.1109/TPEL.2018.2835454.

[8] M. Errouha, S. Motahhir, Q. Combe, and A. Derouich, "Intelligent control of induction motor for photovoltaic water pumping system," *SN Appl. Sci.*, vol. 3, no. 9, pp. 1–14, 2021, doi: 10.1007/s42452-021-04757-4.

[9] M. A. Hannan, J. A. Ali, A. Mohamed, and A. Hussain, "Optimization techniques to enhance the performance of induction motor drives: a review," *Renew. Sustain. Energy Rev.*, vol. 81, pp. 1611–1626, 2018, doi: 10.1016/j. rser.2017.05.240.

[10] B. Fekkak, M. Menaa, B. Boussahoua, and D. Rekioua, "Processor in the loop test for algorithms designed to control power electronics converters used in grid-connected photovoltaic system," *Int. Trans. Electr. Energy Syst.*, vol. 30, no. 2, pp. 1–19, 2020, doi: 10.1002/2050-7038.12227.

[11] M. Ali, A. Talha, and E. madjid Berkouk, "New M5P model tree-based control for doubly fed induction generator in wind energy conversion

system," *Wind Energy*, vol. 23, no. 9, pp. 1831–1845, 2020, doi: 10.1002/we.2519.

[12]   S. Arof, N. H. Diyanah, N. M. Yaakop, P. A. Mawby, and H. Arof, "Processor in the loop for testing series motor four quadrants drive direct current chopper for series motor driven electric car: Part1: chopper operation modes testing," *Adv. Struct. Mater.*, vol. 102, pp. 59–76, 2019, doi: 10.1007/978-3-030-05621-6_5.

[13]   E. Tramacere, S. Luciani, S. Feraco, A. Bonfitto, and N. Amati, "Processor-in-the-loop architecture design and experimental validation for an autonomous racing vehicle," *Appl. Sci.*, vol. 11, no. 16, p. 7225, 2021, doi: 10.3390/app11167225.

[14]   N. Ullah, Z. Farooq, I. Sami, M. S. Chowdhury, K. Techato, and H. Alkhammash, "Industrial grade adaptive control scheme for a micro-grid integrated dual active bridge driven battery storage system," *IEEE Access*, vol. 8, pp. 210435–210451, 2020, doi: 10.1109/ACCESS.2020.3039947.

[15]   Y. Belkhier, N. Ullah, and A. A. Al Alahmadi, "Efficiency maximization of grid-connected tidal stream turbine system: a supervisory energy-based speed control approach with processor in the loop experiment," *Sustainability*, vol. 13, no. 18, p. 10216, 2021, doi: 10.3390/su131810216.

[16]   D. R. Lopez-Flores, J. L. Duran-Gomez, and M. I. Chacon-Murguia, "A mechanical sensorless MPPT algorithm for a wind energy conversion system based on a modular multilayer perceptron and a processor-in-the-loop approach," *Electr. Power Syst. Res.*, vol. 186, p. 106409, 2020, doi: 10.1016/j.epsr.2020.106409.

[17]   S. Motahhir, A. El Ghzizal, S. Sebti, and A. Derouich, "MIL and SIL and PIL tests for MPPT algorithm," *Cogent Eng.*, vol. 4, no. 1, Article 1378475, 2017, doi: 10.1080/23311916.2017.1378475.

[18]   K. K. Prabhakaran, A. Karthikeyan, S. Varsha, B. V. Perumal, and S. Mishra, "Standalone single stage PV-fed reduced switch inverter based PMSM for water pumping application," *IEEE Trans. Ind. Appl.*, vol. 56, no. 6, pp. 6526–6535, 2020, doi: 10.1109/TIA.2020.3023870.

[19]   M. Errouha, S. Motahhir, Q. Combe, and A. Derouich, "Parameters extraction of single diode PV model and application in solar pumping," in *International Conference of Integrated Design and Production (CPI), Oct 14-16,2019 – Fez, Morocco*, 2021, pp. 178–191, doi: 10.1007/978-3-030-62199-5_16.

[20]   M. Errouha, A. Derouich, S. Motahhir, and O. Zamzoum, "Optimal control of induction motor for photovoltaic water pumping system," *Technol. Econ. Smart Grids Sustain. Energy*, vol. 5, no. 1, Article no. 6, 2020, doi: 10.1007/s40866-020-0078-9.

[21]   V. Koneti, K. Mahesh, and T. Anil Kumar, "Improved performance of photovoltaic array for water pumping by fuzzy control in sensorless vector control of induction motor drive," *Int. J. Ambient Energy*, vol. 43, pp. 6598–6607, 2021, doi: 10.1080/01430750.2021.1916589.

[22] S. Rafa, A. Larabi, L. Barazane, M. Manceur, N. Essounbouli, and A. Hamzaoui, "Implementation of a new fuzzy vector control of induction motor," *ISA Trans.*, vol. 53, no. 3, pp. 744–754, 2014, doi: 10.1016/j.isatra.2014.02.005.

[23] H. Parveen, U. Sharma, and B. Singh, "Pole reduction concept for control of SyRM based solar PV water pumping system for improved performance," *IEEE Trans. Ind. Electron.*, vol. 0046, no. c, pp. 1–1, 2020, doi: 10.1109/tie.2020.3000089.

[24] M. Errouha and A. Derouich, "Study and comparison results of the field oriented control for photovoltaic water pumping system applied on two cities in Morocco," *Bull. Electr. Eng. Informatics*, vol. 8, no. 4, pp. 1–7, 2019, doi: 10.11591/eei.v8i4.1301.

[25] M. Errouha, B. Nahid-mobarakeh, S. Motahhir, Q. Combe, and A. Derouich, "Embedded implementation of improved IFOC for solar photovoltaic water pumping system using dSpace," *Green Energy Technol.*, vol. 2021, pp. 435–456, 2021, doi: 10.1007/978-3-030-64565-6_15.

[26] M. Errouha, A. Derouich, N. El Ouanjli, and S. Motahhir, "High-performance standalone photovoltaic water pumping system using induction motor," *Int. J. Photoenergy*, vol. 2020, pp. 1–13, 2020, doi: 10.1155/2020/3872529.

[27] J. Ahmed and Z. Salam, "An improved perturb and observe (P&O) maximum power point tracking (MPPT) algorithm for higher efficiency," *Appl. Energy*, vol. 150, pp. 97–108, 2015, doi: 10.1016/j.apenergy.2015.04.006.

[28] S. Motahhir, A. El Hammoumi, and A. El Ghzizal, "Photovoltaic system with quantitative comparative between an improved MPPT and existing INC and P & O methods under fast varying of solar irradiation," *Energy Reports*, vol. 4, pp. 341–351, 2018, doi: 10.1016/j.egyr.2018.04.003.

Chapter 6

Advanced distributed maximum power point tracking technology

Guanying Chu[1], Huiqing Wen[1] and Yinxiao Zhu[1]

On the worldwide renewables market, photovoltaic (PV) generation systems (PGS) are gaining in popularity. Particularly in areas where energy is difficult to utilize, such as rural villages and mountain villages, the PGS plays a significant role due to its clean, inexpensive, and robust delivery. The PGS demonstrates a high-power conversion efficiency under an ideal condition. However, it exhibits low output efficiency under various non-ideal conditions such as the mismatch among PV modules or submodules. This chapter describes a new distributed maximum power point tracking (DMPPT) system to counter the complex and variable external environment. The low-rated power and simple management of the DC–DC converter in DMPPT technology allow the hardware's small size and low complexity, which is excellent for embedding into the junction-box behind the PV panel to produce a highly integrated module.

6.1 Introduction

Global energy demand continues to increase, at least for a time, as a result of rising wealth and standard of living in developing countries. Existing disparities in energy usage and accessibility persist. The structure of energy development is anticipated to evolve over time as the demand for energy and the need for environmental protection continues to increase. The significance of fossil fuels is falling, while the share of clean, renewable energy, and the role of electricity are growing. The energy consumption by different sources, including oil, natural gas, coal, nuclear, hydro, and renewable energy in the world at different ages is illustrated in Figure 6.1. The data from 2025 to 2050 is predicted by [1] based on a *rapid growth* model. It can be found that the total amount of energy consumption will be stable in the future. However, the proportion of energy types will change sufficiently. The representative of the fossil fuels, such as oil, will gradually decline in proportion, while the proportion of renewable energy that replaces it will increase sufficiently.

[1]Department of Electrical and Electronic Engineering, Xi'an Jiaotong-Liverpool University, China

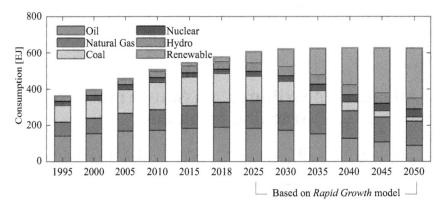

Figure 6.1   Global energy consumption from 1995 to 2050

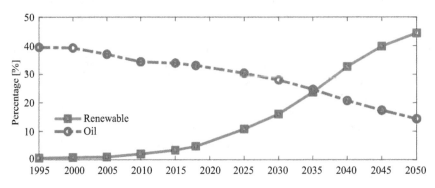

Figure 6.2   The percentage of oil energy and renewable energy consumption

As presented in Figure 6.2, the share of renewable energy will exceed that of oil and may even be close to the 43.7% share of the total energy consumption. Among the many renewable energy sources, PV energy has attracted people's attention and has raised rapidly because of its clean, low cost, and vital distribution characteristics. The consumption of renewable energy in 2015, 2018, and the next 30 years is shown in Figure 6.3. It is noted that the proportion of wind and PV energy is increasing, but the growth rate of PV energy will be much higher than that of wind energy for a period. This equates to a significantly faster rate of expansion in PV energy, supported by a more significant cost decline. Especially, for developing countries such as China and India, where the remote village cannot get utility power, the distributed PV system with a DC grid is a promising solution. According to the *Global Market Outlook* for PVs, China's installed PV capacity only accounts for 13% of the world's total. However, in 2019, this number increased to 31%, installed in nearly one-third of the world's PV capacities, as shown in Figure 6.4.

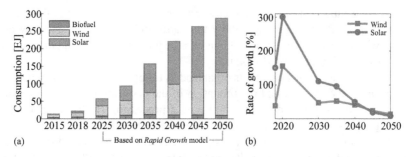

Figure 6.3   (a) Global energy consumption from 1995 to 2050 by renewable and (b) the growth rate of PV and wind

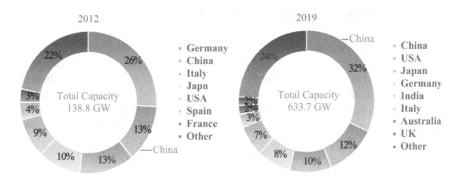

Figure 6.4   Total PV installed shares in 2012 and 2019

According to the International Energy Agency (IEA), the number of individuals without access to electricity has decreased from 1.2 billion in 2020 to 759 million in 2019. Specifically, electrification through localized renewable energy solutions gained traction. Between 2010 and 2019, the number of individuals linked to the microgrid has more than doubled, increasing from 5 million to 11 million. PV microgrids are a feasible option for rural electrification because of their low cost, compact footprint, and dispersed design. It has been discovered that, unlike larger isolated power systems, rural microgrids have a low energy demand since residential and street lighting loads make up the bulk of the loads. Therefore, these microgrids may have a single-phase design. It can be found that PV energy plays a very significant role in balancing energy demand, economic development, and environmental protection. However, what is different from the conventional fossil energy supply is that the energy generation efficiency of the PV system will be affected by the surrounding environment, such as the irradiance, temperature, and cloud location. Therefore, the problem we face is how to extract the maximum power from the PV system in light of a complex and changeable environment. This chapter describes a new DMPPT system to counter the complex and variable external environment. The low-rated power and simple management of the DC–DC converter in DMPPT technology allow the hardware's small size and low complexity, which is excellent for embedding into the junction box behind the PV panel to produce a highly integrated module.

The remainder of this chapter is organized as follows: Section 6.2 discusses the component of PV element and the limitation of the conventional power extracting method. Section 6.3 introduces the full power processing (FPP)-based DMPPT, including topologies, control algorithm, and the inherent flaw. Section 6.4 presents the differential power processing (DPP)-based DMPPT, including topology analysis and control algorithm. Note that this section includes three DPP architectures: PV–PV, PV–bus, and PV–IP. Finally, Section 6.5 presents the conclusion and future work.

6.2   Limitation of the standard MPPT

From top to bottom, the granularity of PGS can be organized as PV array, PV string, PV module, PV submodule, and PV cells. The primary component of the PGS is depicted in Figure 6.5. Numerous PV cells can be found coupled in series to make a PV submodule. Then, a PV module is formed by connecting two to four submodules in series. Finally, multiple PV modules are connected in series and parallel to form a PV array to boost the output voltage and current. This PGS system usually has a high output efficiency but is limited by some non-ideal factors, such as mismatches between PV modules or submodules, which are often the result of partial shading, aging cells, dust collection, heat gradients, or a manufacturing defect [3–11] as shown in Figure 6.6. The characteristic of the internal factor is that it is easy to model over a lifetime. The maintenance engineer can adjust the control plan by evaluating the ageing degree through regular inspection of the PV module. The most common and complex to predict are external factors. The factor of dust gathered can be solved by regularly cleaning the surface of the PV module. The temperature differential is deemed unimportant because

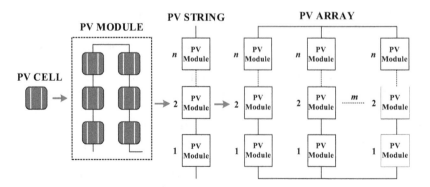

Figure 6.5   Components of the PV array

temperature variations occur so slowly that the temperature value is frequently seen as constant in relation to the fluctuation in irradiance that can occur during the day [12]. Therefore, it may be concluded that partial shading is the most common cause of a mismatch. Figure 6.7 depicts how tree, leaf, shadow, or cloud cover may produce nonuniform illumination of PV modules. This phenomenon is referred to as partial shading (PSCs).

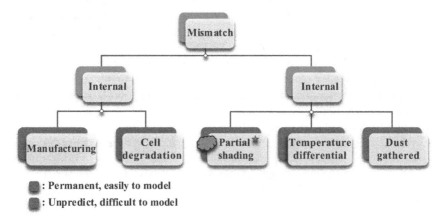

Figure 6.6   The breakdown of the factors that cause a mismatch in PGS

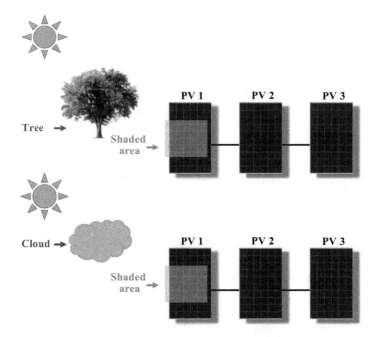

Figure 6.7   Partial shading scenarios in real life

The structure of the conventional centralized PV system is depicted in Figure 6.8. The structure is characterized by its simplicity and low cost in that we only use one centralized DC–DC or DC–AC converter adopting MPPT control to maximize the output of the PV string. Many traditional MPPT algorithms, including perturbation and observation (P&O), are practical for this application. However, the conventional series-string centralized architecture demonstrates poor resistance to non-ideal circumstances. Its characteristic is that a small shading block will limit the output of the entire PV string. In a series-connected PV string, for instance, if one PV module is shaded, as depicted in Figure 6.9, the short-circuit currents of the standard PV modules are higher than the short-circuit of the shaded module. If the PV string current is higher than the short-circuit current of the sha-ded module, the shaded module will be reverse-biased due to the other PV module in the series connection, resulting in a hot-spot effect. It causes local heating of the PV panel to cause further power loss. Cell damage may occur if the temperature is too high [2]. As shown in Figure 6.10, bypass diodes are linked antiparallel with several PV cell strings within PV modules to protect PV cells from hot spots. Standard PV modules are comprised of 60 or 72 PV cells connected in series and separated into two or three submodules by bypass diodes [13]. However, it cannot

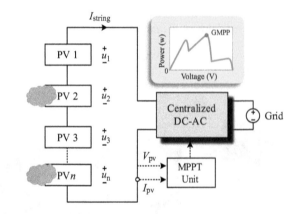

Figure 6.8   Centralized PV system

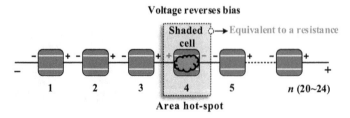

Figure 6.9   PV string under PSC without bypass diode

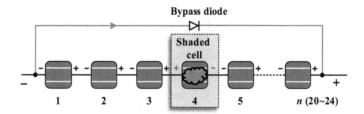

Figure 6.10   PV string under PSC with bypass diode

save the generated power output of the shaded PV submodule; meanwhile, the addition of bypass diodes results in many maxima in the power versus voltage (P–V) curve, which interferes with the standard MPPT algorithm's tracking efficiency.

> Traditional methods to solve the partial shading problem are based on algorithms, i.e., GMPPT algorithm. But this method also faces the problem of possibly being trapped in local maximum power points. Also, the idea based on algorithm cannot recover the energy bypassed by the bypass diode. This is the limitation of the conventional solution for solving the PSC issues.

6.3   DMPPT-full power processing architecture

The DMPPT PV design has been devised to compensate for the power loss caused by the PV module-level mismatch. These methods implement MPPT separately for each PV module. The study investigated the operation and performance of the DC power optimizer (DCPO), which is a distributed PV design based on FPP. It reduces the mismatch power loss by connecting a dedicated DC–DC converter in parallel with each PV module and implementing a separate MPPT [14–22], as shown in Figure 6.11. Furthermore, the partial shading effect on the actual power yield is reduced since any shading of a single module only affects its output power without limiting the performance of other modules in the PGS. The control in FPP architecture can be divided into two parts: local MPPT and global MPPT. Both MPPT algorithm levels are designed to extract and inject maximum power harvesting into the terminal. A simple P&O solution can be implemented for the local MPPT algorithm to drive the converter output voltage to its maximum [31], as shown in Figure 6.12. We can denote the output voltage of ith converter $V_{\text{out},i}$, and the output of the string of PV modules is I_{string}. The global control algorithm is the same; however, the perturbation frequency should be different. This time difference ensures that the two MPPT algorithms do not interfere with each other affecting the system's stability. The author in [31] presented a ratio that the local MPPT frequency has an upper bound of 25 kHz; meanwhile, the global MPPT frequency should be below 2.5 kHz (ten times slower). The two-level synchronous MPPT

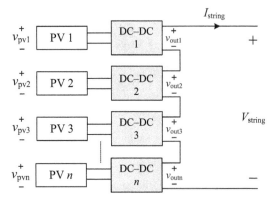

Figure 6.11   One of the architectures of distributed maximum power point tracking: DC power optimized

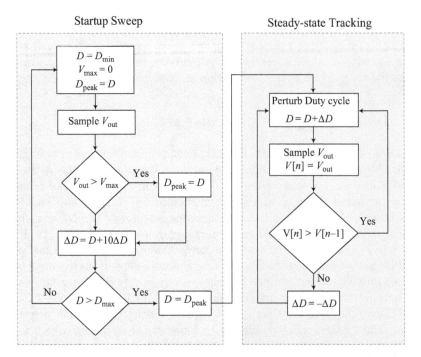

Figure 6.12   Flowchart diagram of the local MPPT algorithm in [17]

algorithm is effective for extracting the maximum power; however, each converter needs to implement real-time MPPT control requiring an independent IC unit that enhances the cost and complexity.

The author in [32] has presented a flexible maximum current point tracking (MCPT) solution, as shown in Figure 6.13. The enable signal clock c equals 1, 0, -1

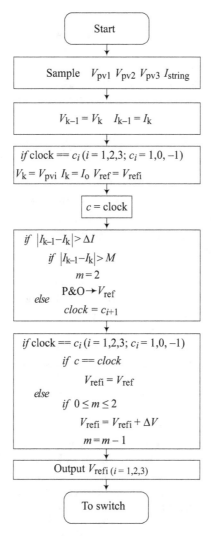

Figure 6.13   Flowchart diagram of the flexible MCPT algorithm in [20]

for PV modules 1, 2, and 3, respectively. At the outset of each algorithm operation, the output voltage of three PV submodules V_{pv1}, V_{pv2}, V_{pv3}, and the total output current I_o are measured. V_{k-1}, V_k, I_{k-1}, and I_k are the common voltage and current variables used for the embedded P&O algorithm. In simple terms, the MCPT control method uses only one IC unit to allow multiple PV modules to achieve MPPT in different time slices. For example, if the signal clock equals 1, the control unit tracks the MPP of PV module 1, while the other two PV modules follow the reference voltage of PV module 1. When the tracking of PV module 1 is finished, switch to the tracking of PV module 2; the whole procedure is cyclic. The motivation for this control method is to reduce hardware costs by reducing the number of IC units, based on the concept of change of irradiance is much slower than the calculation speed of controllers.

The above control solution combined with FPP architecture effectively alleviates the issues of PSC. However, PV module-level FPP techniques exhibit obvious limitations considering practical design and applications, which can be summarized as follows:

1. **System power density**: the power rating of dedicated converters should be set the same as the PV module. Combined with high voltage-ratio transformers and heatsinks, the size of individual converters is large and unfavorable for power density improvement and other power integration designs.
2. **Energy conversion efficiency**: MII typically demands high voltage step-up requirements, resulting in unfavorable transformer power losses. Furthermore, both MII and DCPO techniques can be categorized as FPP architecture since the whole generated power from the PV module has to go through the corresponding DC–DC converter. These characteristics will result in high power losses, which affect the improvement of system conversion efficiency.
3. **Reliability and lifetime**: considering that the electrical stress will largely determine the reliability and lifetime of electrical components, module-level DMPPT techniques face design challenges in reducing electrical stress.
4. **Energy yield performance**: module-level DMPPT techniques cannot address the PV intra-module mismatch problem. However, intra-mismatching is a more common partial shading phenomenon since bird dropping or leaves can only occupy part of PV modules rather than the complete PV module in natural conditions. Thus, the actual energy yield improvement using the module-level DMPPT techniques is limited.
5. **Integration and installation difficulty**: commercial PV panel integration and installation difficulties are challenging for FPP-based DMPPT. Because the PV submodules are serially linked during the lamination process and not via the junction box, complications emerge. The converter cannot function unless the surface lamination structure can be destroyed manually. However, the cost of damaging the surface structure of the PV panel is relatively high and may destroy the entire panel if not appropriately done unless PV panel manufacturers modify the electrical layout of the PV module. Nevertheless, for a mature commodity, it is not easy to do either. Therefore, attempting to embed the converter within the junction box of the FPP-based PV module is structurally impossible.

6.4 DMPPT-differential power processing

From the discussion above, one of the significant limitations of the FPP-based DMPPT is the level of granularity since it can only mitigate the mismatch issues among PV module levels. To solve this issue, the submodule-level DPP technology-based PV architecture that achieves finer-granularity MPPT was introduced in [23–27]. DPP technology is a submodule-level design that tackles the power differential between PV submodules, achieving a finer-granularity MPPT than FPP technology. The characteristic of its structure is to achieve power balance among PV submodules by extracting or injecting a small fraction of differential power through converters. In the case of slight or no mismatch, the converters in DPP architecture can even be turned off to avoid additional losses. However, FPP-based DMPPT architecture cannot achieve this function, and the inside converters must be normally open. Consequently, the hardware size and cost of the DC–DC converter employing DPP technology may be reduced, hence facilitating the design of power integration with PV submodules and even PV cells.

There are three main topologies of the DPP architectures introduced in this chapter, including PV-to-PV (PV–PV), PV-to-bus (PV–bus), and PV-to-isolated-bus (PV–IP), as shown in Figure 6.14(a)–(c), respectively. The commonly used converter type for PV–PV is a bidirectional buck-boost converter. Furthermore, a bidirectional flyback converter (BFC) is commonly used for PV-bus and PV-IP. The primary benefit of the DPP design is that the internal DC–DC converters process only a tiny proportion of the total produced power, hence reducing the power pressure of the DPP converters. Thus, compared with the FPP architecture such as DCPO, the hardware of DPP converters could be designed with a lower power rating to reduce the cost. Furthermore, the overall system efficiency of the DPP-based PV system is higher than that of the FPP-based PV system under a certain PSC. For example, Figure 6.15 shows a PV string with three PV elements. The bottom of the PV element is shaded, and the output power is reduced from 60 W to 30 W. If two solutions with DCPO and PV–bus architecture are adopted to mitigate the PSC, the power stress for each converter is presented in Figure 6.15. It is noticed that both of these two solutions can solve the PSC issues, but the total processed power with PV–bus architecture needs to handle is 150 W less than that using DCPO. Even though the conversion efficiency of the converter in PV–bus architecture is much lower than that in DCPO, the overall system efficiency of PV–bus architecture is still higher, which can be checked in Table 6.1. In addition to efficiency advantages, DPP technologies are well suited to be embedded in junction boxes from highly integrated PV modules thanks to their structural characteristic and low-rated power. The three topologies of DPP architecture own different electrical characteristics resulting in different hardware sizes and difficulty of integration, as shown in Figure 6.16. At the same time, different control strategies may also add or subtract some additional component requirements, such as voltage or current sensors. The following sections present the content of the analysis of topologies and control strategies within DPP architectures of PV–PV, PV–IP, and PV–bus.

6.4.1 PV–PV architecture

6.4.1.1 Topology and structure

The bidirectional buck-boost-based PV–PV architecture is shown in Figure 6.17. In PV–PV architecture, the individual converter number is one less than the PV

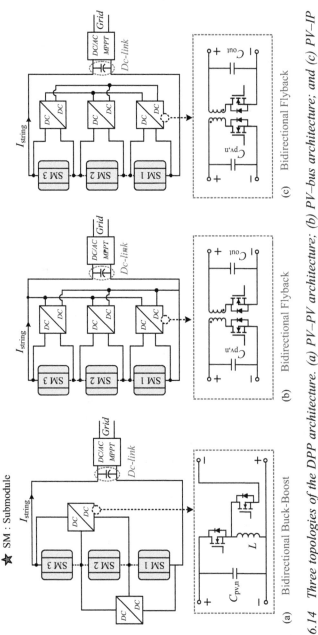

★ SM : Submodule

(a)   Bidirectional Buck-Boost

(b)   Bidirectional Flyback

(c)   Bidirectional Flyback

Figure 6.14   Three topologies of the DPP architecture. (a) PV–PV architecture; (b) PV–bus architecture; and (c) PV–IP architecture

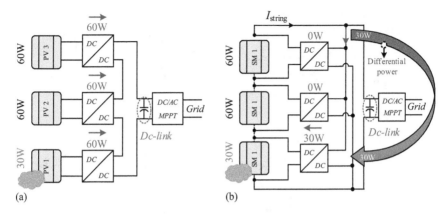

Figure 6.15 Processed powers by DC–DC converter under a certain PSC for (a) DCPO and (b) PV–bus DPP architecture

Table 6.1. Comparison of the efficiency for the shading example in Figure 6.15

Architecture	DCPO	PV–bus architecture
PV-generated power	150 W	150 W
Total processed power by the converter	150 W	30 W
Maximum power stress on a single converter	60 W	30 W
Converter efficiency	95%	90%
Power loss	7.5 W	3 W
Overall system efficiency	95%	98%

submodules, which is a merit of reducing the system cost. The individual bidirectional buck-boost converter is responsible for mismatched compensation among neighbor submodules. A centralized DC–DC converter regulates the string current I_{string} and achieves the function of MPPT through the cooperative operation with the individual sublevel converters. The present works refer to the research on PV–PV that could be found in [28,31,33–51]. The PV–PV DPP architecture owns the advantages of size and integration level. Many papers have presented a good design solution that embeds the DPP converters in the junction box of PV modules. The hardware design of SL-based PV–PV DPP architecture can be found in [28,39,43], and the design of SC-based PV–PV can be found in [50].

The strong coupling characteristics [30], which affect the conversion efficiency and control complexity of the individual submodule converters, pose a design challenge for PV–PV architecture. The degree of coupling signifies the mutual dependency of the system's converters and can also be interpreted as the degree to which the converters can work independently. In PV–PV architecture, if a PV submodule is shaded, the nearby submodules will deliver the compensated power to the shaded submodule through their respective bidirectional converters.

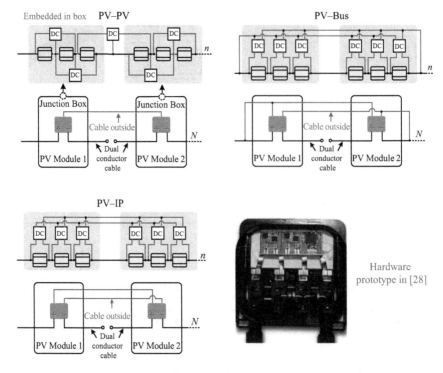

Figure 6.16   Electrical connection diagram for submodule level DPP architecture

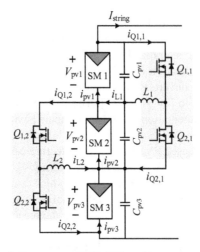

Figure 6.17   Electrical diagrams of PV–PV architecture with bidirectional buck-boost topology

This process results in additional compounded power processing, especially for the case with just one submodule experiencing slight or moderate mismatch. This high-coupling characteristic will challenge achieving MPPT in the PV–PV architecture and increase the power rating of the individual submodule converters, especially for a long PV string. Considering different topologies for individual submodule converters in the PV–PV architecture, including switched-inductor (SL), switched-capacitor (SC) converter, and resonant switched-capacitor (ReSC) converter, the main design, and optimization for each PV–PV category will be presented separately. The PV–PV implementation detail, including topology, hardware, efficiency, and characteristics, is presented in Table 6.2.

If we observe Table 6.2, from the perspective of topology, most of the present works focus on the bidirectional buck-boost converter. The reason is that the control algorithm is more flexible in applying SL-based PV–PV architecture than in SC or ReSC-based architecture. The true MPPT and voltage equalization (VE) can be used in PV–PV architecture introduced in the following part. According to Kirchhoff's current law (KCL), the steady-state inductor current of each DPP converter can be derived as:

$$I_{L,i} = \begin{cases} (I_{pvi} - I_{string})/D_i (i = 1) \\ I_{pvi} - I_{pv,i+1} + D_{i-1}I_{L,i-1} + (1 - D_{i+1})I_{L,i+1} (i > 1) \end{cases} \tag{6.1}$$

where $i_{L,i}$ is the inductor current of the ith DPP converter, $i_{pv,i}$ represents the current of ith PV submodule, and D_i is the duty ratio of the ith DPP converter.

It can be observed (6.1) that the inductor currents of individual converters are highly coupled and dependent. Thus, even if only one PV submodule is slightly shaded, all individual converters must process mismatched power. The power distribution among different individual converters depends on the architecture layout, and more power will be processed for closer DPP converters with respect to the shaded submodule. As shown in Figure 6.17, $n-1$ individual DPP converters are required to regulate the intermediate voltage nodes when there are **n** series-connected PV submodules. The I_{string} can be expressed by:

$$I_{string} = I_{pv1} - (1 - D_1)I_{L1} \tag{6.2}$$

The power coefficient of the total power processed by DPP converters P_{DPP} to the total generated power of all PV submodules P_{GEN} can be expressed by:

$$\beta = \frac{P_{DPP}}{P_{GEN}} = \frac{\sum_{i=1}^{n-1} V_{pv,i}|I_{L,i}|D_{i+1}}{\sum_{i=1}^{n} V_{pv,i}I_{pv,i}} \tag{6.3}$$

When the mismatch among PV submodules becomes severe, the average inductor currents of the SL converter and the power PDPP will increase, which leads to higher power losses. As a result, the system efficiency will decrease with respect to the ratio β.

Figure 6.18(a) shows the diagram of PV–PV architecture with an SC converter, which is widely used for battery voltage balance applications [52,53]. Its basic

Table 6.2 Comparison of PV–PV DPP architecture

Refernces	Topology	Switch number	Submodule control	Current sensing	Report efficiency	Experimental set	Current flow	Integration level
[28]	Bidirectional buck-boost	$2(n-1)$	True MPPT & VE	Yes	• MPPT based: 96.7%** • VE based: 95.7%**	• 3 PV submodules • 2 DPP converters	Bidi	High
[33]	ReSC	$4n$	VE	Yes	• Tracking: >98% • ReSC: >99%*	• 4 PV submodules • 3 DPP converters	Bidi	High
[31]	Bidirectional buck-boost	$2(n-1)$	True MPPT	Yes	• Buck-boost: 95% (peak)*	• 2 PV submodules • 1 DPP converter	Bidi	Medium
[37]	PV equalizer	$2n$	VE (fixed duty cycle)	Yes	• PV equalizer >90%* (SF<0.8) • PV equalizer <85%* (SF=1)	• 4 PV submodules • 4 Equalizers	Uni-dir	Medium
[39]	Bidirectional buck-boost	$2n$	VE (fixed duty cycle)	No	• Buck-boost: 80%–90%* (lower to high load current)	• 18 PV submodules • 18 DPP converters	Bidi	High
[38]	Bidirectional buck-boost	$2(n-1)$	True MPPT (P&O)	-	• Buck-boost: 92% (peak)* • System efficiency 95%**	• 3 PV submodules • 2 DPP converters	Bidi	Medium
[34]	Bidirectional buck-boost	$2(n-1)$	True MPPT	No	• Tracking: 98.4%–99.95% • Buck-boost: 95%(peak)*	• 3 & 6 PV submodules • 2 & 5 DPP converters	Bidi	High
[44]	Bidirectional buck-boost	$2(n-1)$	VE	Yes	• Buck-boost > 90%* • System efficiency >95%**	• 2 PV cells • 1 DPP converter	Bidi	High
[45]	Bidirectional buck-boost	$2(n-1)$	• True MPPT • (TS)	Yes	—	• 3 PV submodules • 2 DPP converters	Bidi	-

Note: *Power conversion efficiency; **Combine both power conversion efficiency and algorithm tracking efficiency.

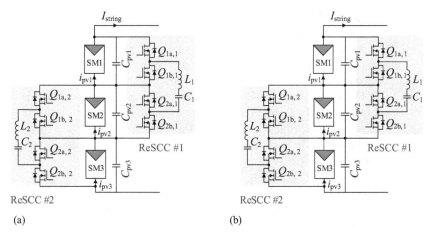

Figure 6.18   *(a) PV–PV architecture implemented with SC converter. (b) PV–PV*
              architecture implemented with ReSC

concept is to regulate the capacitor charges through the turn on/off of switches
according to the principle of VE. Precisely, all PV submodules are controlled to run
at the same output voltage, despite the diversity in their MPP voltages. Compared
to SL converters, SC converters have lower EMI and smaller dimensions [54]. The
research presented in [55] demonstrates that the SC converter is ideally suited for
VE control due to its high efficiency when the input and output voltages are
equivalent or close. In [50], a single flying capacitor was used to create a modular
equalizing SC converter (EQSCC) to equalize the voltages between two nearby PV
submodules. Four switches and one flying capacitor are employed in each EQSCC
unit. About 50% more output power under the worst shading condition is achieved
with the EQSCC-based PV–PV architecture. The switching loss in EQSCC is lower
than that in SL converter due to the soft-switching operation.

Figure 6.18(b) shows the diagram of PV–PV architecture with ReSC, which is
regarded as an extension of SC topology. The ReSC topology can reduce the
intrinsic charge-sharing loss in SC topology by introducing a small inductor. It
allows wide voltage swing sustained on the capacitors, which is beneficial for
power density improvement while ensuring high efficiency simultaneously. ReSC
converters are arranged in a parallel-ladder architecture. Thus, ReSC converters
only handle mismatch power and turn off for the mismatching conditions. Above
99% conversion efficiency was experimentally measured in [40]. The ReSC
topology also shows the merits of near-zero insertion loss, simple co-integration,
and small size due to the elimination of magnetic components.

6.4.1.2   Control and optimization
In addition to the topology, the control algorithm affects the hardware cost and system
efficiency. This section will mainly be introduced three control strategies in PV–PV
architecture, including VE control, true MPPT control, and time-sharing MPPT

control. They have different choices between efficiency, cost, and simplicity, which are discussed in this section. At the same time, these control methods can also be used in PV–IP and PV–bus architectures, which will be discussed in the later section.

Control in the design of the PV–PV DPP can be divided into two levels: submodule control and centralized control. The centralized power stage may employ either a DC–DC converter or a DC–AC inverter. Although the power stage is distinct, it executes the MPPT algorithm to regulate the PV string's total output voltage. In the submodule control, each DPP converter adjusts the duty cycle to regulate the current difference across PV submodules, whereas the centralized control implements the MPPT control.

A.   Voltage equalization (VE)

In theory, if the control of the submodule level performs MPPT for each PV submodule, the power harvesting must be maximized. However, the cost and control complexity of the system will be increased significantly since each PV submodule should be equipped with an MPPT control unit. To overcome these issues, the VE control has been studied and discussed in some works. The control concept is based on the fact that the shift of MPP voltage is not sensitive to the changing of the irradiance; in other words, the MPP voltage under different irradiance is approximately equal. Thus, the VE control can be utilized to keep all PV submodules' voltage fixed at a value of V_{eq}. This control concept reduces control complexity in actual operation and model complexity in theoretical analysis. For example, the actual submodule control is real MPPT in [56]; however, the mathematic model analysis is based on the VE concept to simplify it.

However, it should be noted that VE is not the true MPPT but the suboptimal tracking strategy. The MPP voltage of each PV submodule mainly depends on irradiance and temperature. To test the performance of VE control, the shift of the MPP voltage with respect to the irradiance and temperature tests with 300 silicon-based PV modules, which are randomly selected from the PV database in MATLAB®. The irradiance differences are selected as 700 W/m^2 (SF = 0.3), and 200 W/m^2 (SF = 0.8). The temperature differences are selected as 45 °C and 35 °C. The voltage shift percentage and the power loss percentage are defined by:

$$shift\ of\ V_{mpp}(\%) = \frac{V_{mpp} - V_{eq}}{V_{mpp}} \times 100\% \tag{6.4}$$

$$reduction\ of\ P_{mpp}(\%) = \frac{P_{eq} - P_{mpp}}{P_{mpp}} \times 100\% \tag{6.5}$$

The histogram of the results regarding the difference in irradiance is presented in Figures 6.19 and 6.20, respectively. The x-axis represents the shift percentage of equalized voltage and VE power compared with the MPP voltage and MPP power of each PV element. Furthermore, the y-axis represents the module count. As illustrated in Figure 6.19(a), for 0.3 of the shading factors, around 70% of the PV elements are concentrated in the voltage shift region of −1 to 1%. The power reduction due to voltage shift is slight, and Figure 6.19(b) shows that 94% of the

PV elements have a power reduction of less than 0.5%. For a more severe shading scenario that 0.8 of shading factor, the voltage shift is mainly concentrated in the area of −3 to 3% as shown in Figure 6.20(a); however, the power reduction caused is not apparent that around 96% of PV elements are concentrated in the power reduction area of −1 to 0%, as shown in Figure 6.20(b).

The histogram results of voltage shift and power reduction caused by the different temperatures for VE control are shown in Figures 6.21 and 6.22, respectively. It can be noticed that the MPP voltage shift of the PV element is more sensitive to the temperature gradient. The power reduction is not evident with thermal gradients up to 10 °C (35 °C) around 88% of PV elements are concentrated in power reduction areas of −1 to 0%, as shown in Figure 6.21. The power reduction is more evident under with high thermal gradient of up to 20 °C (45 °C).

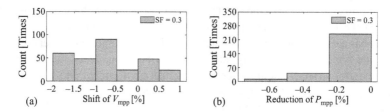

Figure 6.19   Error caused by VE control refers to the shading factor SF = 0.3:
(a) shift of MPP voltage and (b) power reduction

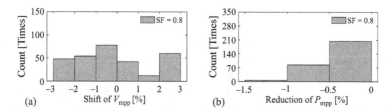

Figure 6.20   Error caused by VE control refers to the shading factor SF = 0.8:
(a) shift of MPP voltage and (b) power reduction

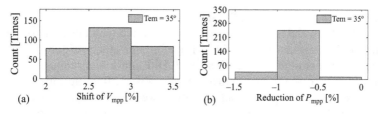

Figure 6.21.   Shift of MPP voltage (S = 1,000 W/m²) refer to the temperature of
(a) T = 35 °C and (b) T = 45 °C

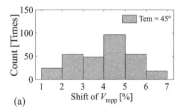 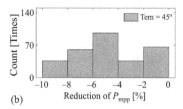

Figure 6.22   Reduction of the power (S = 1,000 W/m²) refer to the temperature of (a) T = 35 °C and (b) T = 45 °C

The MPP voltages of the PV elements with high temperatures are drifting apart from the VE voltage, and the result in power reduction is apparent, as shown in Figure 6.22.

The primary research demonstrates that the irradiance gradient has a negligible effect on the tracking effectiveness of the VE control, whereas the temperature gradient has a discernible effect. According to [57], temperature differences in a PV system are typically produced by non-uniform irradiation. Thus, the utilization of the VE control strategy should refer to the actual environment and the location of the PV system installation to avoid the large temperature gradient among PV panels. As a concluding remark, the author in [57] also made the intriguing observation that the effect of varying temperatures and irradiance may tend to be paired. In other words, these two effects would cancel out the fluctuation in MPP voltage because a decrease in temperature increases MPP voltage while a decrease in irradiance decreases it.

Consequently, the tracking efficiency would be significantly greater in practical PV systems under these conditions. The system efficiency of PV–PV-based DPP architecture using VE and MPPT can be checked in Table 6.2. As mentioned in the previous structure part, the PV–PV architecture mainly uses two topological types, including SL based converter and an SC-based converter. A bidirectional buck-boost converter is the most common topology of SL based on DPP architecture. For the individual bidirectional buck-boost converter, the duty cycle D_i is given by

$$D_i = \frac{V_{pvi}}{V_{pvi} + V_{pv,i+1}} \tag{6.6}$$

where V_{pvi} is the voltage of ith PV submodule.

All PV submodules have the same voltage when the duty cycle of the bidirectional buck-boost converter is 50%. Therefore, the author of [28] has proposed the open-loop control technique whereby all DPP converters can run at a fixed 0.5 duty ratio with open-loop control to accomplish a near MPPT operation. This open-loop VE control technique can significantly lower the DPP system's expense and control complexity. Furthermore, it also decouples the perturbation interface between submodule level control and central level control. However, the author in [44] also expressed concern about control efficiency under large current and long lifetime work reliability. Therefore, the author proposes an integrated circuit (IC) that implements hysteretic current-model control for the inner current control loop of the voltage equalization method. This closed-loop VE control presents better operational reliability and a higher tracking efficiency under severe shading conditions.

The VE strategy is the most common control approach for SC-based converters in the DPP application since the true MPPT is challenging to achieve in this topology. In SC or ReSC DPP architecture, the converter enforces a voltage ratio of 1:1 between adjacent PV submodules, which is equivalent to voltage equalization. However, a significant factor in the voltage drops on equivalent resistance will affect the VE performance. The high switching frequency will minimize this resistance; however, increasing the switching frequency will cause other issues, such as the high current spike that can complicate electromagnetic compatibility [33]. This issue can be mitigated in ReSC converter that achieves a similar equivalent resistance with a much lower switching frequency. A more detailed comparison of the equivalent resistance between SC and ReSC converters can be found in [46].

B. *True MPPT*

Compared with VE control, true MPPT control shows the advantage of higher tracking efficiency under various operating conditions. Currently, the research focus is mainly on reducing the sensing circuits and communication requirements between individual DPP converters and the central converter while simultaneously improving the convergence speed of MPPT algorithms for all PV submodules. True MPPT can be performed without needing local current sensing at each submodule. According to [58], the generation control circuit achieves the correct MPPT for each PV submodule without monitoring the local current. However, communication between all individual converters and a centralized controller must be employed. Considering a system with N individual DPP converters, the centralized controller must implement $2N$ duty ratio perturbations during each tracking step to realize the true MPPT operation of each submodule, which is not feasible for the practical implementation of an extensive PV system. To circumvent the constraint, a multilevel control and MPPT method for variable conversion ratio SL-based PV–PV DPP architecture [42] was developed. Regardless of the number of PV submodules, the real MPPT can be attained for each submodule with a rapid convergence rate. Nevertheless, this approach involves communication between individual DPP converters and a centralized controller. In [31], a distribution MPPT approach for the PV–PV architecture has been presented, significantly reducing communication requirements. The convergence time is reduced even for a large number of series-connected stages. However, this method necessitates sensing the current of each PV submodule, which incurs additional hardware costs. The distributed true MPPT system is described in [43], where a slow-changing current-based P&O control is implemented in the central control, and a rapid control loop is established in the submodule control. In addition, submodule control required two perturbations and information exchanges between adjacent DPP converters. Both simulation and experimental findings for the DPP system with five submodules are shown.

6.4.2   PV–bus architecture

In contrast to the PV–PV architecture, the PV–bus architecture incorporates mismatch compensation between the PV submodule and the power bus. For grid-connected PV systems, the DC-link voltage is often substantially higher than that of each PV submodule to satisfy the PV inverter's voltage requirement. In the PV–bus architecture, the voltage gain of the individual DPP converters from PV submodules to the DC-link is thus extraordinarily high. Typically, significant voltage gain is achieved using high-frequency transformers. In addition, the ground nodes of the PV submodules differ, which is another reason for the galvanic insolation design. The current works reference to research on PV–PV that can be found in [23,29,56,59–70]. Table 6.3 summarizes the primary characteristics and implementation details of the PV–bus DPP architecture.

6.4.2.1   Topology and structure

Several isolated converter topologies can be implemented with this architecture, including a bidirectional flyback converter (BFC) and dual active bridge (DAB) converter. According to the interconnection configure ratio, two kinds of PV–bus architectures can be classified. Figure 6.23 shows the electrical diagram of PV–bus architecture one, where each DPP converter is paralleled with the PV submodule, and the output of DPP converters is paralleled with the DC–bus and connected to the input of the centralized inverter.

In PV–bus Architecture One, the generated power from PV submodules will be fed to the AC grid in two stages: submodule converters and a centralized inverter. In this PV–bus architecture, the number of individual DPP converters is the same as that of submodules. Thus, there is an $n +1$ control variable, including the string current I_{string}, for n control objectives in an *n-submodule*-based PV–bus architecture. Therefore, the additional degree of freedom can reduce the overall power processed by DPP converters. In order to minimize the total power loss, particularly the total power processed by DPP converters, [35] and [30] explore the operation optimization for the individual DPP converters. However, the control of the string current I_{string} must be implemented in the centralized inverter, which affects the decoupling effect between DPP converters and centralized inverter. Furthermore, the currents of the secondary side of the DPP converters will flow into the DC–bus together with the string current, making it difficult to control the string current directly. More specific control analysis of PV–bus Architecture One can be found in [59].

In order to improve the decoupling effect, the PV–bus architecture two has been discussed by adding a centralized DC–DC converter between the PV string and DC-link [29,56,62,68,69]. Figure 6.24 shows the electrical diagram of PV–bus architecture two, where DPP converters and centralized DC–DC converters are paralleled between PV elements and DC-Link. However, only DPP converters are arranged between the PV elements and DC-link for the PV–bus Architecture One. Although the PV–bus Architecture Two may add power losses. However, there are several advantages for the PV–bus Architecture Two compared with Architecture One:

Table 6.3 Comparison of PV-bus DPP architecture

References	Topology	Switch number	Submodule control	Central control	Current sensing	Report efficiency	Experimental set	Current flow	Integration level
[57]	BFCs (DCM)	$2n$	VE	–	No	VE based: 90–100%** $(0 \leq SF \leq 1)$	3 PV submodules 3 DPP converters	Bidi	Low
[61]	Multi-winding flyback	n	True MPPT	–	Yes	–	2 PV submodules 2 DPP converters	Uni-dir	Low
[29]	BFC (CCM)	$2n$	True MPPT	LPPT	Yes	Flyback: 91.7–92%* Boost:96.6–97%* LPPT&MPPT based: 95.7% (peak)*	2 PV submodules 2 DPP converters	Bidi	Medium
[62]	Flyback (CCM)	n	True MPPT	IMPP	Yes	LPPT&IMPP based: >95.7% **	4 TEG submodules 4 DPP converters	Uni-dir	Low
[56]	BFC (CCM)	$2n$	True MPPT	UM-LPPT	Yes	Flyback: 89–92%* Boost:96.6–97%* MPPT&UM-LPPT based: 96.5% (peak)**	4 PV submodules 4 DPP converters	Bidi	Low
[65]	Flyback (DCM)	n	True MPPT	–	No	Flyback: >96.7%* Tracking eff = 94.4%	2 PV submodules 2 DPP converters	Uni-dir	Low
[67]	Flyback (DCM)	n	True MPPT	IMPP	No	–	2 PV submodules 2 DPP converters	Uni-dir	–
[68]	BFC (CCM)	$2n$	True MPPT (TS)	TMPP	Yes	Flyback: 87–92.2%* Boost:94.5–97.4%* LPPT&TMPP based: 95.5% (peak) **	3 PV submodules 3 DPP converters	Bidi	Medium
[69]	BFC (CCM)	$2n$	VE	TMPP	Yes	Flyback: 84–92%* Boost:95–97.4%* MPPT&TMPP based >91.4% **	3 PV submodules 3 DPP converters	Bidi	–

Note: *Power conversion efficiency; **Combine both power conversion efficiency and algorithm tracking efficiency.

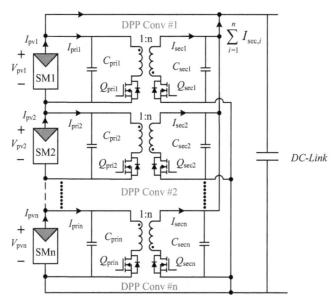

Figure 6.23 PV–bus Architecture One with bidirectional flyback converters

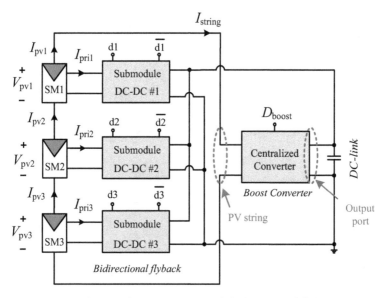

Figure 6.24 PV–bus Architecture Two with bidirectional flyback converters

- **Improved decoupling effect**: the string current I_{string} can be easily controlled independently by the added centralized DC–DC converter without affecting the decoupling effect between DC–DC power conversion for mismatched compensation and centralized DC–AC power conversion for grid connection.

- **Efficiency improvement**: in contrast to the first PV–bus architecture, the second PV–bus architecture allows the string current to be directly regulated by a centralized DC–DC converter to optimize the DPP converter's processing of the total minimum power while minimizing power loss. Moreover, in the DC microgrid application, the differential power in the previous PV–bus Architecture One must be handled twice, i.e., by DPP converters and a centralized converter. Consequently, more losses are incurred. However, with PV–bus design two, mismatched power will only be processed by DPP converters, which is advantageous for efficiency.
- **Voltage gain**: a problematic aspect of designing the PV–bus Architecture One is the high voltage step-up ratio from primary to secondary [30]. The DC-link voltage for a PV string with n PV submodules is close to *n* times that of individual PVs. While this is OK for a small number of PV submodules, the voltage strain on secondary side switches becomes problematic for lengthy strings of PV submodules. Nevertheless, the DC-link voltage is independent of the PV string voltage, allowing for greater design flexibility.
- **Scalability**: scalability is another design challenge. Once the converter is designed and optimized for a set voltage ratio, the system cannot significantly increase the number of PV elements without redesigning the DPP converter. It lacks the flexibility to interface the PV module voltage to the grid requirement since the transformer-less (TL) inverter is currently popular. The new type of PV–bus architecture provides freedom in selecting the number of PV modules. The centralized boost converter can step up voltage, and the submodule-level bidirectional flyback converter can quickly step up/down voltage by adjusting the duty cycle. Therefore, a new type of architecture can make the system quickly meet the grid voltage requirement.
- In summary, PV–bus Architecture Two provides freedom in selecting the number of PV modules. As shown in Figure 6.24, the centralized boost converter can step-up voltage, and the submodule-level bidirectional flyback converter can quickly step up/down voltage by adjusting the duty cycle. Therefore, PV–bus Architecture Two can quickly meet the grid connection and integration since it can provide more design and control flexibility. Thus, the PV–bus Architecture Two is preferred for DC microgrid applications, considering the fast development of renewable energy and energy storage systems.

6.4.2.2 Control and optimization

A. *Least power point tracking control*

Considering the advantage of PV–bus Architecture Two, control and optimization mainly focus on this architecture. The control typically covers two levels: achieving the submodule level distributed MPPT control and realizing the centralized system control, such as the minimum power loss by adjusting I_{string}. As mentioned before, the power flow in each DPP converter depends on changing the string current I_{string}. Furthermore, due to the power extraction and injection requirement, the DPP converter should have bidirectional characteristics.

In this section, we will analyze the bidirectional flow of the power in DPP converters concerning the different values of string current. For example, in a PV system with three submodules, as shown in Figure 6.25. The PV submodule 1 is normally working; however, the PV submodule 2 and submodule 3 are shaded. The corresponding currents are 3.5 A, 2.8 A, and 1.755 A, respectively. The VE method controls each PV submodule, and the equalized voltage is V_{eq}. If we divided the string current into different regions, the following cases could be derived:

1. Case I: when "$I_{string} \leq I_{pv3}$", The distribution of current is depicted in Figure 6.25(a). In this case, string current is smaller than the output current of all PV submodules. According to KCL, DPP converters extract energy from the corresponding PV submodules.

2. Case II: when "$I_{pv3} \leq I_{string} \leq I_{pv2}$", for example, I_{string} =2.3A, the current distribution is depicted in Figure 6.25(b). In this case, the energy no longer flows in a single direction. According to KCL, DPP converters 1 and 2 are extracting surplus energy from PV submodules 1 and 2. Then part of the extracted energy flows to DPP converter 3 through the secondary circuit.

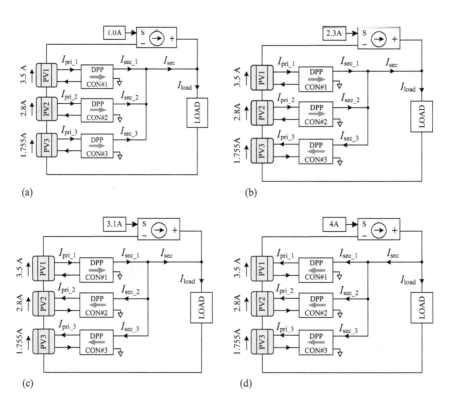

Figure 6.25   Four different operation cases with respect to the string current value: (a) Case I: $I_{string} \leq I_{pv3}$; (b) Case II: $I_{pv3} \leq I_{string} \leq I_{pv2}$; (c) Case III: $I_{pv2} \leq I_{string} \leq I_{pv1}$; (d) Case IV: $I_{string} \geq I_{pv1}$

3. Case III: when "$I_{pv2} \leq I_{string} \leq I_{pv1}$", the current distribution is depicted in Figure 6.25(c). Similar to case 2, the energy flow direction is not consistent. According to KCL, DPP converter 1 extracts surplus energy from PV submodule 1. Then part of the extracted energy flows to DPP converters 2 and 3 through the secondary circuit.

4. Case IV: when "$I_{string} \geq I_{pv1}$", Figure 6.25(d) depicts the current distributions. In this case, string current is greater than the output current of all PV submodules. According to KCL, DPP converters extract energy from the bus and injecting to all PV submodules.

Note that each PV submodule's MPP will be controlled in these modes. Nevertheless, the power allocation among DPP converters fluctuates continuously with the string current. For every operating mode, the total power delivered by DPP converters for a three PV submodule layout may be expressed as:

$$P_{DPP} = V_{eq}(|I_{string} - I_{pv1}| + |I_{string} - I_{pv2}| + |I_{string} - I_{pv3}|) \tag{6.7}$$

Assume the same shading situation as shown in Figure 6.25. As depicted in Figure 6.26, the relationship between the total power processed by DPP converters and the string current in this shading state may be determined using (6.7). The *x*-axis represents the string current I_{string}, while the *y*-axis represents the overall power handled by DPP converters P_{DPP}. At point A, the whole generated power by PV submodules is processed by the DPP converters. The *x*-axis values of points B, C, and D reflect the MPP current of PV3, PV2, and PV1.

According to (6.7), the relation between total processing power by DPP converter and string current can be expressed in Figure 6.26. The current value in points B, C, and D is the MPP current of PV 3, PV 2 and PV 1, respectively. The part from points A to B consists of two lines; although the slope of the two lines is not the same, the total power is decreased with string current. When the string current value passes by point C, the total power constantly increases with string

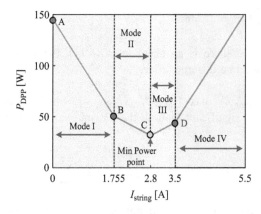

Figure 6.26 Total processed power versus string current under a certain PSC

current. Thus, point C is a critical point, and the current value in point C is 2.8 A which can minimize the power processed through the DPP converter.

It is noticed that there exists a string current value that minimizes the overall power processed by the DPP converter, and this value corresponds to the current of one of the PV submodules. Next, the condition of *n* PV submodules will be considered. The aim is to find the generalized relationship between P_{DPP} and I_{string} for *n* PV submodule DPP converter. The generalized equation for optimal I_{string} at minimum point will be derived. In the ideal model, we assume the conversion power loss in the DC–DC converter and MPP tracking error are ignored. The total power processed by DPP converters is the sum of the power flowing through each DPP converter:

$$P_{DPP} = \sum_{i=1}^{n} P_{dpp,i} \tag{6.8}$$

If we ignore the power loss in the DC–DC converter, the power flowing in each DPP converter can be calculated by

$$P_{dpp,i} = V_{pvi}|I_{pvi} - I_{string}| \tag{6.9}$$

where V_{pvi} is the voltage of *i*th PV submodule, I_{pvi} is the current of *i*th PV submodule, and I_{string} is the string current. Then, the total power processed by the DPP converter can be expressed by

$$P_{DPP} = \sum_{i=1}^{n} V_{pvi}|I_{pvi} - I_{string}| \tag{6.10}$$

In order to simplify the analysis, we assume that the submodule-level control adopts VE control. Then, the voltages of each PV submodule are equalized with the same value V_{eq}

$$P_{DPP} = \sum_{i=1}^{n} V_{eq}|I_{pvi} - I_{string}| \tag{6.11}$$

Considering that there is an absolute in (6.11), the value of string current should be discussed in a different zone.

1. Mode I: If $I_{string} \geq \max \{I_{pv1}, I_{pv2} ..., I_{pvn}\}$, which means that the string current is more significant than all the PV submodule current. Then (6.11) is rewritten as

$$P_{DPP} = \sum_{i=1}^{n} V_{eq}(I_{string} - I_{pvi}) = V_{eq}I_{string} + V_{eq}\sum_{i=1}^{n} I_{pvi} \tag{6.12}$$

It is noted that the slope of the power curve will be positive since the voltages of the PV submodule are positive.

2. Mode II: If $I_{string} \leq \min \{I_{pv1}, I_{pv2} \ldots, I_{pvn}\}$, the string current is lower than all the PV submodule currents. Then (6.11) is rewritten as

$$
\begin{aligned}
P_{DPP} &= \sum_{i=1}^{n} V_{eq}(I_{pvi} - I_{string}) \\
&= -V_{eq}I_{string} + V_{eq}\sum_{i=1}^{n} I_{pvi}
\end{aligned}
\tag{6.13}
$$

3. Mode III: The last case is the string current value between the lowest and the highest PV submodule current. If $\min \{a_1, a_2 \ldots, a_n\} < I_{string} < \max \{a_1, a_2 \ldots, a_n\}$, combined with $a_m < I_{string} < a_{m+1}$, $m \in [1, n]$, thus

$$
P_{DPP} = \sum_{i=1}^{n} V_{pvi} \left[m \times I_{string} - \sum_{i=1}^{n} I_{pvi} \right] - V_{pvi} \left[(n - m) \times I_{string} - \sum_{i=m+1}^{n} I_{pvi} \right]
\tag{6.14}
$$

It is noted that if $2m > n$ then $dy/dx > 0$, which means that the slope of the power curve is positive. If $2m < n$ then $dy/dx < 0$, which means that the slope of the power curve is negative. If $2m+n = 0$ then $dy/dx = 0$, which means that the slope of the power curve is zero at this point.

The slope sign will change from negative to positive when the string current increases, as shown by the slope analysis of the power curve. Consequently, the curve of processed power versus string current in PV–bus DPP architecture is convex such that there is a single point or a unique collection of points that minimizes the power processed by the DPP converter; this point is designated as TMPP. Moreover, this string current is equal to one of the output currents of PV submodules, which means $I_{string} \in \{I_{pv1}, I_{pv1} \ldots, I_{pvn}\}$.

As mentioned, BFC is the optimal topology for the individual DPP converter due to its galvanic isolation, high voltage gain, and simple control implementation. The boost converter is selected as the preferred topology for the centralized DC–DC converter due to efficiency improvement. Expressly, the submodule-level control adopts the conventional MPPT method with direct duty cycle regulation. In contrast, the system-level control regulates the string current using the standard proportional–integral (PI) closed-loop control for the boost converter. In terms of power distribution, more power should be distributed to the boost converter considering the higher power conversion efficiency of the non-isolated boost converter compared with isolated BFC [62]. This real-time control algorithm features the simultaneous implementation of submodule-level MPPT and system-level least power point tracking control (LPPT), effectively maximizing the power yield. However, the work in [56] also shows apparent limitations:

1. **Uneven power distribution**: Due to the complex partial shading circumstances, the standard TMPP may concentrate power on a single DPP converter, which increases the power rating and cost of DPP converters.

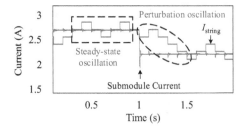

Figure 6.27   Perturbation oscillation by using LPPT control

2. **Distinct oscillation**: There is a two-level oscillation using TMPP control that results in an additional power loss of the system. As seen in Figure 6.27, the LPPT algorithm is responsible for the steady-state oscillation, while the LPPT and MPPT algorithms are responsible for the dynamic oscillation. This feature is not conducive to the system in the complex external environment.

3. **High requirement of coordination**: Due to the fact that two extremum-seeking controls are concurrently executed, the perturb stages must be well-coordinated. Multiple LPPT steps should be conducted within a single MPPT step to control the string current to the new steady-state optimum point. To guarantee steady functioning during an experimental test of [31], the MPPT perturbation step is raised to 10 s. This cooperation will complicate the algorithm's implementation and result in tracking failure in some PSCs with rapid change. This drawback can also be reflected in Figure 6.27. The blue waveform indicates the output current of each PV submodule, and the red waveform indicates the string current at the central level. It can be found that the perturbation frequency of the submodule level is higher than that of the central level. The MPPT algorithm of the submodule level is perturbed every 1s, and the LPPT algorithm of the central level is perturbed every 0.1s. This time difference ensures that the LPPT algorithm can have sufficient time to find the working point. However, the steady-state oscillations are heavily dependent on the perturbations of the LPPT algorithm, whereas the dynamic oscillations depend on the perturbations of both extremum-seeking algorithms.

B. *Total minimum power point control*

To address the above issues, the fixed-duty cycle control was implemented in [28] to eliminate the steady-state and dynamic perturbation for MPPT while reducing the cost and control complexity for the whole system. To eliminate the perturbation caused by the LPPT algorithm, the direct TMPPT without submodule perturbations is presented in [68,69], which seems to be a suitable solution for solving the limitations of the LPPT-based two-extremum-seeking algorithm. The VE algorithm in BFCs controls the PV submodules. The basic concept of TMPP control is to seek the minimum power proceeded by DPP converters by setting the string current at one of the submodules' MPP currents. The TMPP concerning string current is:

$$I_{stringTMPP} = \underbrace{\arg\min}_{I_s \in \{i_{pv1}, i_{pv2}, \ldots, i_{pv3}\}} \left(\sum_{i=1}^{n} V_{eq} |I_{pvi} - I_s| \right) \tag{6.15}$$

where V_{eq} represents the equalized voltage for PV submodules, and I_{pvi} represents the current of ith PV submodule. Furthermore, for different numbers of PV submodules, the optimal I_{string} by using TMPP can be simply expressed by:

$$I_{string,TMPP} = \begin{cases} I_{pv,\frac{n+1}{2}} & \text{when } n \text{ is odd} \\ \in \left[I_{pv,\frac{n}{2}}, I_{pv,\frac{n}{2}+1} \right] & \text{when } n \text{ is even} \end{cases} \tag{6.16}$$

It can be noticed that there is no need for perturbations by using the TMPP algorithm because the optimal string current value can be calculated directly by sensing the current of each submodule and the equalized voltage.

Additional efforts have been undertaken to increase the model accuracy of the direct TMPP method, which is often derived from an ideal power loss model without taking into account the power losses in the individual DPP converters. The corresponding total processed power by the DPP converter under ideal conditions can be expressed by

$$P_{total} = \sum_{i=1}^{n} V_{eq} |I_{pv,i} - I_{string}| \tag{6.17}$$

Another factor that should be considered to improve the model accuracy is the power flow direction since the power flow direction for DPP converters is bidirectional. The expression shown in (6.17) is only held when the power flows from the PV submodule to the DC-link side. As shown in Figure 6.28(a), the power losses of DPP converters will not be considered in the model under this condition. However, the power loss during the power transformation process should be considered when the power flows from the DC-link side to the PV submodule, as shown in Figure 6.28(b).

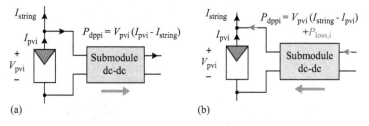

Figure 6.28   The current direction in the DPP converter: (a) flowing from PV submodule side to the secondary side and (b) flowing from the secondary side to the PV submodule side

Therefore, the actual power model of the DPP converter concerning the power loss and power flow direction can be expressed as:

$$P_{total} = \underbrace{\sum_{i \in P} V_{eq}(I_{pv,i} - I_{string})}_{\substack{\text{power from PV side to bus side} \\ I_{pvi} > I_{string}}} +$$

$$\underbrace{\sum_{i \in K} \left[V_{eq}(I_{string} - I_{pv,i}) + P_{loss,i} \right]}_{\substack{\text{power from bus side to PV side} \\ I_{pvi} < I_{string}}} \tag{6.18}$$

where P is the set of positive currents in DPP converters, K is the set of negative currents in DPP converters, and $P_{loss,i}$ is the power loss of ith DPP converter in a set of K, which can be expressed by:

$$P_{loss,i} \approx a(I_{string} - I_{pv,i})^2 + b(I_{string} - I_{pv,i}) + c, \forall i \in K \tag{6.19}$$

where a, b, and c represent the factors of the power loss model

The study in [69] shows that the TMPP algorithm based on (2.8) and (2.9) is still held once the following condition can be fulfilled:

$$\max \left\{ 2a \left(kI_{string} - \sum_{i \in K} I_{pv,i} \right) + bk \right\} \leq V_{eq}, I_{string} \in \left[I_{pv,i}, I_{pv,\frac{n+1}{2}} \right] \tag{6.20}$$

Considering the practical specifications of PV modules and power devices, the condition in (6.20) is frequently satisfied. Therefore, (6.15) and (6.16) can be adopted to implement the TMPP algorithm

This TMPPT algorithm can mitigate the computation burden of the central-level algorithm. However, the PV submodule level still needs individual MPPT control for each unit, which increases the system's cost and complexity. As shown in Figure 6.29, considering that each PV submodule should realize the true MPPT simultaneously, each DPP converter must be equipped with a dedicated microcontroller. In an

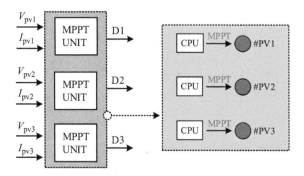

Figure 6.29   The conventional MPPT control scheme in the DPP system

n-submodule-based DPP system, n microcontrollers are required for PV–bus and PV–IP architecture, while "$n-1$" microcontrollers are required in PV–PV architecture. Even though the microcontroller number is one less than that in the other architectures, the controller cost still accounts for a high ratio in the total system, especially for a long-string PV system. However, the controller cost accounts for a high ratio of the total system cost, especially for a long-string PV system.

To further consummate the whole TMPPT control, a submodule-level time-sharing control is a promising solution. Different from the conventional MPPT control in DPP architecture, the time-sharing MPPT strategy is to achieve the MPPT for each PV submodule in a different time slice, which means that the MPPT can only be adopted for one PV submodule in a period. This method is based on the fact that the irradiation change is much slower than the calculation speed of the microcontroller. Consequently, the MPP voltage reference from the MPPT controller can be regarded as a constant most of the time. It indicates that it is unnecessary to equip an MPPT controller for each PV submodule. Thus, for the time-sharing-based MPPT control strategy, the hardware can only use one microcontroller to handle a number of PV submodules, for example, three PV submodules, as shown in Figure 6.30.

Simultaneously, the amount of the PV submodules that the controller can handle is not limited to just three but can also extend to four, five, or more. However, a balance should be considered between the number of PV submodules with the time-sharing control and the stability of the PV–PV architecture, especially under complex irradiance environmental conditions. For the PV–bus architecture with three PV submodules, the time-sharing signal of "1," "0," and "−1" can be used to regulate the MPPT order for PV submodules 1, 2, and 3. The block of the time-sharing signal is presented in Figure 6.31. It can be found that there are two states of each PV submodule, including perturbation and fixed. If one PV submodule is at the state of perturbation, the MPPT is achieved, and others follow the voltage reference of this PV submodule. The basic principle is presented here:

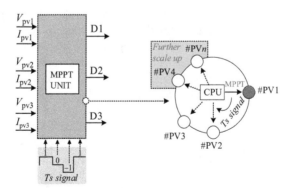

Figure 6.30   *The MPPT control scheme in the DPP system. (a) Simultaneous n PV submodule for n MPPT unit. (b) Time-sharing MPPT, several PV submodules using one MPPT unit, achieving MPP operation in different time-slice*

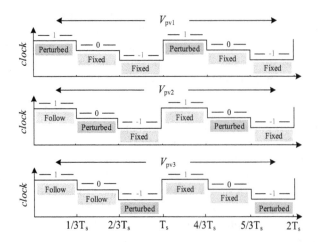

Figure 6.31   The clock single of the time-sharing control

when the time-sharing signal is "1," the MPPT controller will track the MPP of PV submodule 1 corresponding to the perturbation state. Meanwhile, the PV sub-modules 2 and 3 follow the MPP voltage reference of the PV submodule corresponding to the following state. When the MPP of PV submodule 1 is tracked, the duty cycle of DPP converter 1 will be fixed to maintain the last value. Meanwhile, the time-sharing signal is changed from "1" to "0," and the PV submodule is starting to track its MPP. At the same time, the following aim for PV submodule 3 is changed to PV submodule 2. Similar to the previous introduction, the duty cycle of the DPP converter will be fixed when the MPP of the PV submodule is tracked. Then, the time-sharing signal is changed from "0" to "−1," and the PV submodule 3 starts to track its MPP.

In conjunction with a time-sharing solution, the system features two control parts for a submodule and a centralized converter. In the PV–bus DPP design, a centralized boost converter modulates the string current to accomplish real-time TMPPT. Figure 6.32 depicts the control strategy for the whole system.If there is no mismatch between PV submodules, the string current reference for the central boost converter is set equal to the MPP current of PV submodules, which means the expression of "$I_{string,ref} = I_{pv2,ref} = I_{pv3,ref}$." According to the time-sharing signal sent to the controller, the MPPT is achieved for each PV submodule in a different time slice. In addition, the whole processed power by the DPP converter is adjusted by central control by adjusting the value of the string current. Each PV sub-module's output current is sent to the TMPPT control part, and the optimal string current reference value is then sent to the PI controller. The system can run in an optimal state through two levels of coordinated control.

To analyze the performance of the proposed method, three PV modules were linked in series on the experimental platform to examine the submodule-level performance. Figure 6.33 depicts the experimental prototype, while Table 6.4 lists the primary characteristics of the constructed converter.

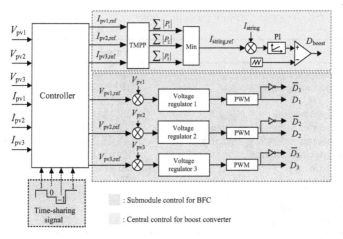

Figure 6.32   System control diagram

Figure 6.33   Experimental setup

Table 6.4   Hardware component specifications

Parameters	Values	Qty
Switch for flyback converter	IRFP460	6
Switch for boost converter	IRFP250N	1
Inductor	KS1571-0.5 mH 8 A	1
Transformer framework	EC28	3
Flyback converter primary capacitance	220 μF 50 V	3
Flyback converter secondary capacitance	47 μF 250 V	3
Boost converter capacitance	47 μF 250 V	1
Switching frequency	20 kHz	–
Magnetic inductance	300 μH	–
Flyback converter input voltage range	16–19 V	–
Boost converter power rating	48–57 V	–
MPPT perturbation time	0.5 s	–

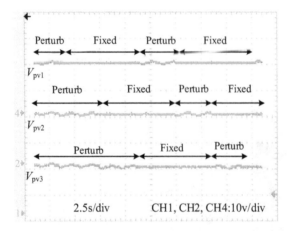

Figure 6.34   Experimental submodule output voltage

The steady-state shading condition is a test to verify the performance of the proposed control algorithm. The PV submodule 1 is normally working, and the corresponding irradiation is 1,000W/m^2. The PV submodules 2 and 3 are shaded, and the corresponding irradiation is 400 W/m^2 and 200 W/m^2, respectively. The theoretical MPP voltage and current is: V_{mpp1} = 18.5 V and I_{mpp1} = 5.4 A, V_{mpp2} = 18.1 V and I_{mpp2} = 2.3 A, and V_{mpp3} = 17.5 V and I_{mpp3} = 1.1 A. The MPPT perturbation frequency is 2 Hz, and the corresponding step time is 0.5 s. The measured output voltage of each PV submodule is shown in Figure 6.34.

Each PV submodule is found to be controlled at its MPP, matching the time-sharing signal. Presented is a three-level steady-state oscillation in which the output voltage is stepped above and below the MPP voltage. It signifies the correct operation of the P&O-based time-sharing MPPT algorithm. Figure 6.35 depicts the experimental PV submodule current and the string current. It is noted that all the

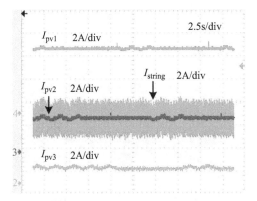

Figure 6.35   Experimental submodule current and string current

DPP converters are working on CCM. The string current waveform exhibits a more significant ripple than the PV submodule current, which can be optimized in the future to increase the switching frequency or increase the inductance value in the boost converter. The three-level steady-state oscillations in I_{pv2} show the experiments' 0.5 s MPPT perturbation step time. The theoretical optimal string current value at TMPP is 2.3 A, which is the MPP current for PV submodule 2. As illustrated in Figure 6.35, the string current is following the current of PV submodule 2, which means the TMPPT control algorithm is working.

The irradiance-change condition is tested to verify the proposed algorithm's tracking speed. Assume PV submodule 1 is not shaded, and the corresponding irradiation is 1,000 W/m² at all times. The PV submodules 2 and 3 are shaded, and the corresponding irradiation is 400 W/m², and 200 W/m² in the beginning, respectively. At time t_1, the irradiation of the PV submodule is reduced to 300 W/m², and the corresponding MPP current changes from 2.3 A to 1.64 A. At the same time, the irradiation for the PV submodule is increased to 300 W/m², and the corresponding MPP current is increased from 1.1 A to 2.75 A. The dynamic results, including PV submodule current and string current, are shown in Figure 6.36. PV submodule 1 is not affected by the changing irradiance. PV submodule 2 and PV submodule 3 were also quickly tracked to the vicinity of the MPP after the irradiance changed. Furthermore, the string current value at TMPP is also changed from 0.9 A to 1.3 A regarding the irradiance change. It can be found that the string current tracks and works at the correct TMPP quickly. The convergence time is measured at less than 0.1 s.

C.   *Unit minimum LPPT control*

The LPPT and TMPP algorithm minimizes the total proceeded power by DPP converters. However, the power distribution among individual DPP converters has not been optimized, which will affect the power rating of individual DPP converters. Typically, the rated power of DPP converters is determined according to the worst-shading scenario since the modular design will be adopted for submodule

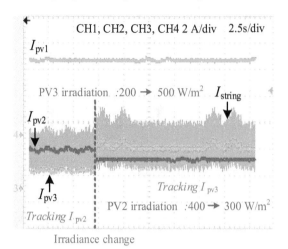

Figure 6.36   Experimental submodule and string current for dynamics

DPP architecture. For instance, according to the worst-shading scenario analysis, the rated power for DPP converters for 83 W PV submodules will be 55 W. However, the actual proceeded power by most DPP converters under most conditions is far below 55 W [69]. For PV–PV and PV–IP architecture, the processed power of each DPP converter is almost unchanged when the shading scenario is specific. Thus, optimizing the power rating of individual DPP converters for the two architectures is difficult. However, for PV–bus architecture, it is possible to optimize the rated power of DPP converters by regulating the power distribution among DPP converters even to further reduce the size and cost of modular DPP converters and improve the power density.

In order to solve this issue, a unit-minimum LPPT (UM-LPPT) control was proposed in [56]. Similar to the LPPT, the UM-LPPT algorithm is implemented by the centralized boost converter by using the classical P&O to search for the optimum point. The difference between the LPPT and TMPP, the optimization objective is the rated power for each DPP converter rather than the total proceeded power. Thus, the UM-LPPT ensures a more even power distribution among DPP converters than that using LPPT or TMPP control. The MSX60W PV module controller from TMPP has a DPP converter power stress of 60 W. Using PBP, however, the worst-case power a DPP converter must handle is 30 W, meaning that the rated power of PBP control can be reduced by 50% compared to TMPP. Considering some design margins, the rated power of the DPP converter can be set to 35 W in this work. Table 6.5 lists the power distribution under different shading with case #1, case #2, case #3, and case #4. By using TMPPT control, it can be seen that DPP converter 3 has the most significant power in most cases. Significantly for case #2, DPP converter 3 increases the total power by 90%, which causes the operating temperature of DPP converter 3 to be higher than other DPP converters, as shown in Figure 6.37(a). Increasing temperatures will cause an

Table 6.5 Four shading cases by using TMPP and PBB control

Scenario	I_{pv1}	I_{pv2}	I_{pv3}	TMPP	P_{DPP}	Distribution	PBP	P_{DPP}	Distribution
Case #1	3.55 A	3 A	0.4 A	3 A	53.55 W	• P_{dpp1} = 44.2 W • P_{dpp2} = 0 W • P_{dpp3} = 9.35 W	1.975 A	70.97 W	• P_{dpp1} = 26.7 W • P_{dpp2} = 19.9 W • P_{dpp3} = 26.7 W
Case #2	3.55 A	3.2 A	0 A	3.2 A	60.35 W	• P_{dpp1} = 54.4 W • P_{dpp2} = 0 W • P_{dpp3} = 5.95 W	1.775 A	84.57 W	• P_{dpp1} = 30.1 W • P_{dpp2} = 24.2 W • P_{dpp3} = 30.1 W
Case #3	0.7 A	3 A	0.9 A	0.9 A	39.1 W	• P_{dpp1} = 3.4 W • P_{dpp2} = 35.7 W • P_{dpp3} = 0W	1.85 A	55.25 W	• P_{dpp1} = 19.5 W • P_{dpp2} = 19.5 W • P_{dpp3} = 16.1W
Case #4	0.4 A	0.5 A	3.55 A	0.5 A	53.55 W	• P_{dpp1} = 1.7 W • P_{dpp2} = 0 W • P_{dpp3} = 51.8 W	1.975 A	78.63 W	• P_{dpp1} = 26.7 W • P_{dpp2} = 25.0 W • P_{dpp3} = 26.7 W

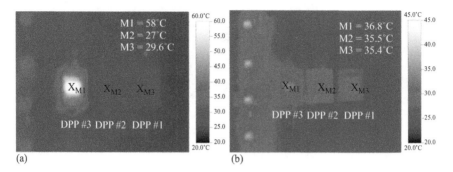

Figure 6.37   The operation temperature by using (a) TMPP control and (b) PBP control

increase in the failure rate of switching devices (such as MOSFETS) [45]. Consequently, the mean time between failures (MTBF) of DPP converter 3 is diminished [46]. Therefore, PBP control is essential for balancing power stress and achieving uniform temperature distribution among DPP converters. Figure 6.37(b) depicts the temperature distribution utilizing PBP control, which demonstrates that a uniform temperature distribution has been obtained. As demonstrated in Table 6.5, the utilization of PBP control reduced the maximum power stress of the DPP converter to around 30 W, a reduction of 50% compared to the TMPP control. The results demonstrate that when the total power is significantly raised, the power stress can be lowered by 32% and 40% in cases #1 and #2, respectively. Consequently, the mode must be switched between TMPP and PBP based on the actual output power of each DPP converter.

According to the above analysis, the DPP converter power distribution affects the PV–bus architecture operation mode. The TMPP control is implemented to ensure the overall system efficiency if the power stress on each DPP converter falls below the predetermined threshold. Otherwise, the system is switching to PBP control to adjust the power distribution to maintain the DPP converter's safe operation. As illustrated in Figure 6.38(a), in the shading cases #3 and #4 from Table 6.4, the most significant power pressure on the DPP converter is lower than the designed rated power by us neither TMPPT nor PBP control. Nonetheless, as shown in Figure 6.38(b), the total power processed by the DPP converter when PBP is employed is more than when TMPPT is employed. Figure 6.39(a) and (b) exhibits the BFC and centralized boost conversion efficiencies, respectively. Figure 6.39(c) illustrates that the power loss calculated using TMPPT and PBP for these four shading instances may be found. It demonstrates that TMPPT can decrease the additional power losses in DPP converters compared to PBP.

In [56], the rated power of each DPP converter can reduce to approximately half of that using LPPT or TMPP control. Although the total proceeded power using the UM-LPPT algorithm is slightly higher than that of the LPPT or TMPP algorithm, a lower power rating of modular DPP converters and even power distribution among DPP converters have been achieved. This characteristic is

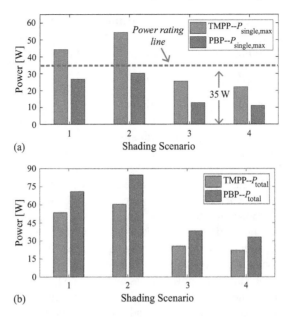

(a)

(b)

Figure 6.38 Comparison of the power distribution for TMPP and PBP: (a) maximum power stress and (b) total processed power

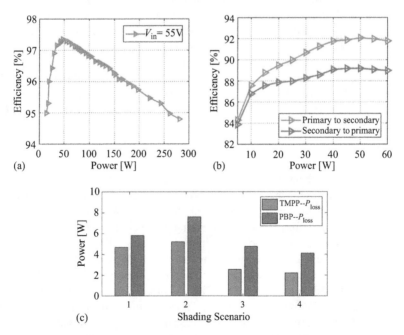

(a)

(b)

(c)

Figure 6.39 (a) The efficiency of BFC. (b) The efficiency of boost converter. (c) Power loss caused by the control method

beneficial for cost reduction and system lifetime improvement. Further optimization of the UM-LPPT algorithm reduces the convergence time, minimizes the steady-state oscillations, and lowers the algorithm complexity and coupling effect.

D.   Hybrid power-balancing-point control

In order to combine the merits of TMPP and the UM-LPPT algorithm, a low-complexity hybrid power-balancing-point (PBP) tracking algorithm was discussed in [69] to find the optimal I_{string} with high system efficiency and low-power-rating modular design for DPP converters. For the submodule-level control of the PBP algorithm, the VE algorithm is employed to reduce control complexity, eliminate costly communication devices, and guarantee a high tracking speed. The suggested hybrid method is implemented for system-level control by switching between UM-LPPT and TMPP control to allocate the optimal string current precisely, rapidly, and with reduced oscillation. The switch between the UM-LPPT and TMPP depends on the power stress of DPP converters under changing environmental conditions. The boundary is typically set as half of the power rating by using the TMPP control. If the maximum power stress on a single DPP converter exceeds the predetermined limit, the UM-LPPT is implemented to actualize the standardized modular design for the DPP converter and guarantee the healthy functioning of DPP converters. If the highest power stress on a particular single DPP converter exceeds the predetermined limit, the TMPP is implemented to seek the lowest proceeded power and enhance the power conversion efficiency.

The hybrid PBP algorithm is implemented through direct calculation because the PBP is only determined by the minimum and the maximum PV submodule current in the PV–bus Architecture Two, no matter the number of PV submodules in the system. Assume the output current of PV submodules meets the relationship of "$I_{pv1} < I_{pv2} \ldots < I_{pvn}$," where the I_{pv1}, I_{pv2}, and I_{pvn} represent the PV submodule currents, the I_{string} at PBP. If the efficiency of the DPP converters is considered, it can be expressed as

$$I_{stringPBP} = \frac{I_{pv1} + \eta_1 I_{pvn}}{1 + \eta_1} \tag{6.21}$$

where $I_{stringPBP}$ is the string current at PBP, and η_1 is the conversion efficiency of the DPP converter with the lowest submodule current. Although the conversion efficiency η_1 is changing for different environmental and load conditions, the expression still holds for practical engineering applications.

The flowchart of the hybrid algorithm combining UM-LPPT and TMPP is presented in Figure 6.40. The output currents of each PV submodule and the string current are measured first. The algorithm must then determine whether the current of each PV submodule exceeds 0.1 A. The PV submodule is severely shadowed if the output current is lower than 0.10 A. The algorithm will turn off the relevant DPP converter and activate the internal bypass diode to bypass this PV submodule. Then, the power stress for each DPP converter under TMPP is computed and compared to the rated power $P_{PBP,worst}$ at the design boundary. If $\{P_{dpp1}, P_{dpp2}, \ldots,$

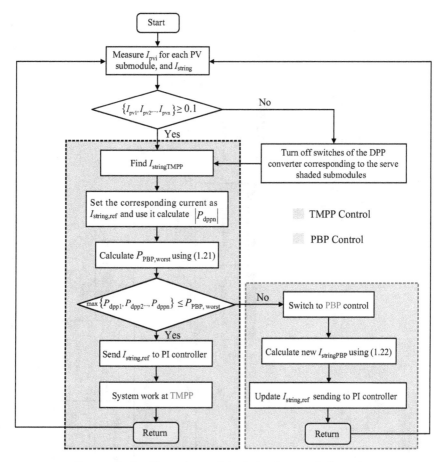

Figure 6.40 Flow chart of the hybrid control algorithm

$P_{dpp3}\} \leq P_{PBP,worst}$, the system will operate by using TMPP. Alternatively, the system will switch to PBP control to keep the DPP converter operating within a safe power range. The string current value at PBP can be calculated by (6.21) directly.

Three PV submodules under changing irradiance conditions are tested to verify the proposed control algorithm's performance. The partial shading pattern for each PV submodule is shown in Figure 6.41(a). The value of E_1, E_2, and E_3 correspond to the irradiation for PV submodules 1, 2, and 3, respectively. For PV submodule 1, the irradiation is increasing from 800 W/m² to 1,000 W/m², and then decrease to 900 W/m², at $t = 0.3$ s and $t = 0.6$ s, respectively. For PV submodule 2, the irradiation is decreasing from 600 W/m² to 300 W/m², and then increases to 500 W/m², at = 0.3 s and $t = 0.6$ s, respectively. Similarly, for PV submodule 3, the irradiation is decreasing from 400 W/m² to 200 W/m², and then decrease to 0 W/m², at = 0.3 s and $t = 0.6$ s, respectively.

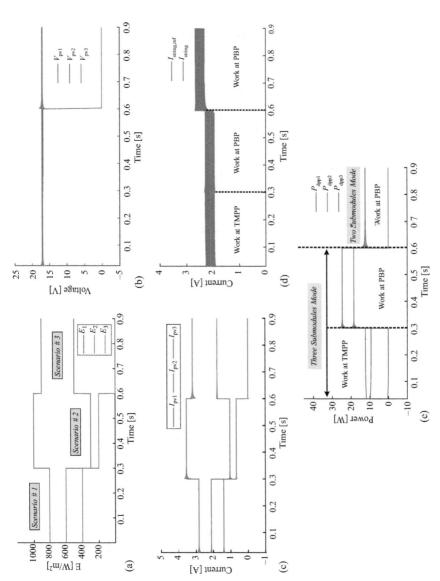

Figure 6.41   The simulation results: (a) Changing irradiance; (b) output voltage of each PV submodule; (c) output current of each PV submodule; (d) string current and reference string current; (e) power stress for each DPP converter

The simulation results are presented to verify the performance of the hybrid control algorithm. Due to the VE control, each PV submodule has the same voltage in scenarios #1 and #2. It is observed that the voltage of PV submodule 3 drops to zero since it is completely shaded and the related DPP converter is turning off to reduce additional power loss, as shown in Figure 6.41(b). The PV submodule currents are shown in Figure 6.41(c). In the first shading scenario, the total minimum power point value is 2.12 A; bring this value into the equation $P_{\mathrm{dpp,n}} = V_{\mathrm{eq}}\,|I_{\mathrm{string}}-I_{\mathrm{pv,n}}|$; bring this value into the equation $P_{\mathrm{dpp,n}} = V_{\mathrm{eq}}\,|I_{\mathrm{string}}-I_{\mathrm{pv,n}}|$, then the power processed by each DPP converter is $\{P_{\mathrm{dpp1}} = 12.3\text{ W}, P_{\mathrm{dpp2}} = 12.3\text{ W}, P_{\mathrm{dpp3}} = 0\text{ W}\}$ satisfying the requirement that each converter's processed power is less than the designed power rating. In the second scenario, the total minimum power point value is 1.03 A, bring this value into the equation $P_{\mathrm{dpp,n}} = V_{\mathrm{eq}}\,|I_{\mathrm{string}}-I_{\mathrm{pv,n}}|$, then the power distribution among each converter is $\{P_{\mathrm{dpp1}} = 43\text{ W}, P_{\mathrm{dpp2}} = 0\text{ W}, P_{\mathrm{dpp3}} = 5.6\text{ W}\}$. The power flowing into submodule DPP converter 1 is 43 W, which is greater than the rated power. The suggested technique, therefore, shifts the string current from TMPP to PBP. In the third scenario, submodule 3 is entirely shaded without output power harvesting; meanwhile, according to the proposed algorithm, the switches of the relevant DPP converter are turned off to increase the system's efficiency. Next, the system switches to the mode with two submodules, and in this mode, the string current always performs power balancing. The above description can be proved by Figure 6.41(d) and (e). In scenario #1, the value of string current on TMPP is 2.12 A, and the processing power of DPP converters 1 and 2 are 12.3 W, respectively. In scenario #2, the value of the string current at the PBP is 2.11 A, and the DPP converters 1, 2, and 3 handle 24.5 W, 24.5 W, and 18.5 W power, respectively. In scenario #3, the value of the string current at PBP is 2.48 A, and the power handled by DPP converters 1 and 2 is the same, which is 12.2 W. At the same time, due to the shutdown, the DPP converter 3 does not process power.

A prototype BFC was constructed for further evaluation of the performance of the BFC-based PV–bus architecture in PV system applications, as seen in Figure 6.42. Theoretically, one BFC may process up to 9.8 W of electricity utilizing

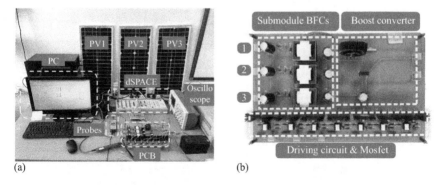

(a) (b)

Figure 6.42 Photograph of indoor experimental test. (a) Experimental test bench. (b) PCB hardware

PBPT under the most severe shadowing conditions. Consequently, the prototype converter is rated at 10 W in this study. IRF740 was chosen as the switching device, and 100 kHz is the switching frequency. The primary and secondary of the transformer were each constructed using 10 twists of 18-gauge Litz wire. In Table 6.6, the precise BFC values are displayed.

In the experiment test, the platform of the dSPACE DS-1104 completes the control portion. This platform has a strong MATLAB® interface link. The execution duration of the code in dSPACE was measured at 5.186 ms; therefore, the microcontroller's calculating load may be disregarded. The purpose of the experiment is to evaluate the performance of the suggested hybrid control algorithm in terms of tracking. Figure 6.43 depicts a partial shade pattern similar to the simulation portion test. It is noted that PV submodule 1 always receives uniform illumination. The PV submodules 2 and 3 are shaded, and the irradiance at each time slice is distinct.

Table 6.6   *Specification of the designed BFCs*

Item	Parameter
Input voltage	4–21 V
MOSFET	IRF740
Transformer turn ratio	1:1
Magnetizing inductance	50 μH
Power rating	10 W
Capacitor	240 μF × 6
Converter peak efficiency	92%

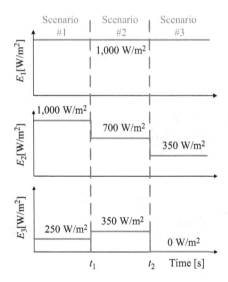

Figure 6.43   *Partial shading scenario for PV1, PV2, and PV3 (from top to bottom)*

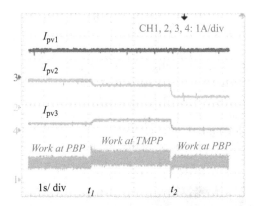

Figure 6.44   Indoor experimental test results under three defined partial shading patterns

The experimental results of the indoor test refer to the three-shading scenario, which is presented in Figure 6.44. For shading scenario #1, the string current value at TMPP is 1.1 A. The power stress for each DPP converter under TMPPT control is P_{dpp1} = 5 W, P_{dpp2} = 8.7 W, and P_{dpp3} = 14.45 W. It can be seen that the power stress for DPP converter 3 is larger than the design rate power if we use TMPPT control. Thus, the system should work at PBP for shading scenario #1. Then, the irradiation of PV submodules 2 and 3 changed at t_1; refer to shading scenario #2 in Figure 6.44. For this shading scenario, the string current at TMPP is 0.9 A. The power stress for each DPP converter under TMPPT control is P_{dpp1} = 7.3 W, P_{dpp2} = 6.1 W, and P_{dpp3} = 8.5 W. It can be found that all DPP converters' power stress is less than the designed rate of power. Thus, the system is working TMPP to ensure the whole DPP converter process has the minimum power, improving system efficiency. For the last shading scenario #3, the irradiation of PV submodule 3 drops to 0 W/m^2. The corresponding DPP converter is switched off, and the output voltage and current are 0. As a consequence, the number of PV submodules is becoming two, and the system is always working on PBP as introduced before. The convergence speed in the experimental results is around 150 ms, which presents 260 times faster than that in [29,56].

Figure 6.45 demonstrates the outside experimental prototype. Three PV panels were linked in series and put with test equipment on the north-facing flat roof of a building on the school's campus. In addition to no longer using a DC-power source to connect the PV panel to simulate its photo-diode current, the remaining parameters are consistent with indoor testing. The PV submodule 1 is shaded by a fixed leaf to simulate a setting with fixed shading. The second PV submodule is regularly operating without any shading. For PV submodule 3, a moving garbage can is used to simulate the various shade levels. According to the two distinct locations of the trash can, the two shading possibilities corresponding to those positions are simulated. In the first case with shading, the string current at TMPP is 0.6 A. The maximum power stress on a single DPP converter is around 8.5 W, less than the rated power. As a result, TMPP is implemented to increase system efficiency, and the string current value should be around 0.6 A, as shown in Figure 6.46. Then, the movable shading block is slightly adjusted to the north at time t_1; meanwhile, the shading of PV submodule 3 is mitigated. At the same, under the second scenario, which corresponds to the results reported after the time instant t_1, PV submodule 1 and PV submodules are still operating in their original operational states. For the second shading scenario, the current of I_{pv3} is increasing from 0.6 A to 1 A, and the power distribution among DPP converters by using TMPP is $P_{dpp1} = 12$ W, $P_{dpp2} = 0$ W, and $P_{dpp3} = 0$ W. It can be found that the power stress for DPP converter 1 is

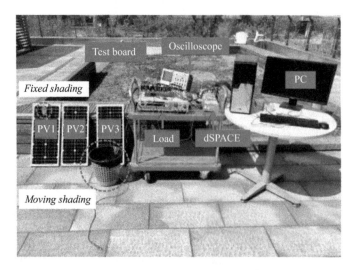

Figure 6.45   Outdoor experimental setup

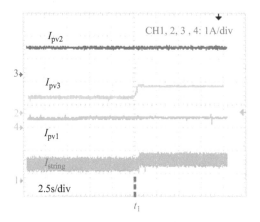

Figure 6.46   Outdoor experimental results

higher than the designed power rating if we use TMPPT control. Therefore, the algorithm has switched the operation point of the string current from TMPP to PBP to keep all DPP converters working within a safe power range.

6.4.3 PV–IP architecture

6.4.3.1 Topology and structure

Figure 6.47 shows the diagram of the PV–IP DPP architecture. Similar to the PV–bus DPP architecture, bidirectional power converters such as BFC and DAB can be used for both architectures. The present works refer to the research on PV–PV that could be found in [57,71–80]. The specific comparisons of PV–IP architectures are presented in Table 6.7.

The distinction between the PV–bus and PV–IP DPP architectures is whether the secondary side of the DPP converters is connected to the DC-link side. As illustrated in Figure 6.47, the secondary sides of the converters in the PV–IP DPP architecture are paralleled and connected to a port independent of the DC link. Individual DPP converters typically utilize suboptimal control schemes such as VE control rather than true MPPT to limit the number of sensors and communications. Compared to the PV–bus DPP architecture, the PV–IP DPP architecture has the following advantages:

1. **Low cost**: since the secondary side of the PV–IP architecture is an isolated port rather than the DC link, the voltage can be selected independently from either

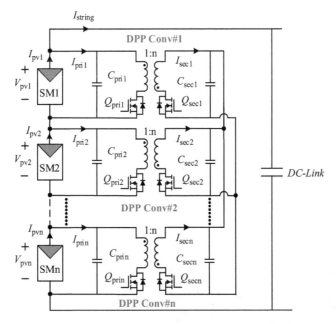

Figure 6.47 PV–IP DPP architecture with bidirectional flyback converters

Table 6.7 Comparison of PV–IP DPP architecture

Ref.	Topology	Switch number	Submodule control	Central control	Current sensing	Current sensing	Report efficiency	Experimental set	Current flow	Integration level
[54]	BFCs (DCM)	$2n$	VE	True MPPT	No	No	VE based: 96.5%** (significant shading)	• 3 PV submodules • 3 DPP converters	Bidi	Medium
[55]	BFCs (DCM)	$2n$	VE	–	No	No	VE based: >99%** (significant shading)	• 3 PV submodules • 3 DPP converters	Bidi	High
[56]	BFCs (DCM)	$2n$	VE	Load sweep	No	No	VE tracking > 98%	• 3 PV submodules • 3 DPP converters	Bidi	–
[57]	BFCs (CCM)	$2n$	True MPPT	MPP sweep	Yes	Yes	• Flyback: 85-94.5%* • System:>97.42%**	• 3 PV submodules • 3 DPP converters	Bidi	Medium
[58]	DAB (CCM)	$8n$	VE	True MPPT	Yes	Yes	• DAB: 85-95%* • System:>98.7%**	• 4 Server submodules • 4 DPP converters	Bidi	High
[59]	Flybac (CCM)	$2n$	VE	True MPPT	No	No	• Flyback: >96.7%* • Tracking eff = 94.4%	• 3 PV submodules • 3 DPP converters	Bidi	Medium
[64]	Flybac (CCM)	$2n$	VE	Load Swee-p	Yes	Yes	–	• 3 PV submodules • 3 DPP converters	Bidi	Medium

the PV string voltage or the DC link voltage. Thus, both primary and secondary side devices can use low-voltage semiconductor devices, benefiting the system's cost reduction and efficiency improvement.

2. **Flexible voltage gain**: the voltage of the isolated port in the PV–IP DPP architecture can be regarded as a control variable independent of the PV submodule voltage and bus voltage. Thus, the conversion ratio can also be selected independently from the number of PV submodules [59]. The flexibility in selecting the voltage gain also benefits the reduction in the magnetic size.

However, compared with PV–bus DPP architecture, one additional constraint must be fulfilled in the control of PV–IP DPP architecture, which is expressed by:

$$P_{\text{dpp.net}} = \sum_{i=1}^{n} V_{\text{pv},i} I_{\text{dpp},i} = 0 \tag{6.22}$$

The constraint (6.22) lies in the interconnection configuring the ratio of PV–IP DPP architecture. Considering the isolated port in PV–IP DPP architecture, the sum of the primary and secondary side power must be zero. The injected power from DPP converters will be the same as the extracted power from submodules. Thus, the net power for the PV–IP DPP architecture is always zero. With the constraint, the MPP for each PV submodule may not always be achievable for any string current. The only string current that is capable of extracting the maximum output power from the PV submodule can be expressed as:

$$I_{\text{string}} = \frac{\sum_{i=1}^{n} I_{\text{pv},i}}{n} \tag{6.23}$$

where I_{pvi} is the current for each PV submodule, and n is the number of submodules.

Due to the power constraint of (6.22), the total processed power for DPP converters in PV–IP DPP architecture is higher than that in PV–bus DPP architecture using LPPT or TMPP control. The study in [59] indicates that the power processed by the DPP converters in PV–IP architecture under the worst scenario is 33% higher than that in PV–bus architecture.

The two structures have some similarities in hardware integration, i.e., the integration difficulty is greater than that of the PV–PV. The BFCs and DAB in PV–IP and PV–bus DPP architecture present a low advantage in size and integration level due to the isolated transformer, which makes the DPP converters in these two-architecture difficult to fit into the existing junction box of the PV module. However, some excellent IC designs still put the hardware of these two architectures in the PV junction box, such as the BFC-based PV–IP architecture in [72]. The dimension of the board of the work in [72] is of 70 mm by 110 mm with 14 mm height with a volume of 113 cm³, which is an excellent design in present DPP works. However, it is still worth noting that its volume is still 2.46 times larger than SC-based PV–PV in [40], and 6.5 times larger than SL-based PV–PV in [28]. Furthermore, the volume in [72] is based on the converter's power rating and is not designed according to the worst shading scenario. Therefore, compared with PV–PV architecture, the hardware prototype in PV–bus and PV–IP is still more challenging to design to be small enough to fit into most of the PV junction boxes on the market. The issues can be solved by manufacturing by increasing the junction box size, or we can add another box behind the PV module to place the DPP converters. However, both two methods will increase the cost. In a high-frequency mode, the size of the magnetic components can be significantly reduced, but the operating frequency of silicon-based switches is limited. Therefore, combining the gallium nitride (GaN) switches with better high-frequency characteristics can significantly reduce the size of the DPP converter for PV–IP and PV–bus DPP architecture.

6.4.3.2   Control and optimization

A.   *VE control*

In PV–IP DPP architecture, the true MPPT cannot be guaranteed under various environmental conditions due to an isolated port. Thus, the VE control is widely used as sub-optimal control. The VE control based on DCM flyback converters was first introduced in [59]. The algorithm for the local controller of each flyback converter with the aim to balance the voltages of each PV module with respect to the isolated port voltage can be expressed as [57,71,72,79]

$$I_{pri,i} = K(V_{pv,i} - V_{sec}) \tag{6.24}$$

The DCM BFCs inject or extract current parallel to the PV submodule to equalize their voltage. When the PV submodule voltage V_{pvi} is higher than the V_{sec}, the current $I_{pri,i}$ is subtracted from the PV submodule and injected into the isolated port. When the PV submodule voltage V_{pvi} is lower than the V_{sec}, the current $I_{pri,i}$ is injected into the PV submodule and is subtracted from the isolated port. While the DPP converters will balance the PV submodule voltage, they cannot regulate the precise output voltage value. Therefore, an external central stage should be utilized to optimize the voltage of the module. The entire control mechanism can be summed up as follows: the internal DPP converter equalizes the voltage of each PV submodule, ensuring that the operating point is on a straight line. The central DC–DC converter or DC–AC inverter can then utilize the MPPT algorithm to maintain the maximum power harvest.

In [57], the effect of parameter K in (6.24) on the system efficiency has been analyzed, which indicates that the selection of K should be the trade-off between the power conversion efficiency and the regulation performance of the output voltage. In [71], a non-linear control was introduced, and the maximum difference between two PV submodule voltages is bounded by the constant C and the controller gain K, which can be expressed by

$$\left| V_{pv,i} - V_{pv,j} \right| \leq \frac{C}{K}, \ \forall i,j = 1, 2, ..., n, i \neq j \tag{6.25}$$

The bound (6.25) hold for any submodule current difference, and the controller gain K should be set higher than a minimum boundary to ensure the control performance.

In order to reduce the voltage sensors and communications for individual DPP converters that are required for the VE control in (6.25), a fixed-duty cycle VE control was proposed in [75], and the mathematical expression is given by:

$$V_{sec} = V_{pv1} \frac{D}{1-D} = ... = V_{pvn} \frac{D}{1-D} \tag{6.26}$$

where D is the duty cycle of the BFCs.

It is noticed that if we let duty cycle in (6.26) equal to 50%, and then $V_o = V_{in\_1} = V_{in\_2} = V_{in\_3}$. Therefore, the voltage output voltage of each PV submodule is equalized. In other words, the submodule voltages are controlled to reside on a straight line. The operation principle of the BFCs is presented in Figure 6.48. If the power transfers from the primary side to the secondary side of the BFCs, the primary side switch is switched on. Meanwhile, the secondary side switch is switched off, and the built-in diode in Mosfet is working as the inversed diode. If the architecture needs to transfer the power from the secondary side to the primary side to compensate for the shaded submodule, the secondary side switch is switched on. Meanwhile, the primary side switch is switched off.

The conventional PV system uses a bypass diode against mismatched conditions. However, the drawbacks are limited save power, and the output P–V curve shows multi-peaks characteristic under PSC. Using the VE control method, we have compared the power improvement of the PV–IP architecture with the bypass diode method under three shading scenarios, including severe mismatch condition, moderate mismatch condition, and slight mismatch condition. For the severe shading condition, the normalized illumination value for PV 1, PV 2, and PV 3 are

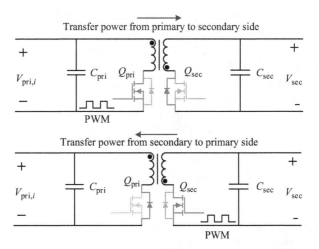

Figure 6.48 Power transformation in the bidirectional flyback converter

100%, 50%, and 25%, respectively. The normalized illumination values for PV 1, PV 2, and PV 3 under mild shadowing conditions are 100%, 75%, and 50%, respectively. The normalized illumination values for PV 1, PV 2, and PV 3 under the situation of minor shadowing are 100%, 90%, and 80%, respectively. The electronic load is used to sweep the P–V and I–V curves to check the maximum output power by using the conventional bypass diode and the PV–IP methods for these different shading scenarios. Figure 6.49(a) and (b) shows the slight shading scenario results. The bypass diode method has a limited influence on the power reduction under PSC, and the power improvement with the PV–IP is slight (5.32%). However, if we increase the shading level, the power improvement with PV–IP architecture becomes apparent. It is noticed that there is a 33.91% power improvement under moderate PSC, and 56.25% power improvement under severe PSC, as shown in Figure 6.49(c) and (e). For these two shading scenarios, the string current is primarily constrained by the severely shaded PV submodule, and a significant portion of the power supplied by the normally functioning PV submodule cannot be output using the standard bypass diode technique.

To further research the power improvement led by PV–IP architecture compared with the bypass diode in more shading conditions, a test was made under various irradiance conditions. In this test, PV submodule 1 is fully shaded at all times, and the normalized insolation is set to 100%. Normalized insolation is set to 25%, 50%, 75%, and 100% for PV submodule 2. The normalized insolation of PV submodule 3 is varied between 10% and 100%. The power enhancement outcomes are depicted in Figure 6.50. It has been determined that the power improvement is as great as 86% for the shading scenarios PV 1: 100%, PV 2: 25%, and PV 3: 40%. Using a PV–IP design can significantly increase power, particularly under a more severe PCS.

B.   *True MPPT*

With the change of PV–IP DPP architecture, the true MPPT can be implemented under some environmental conditions [73]. Specifically, a large bus capacitor is connected to the isolated port. In [73], the added capacitance is high up to 50 mF, which is big and accounts for a large part of the experimental prototype. The main function of the isolated-port capacitor is to mitigate the perturbation coupling and act as a large reservoir to store energy to ensure each PV submodule can reach its respective MPPs. The selection of the isolated-port capacitance should be made carefully to main the system stability. Typically, two modes will be adopted to achieve the function of MPPT, namely bus charging mode and central MPP sweeping mode. The isolated-port capacitor will be charged to the reference PV submodule voltage within the bus charging mode. In the central MPP sweep mode, the central DC–DC converter will regulate the string current to maximize the output power from the whole PV system. It should be noted that the PV submodules operate at the VE point rather than the true MPPs. The two modes require a time of 40 s and 20 s, respectively, before the true MPPT algorithm is implemented for PV submodules.

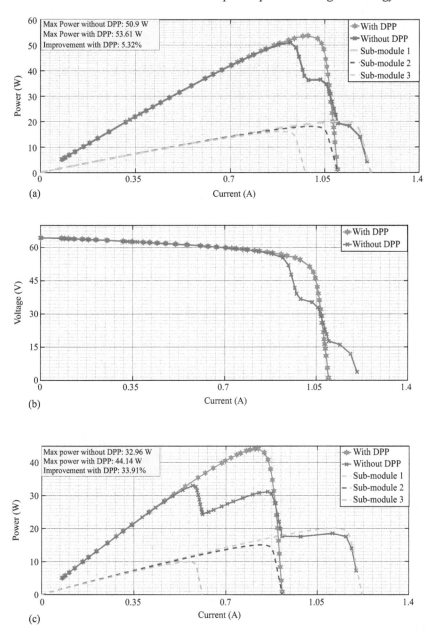

*Figure 6.49   Output P–V and I–V curves by using PV–IP architecture and
conventional bypass diode under different PSCs*

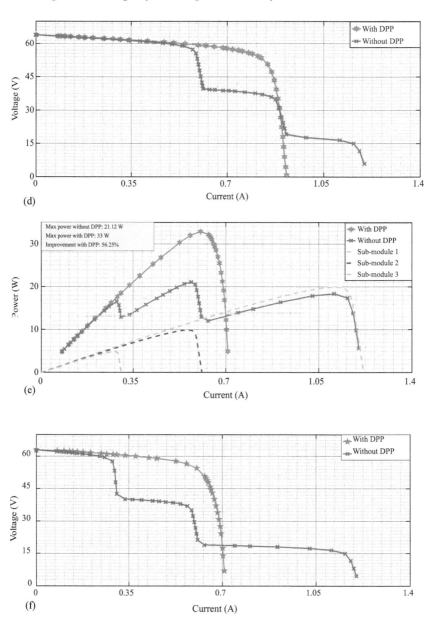

Figure 6.49   (Continued)

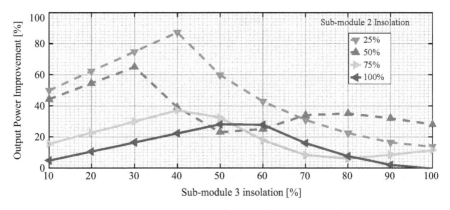

Figure 6.50   Output power improvement under various insolation conditions

6.5   Conclusion and future work

The content in this chapter has discussed the topologies and control strategies of the advanced submodule DMPPT technologies named DPP architecture. The primary advantage of the DPP architecture is that DC–DC converters process just a tiny portion of the total produced power, reducing the power pressure of the DPP converters. This characteristic results in high overall output efficiency and small hardware size. The DPP design may be placed in the junction box behind PV modules without incurring extra surface lamination damage. Especially for PV–PV architecture, the merit of not having a magnetic core original makes the overall hardware smaller than PV–bus and PV–IP architectures. However, the issues of high coupling and high processed power in PV–PV architecture are not solved yet. It is an issue caused by the topology structure's characteristics, which cannot be solved only by optimizing the control algorithm. The PV–bus and PV–IP have a similar structure, and the only difference is that the secondary side of the PV–IP is isolated. The advantages of the insolation are low coupling, simple structure, and low voltage pressure on secondary side switches. However, on the other hand, the control strategies available for PV–IP are much less than for the PV–bus architecture, such as the LPPT and TMPPT control for the centralized converter.The DPP architecture is still relatively modern and needs to be improved. Future work can focus on the following aspects:

- **Integration**: The BFCs and DAB in PV–IP and PV–bus DPP architecture present a small size and integration level advantage due to the isolated transformer, which makes the DPP converters in these two-architecture difficult to fit into the existing junction box of the PV module. The issues can be solved by manufacturing by increasing the junction box size, or we can add another box behind the PV module to place the DPP converters. Another solution is to optimize the control method with increasing switch frequency. In a high-frequency mode, the size of the magnetic components can be significantly reduced, but the operating frequency of silicon-based switches is limited. Therefore, combining the GaN switches with better high-frequency characteristics can significantly reduce the size of the DPP converter for PV–IP and PV–bus DPP architecture.

- **Cable**: This is an issue that many researchers overlook. However, the cable stress caused by DPP technology is considerable. The high cable number will increase the cost and the difficulty of installation. The works in [59] have given a wiring diagram of a PV module with three DPP converters, and we can see the complexity and difficulty of the whole system. Therefore, reducing the cables will become an essential optimization direction. For example, we can use the main circuit wire to achieve communication purposes and reduce cable numbers.

- **Reliability and fault-tolerant control:** the DPP-based PGS significantly increases the number of power DC–DC converters compared with the conventional series PGS, which increases the cost and influences the lifetime of the system. In a PGS, the PV module's viability is bolstered by relatively long panel lives and often 25-year or more extended warranties. In the past, the power electronics in PGS often had significantly poorer dependability than PV modules. According to reliability research, switching devices are identified as the most vulnerable element contributing to 69.24% of the overall failure rate. Therefore, the power electronics fault occurs more often in DPP-based PV systems than in the conventional series PV system. The failure of the power electronics reduces the power harvesting and may lead to instability of the whole system. In light of this, the fault diagnosis and fault-tolerant control mechanism for submodule-level DPP architecture in PGS are highly significant. However, there is currently a dearth of relevant literature research.

References

[1] BP, "BP energy outlook 2020 edition," 2020. Available: https://bp.com/en/global/corporate/energy-economics/energy-outlook.html.

[2] Y. Hu, W. Cao, J. Ma, S. J. Finney, and D. Li, "Identifying PV module mismatch faults by a thermography-based temperature distribution analysis," *IEEE Transactions on Device and Materials Reliability*, vol. 14, no. 4, pp. 951–960, 2014, doi:10.1109/TDMR.2014.2348195.

[3] G. Velasco-Quesada, F. Guinjoan-Gispert, R. Pique-Lopez, M. Roman-Lumbreras, and A. Conesa-Roca, "Electrical PV array reconfiguration strategy for energy extraction improvement in grid-connected PV systems," *IEEE Transactions on Industrial Electronics*, vol. 56, no. 11, pp. 4319–4331, 2009.

[4] S. Motahhir, A. El Ghzizal, S. Sebti, and A. Derouich, "Shading effect to energy withdrawn from the photovoltaic panel and implementation of DMPPT using C language," *International Review of Automatic Control*, vol. 9, no. 2, pp. 88–94, 2016.

[5] F. A. Silva, "Power electronics and control techniques for maximum energy harvesting in photovoltaic systems (Femia, N. *et al*; 2013) [Book News]," *IEEE Industrial Electronics Magazine*, vol. 7, no. 3, pp. 66–67, 2013.

[6] Y. A. Mahmoud, W. Xiao, and H. H. Zeineldin, "A parameterization approach for enhancing PV model accuracy," *IEEE Transactions on Industrial Electronics*, vol. 60, no. 12, pp. 5708–5716, 2013, doi:10.1109/TIE.2012.2230606.

[7] Y. Liu, S. Huang, J. Huang, and W. Liang, "A particle swarm optimization-based maximum power point tracking algorithm for PV systems operating under partially shaded conditions," *IEEE Transactions on Energy Conversion*, vol. 27, no. 4, pp. 1027–1035, 2012, doi:10.1109/TEC.2012.2219533.

[8] K. Ding, X. Bian, H. Liu, and T. Peng, "A MATLAB-Simulink-based PV module model and its application under conditions of nonuniform irradiance," *IEEE Transactions on Energy Conversion*, vol. 27, no. 4, pp. 864–872, 2012, doi:10.1109/TEC.2012.2216529.

[9] A. Mäki and S. Valkealahti, "Power losses in long string and parallel-connected short strings of series-connected silicon-based photovoltaic modules due to partial shading conditions," *IEEE Transactions on Energy Conversion*, vol. 27, no. 1, pp. 173–183, 2012, doi:10.1109/TEC.2011.2175928.

[10] T. Takashima, J. Yamaguchi, K. Otani, T. Oozeki, K. Kato, and M. Ishida, "Experimental studies of fault location in PV module strings," *Solar Energy Materials and Solar Cells*, vol. 93, no. 6, pp. 1079–1082, 2009/06/01/ 2009, doi: https://doi.org/10.1016/j.solmat.2008.11.060.

[11] H. Patel and V. Agarwal, "MATLAB-based modeling to study the effects of partial shading on PV array characteristics," *IEEE Transactions on Energy Conversion*, vol. 23, no. 1, pp. 302–310, 2008, doi:10.1109/TEC.2007.914308.

[12] N. Femia, G. Petrone, G. Spagnuolo, and M. Vitelli, *Power Electronics and Control Techniques for Maximum Energy Harvesting in Photovoltaic Systems*, London: CRC Press, 2013. Available: https://doi.org/10.1201/b14303.

[13] O. Khan and W. Xiao, "An efficient modeling technique to simulate and control submodule-integrated PV system for single-phase grid connection," *IEEE Transactions on Sustainable Energy*, vol. 7, no. 1, pp. 96–107, 2016.

[14] G. R. Walker and P. C. Sernia, "Cascaded DC-DC converter connection of photovoltaic modules," in *2002 IEEE 33rd Annual IEEE Power Electronics Specialists Conference. Proceedings (Cat. No.02CH37289)*, 23–27 June 2002, vol. 1, pp. 24–29.

[15] L. Linares, R. W. Erickson, S. MacAlpine, and M. Brandemuehl, "Improved energy capture in series string photovoltaics via smart distributed power electronics," in *2009 Twenty-Fourth Annual IEEE Applied Power Electronics Conference and Exposition*, 15–19 Feb. 2009, pp. 904–910.

[16] A. I. Bratcu, I. Munteanu, S. Bacha, D. Picault, and B. Raison, "Cascaded DC–DC converter photovoltaic systems: power optimization issues," *IEEE Transactions on Industrial Electronics*, vol. 58, no. 2, pp. 403–411, 2011.

[17] R. C. N. Pilawa-Podgurski and D. J. Perreault, "Submodule integrated distributed maximum power point tracking for solar photovoltaic applications," *IEEE Transactions on Power Electronics*, vol. 28, no. 6, pp. 2957–2967, 2013.

[18] O. Khan, W. Xiao, and H. H. Zeineldin, "Gallium-nitride-based submodule integrated converters for high-efficiency distributed maximum power point tracking PV applications," *IEEE Transactions on Industrial Electronics*, vol. 63, no. 2, pp. 966–975, 2016.

[19] C. Deline and S. MacAlpine, "Use conditions and efficiency measurements of DC power optimizers for photovoltaic systems," in *2013 IEEE Energy Conversion Congress and Exposition*, 15-19 Sept. 2013 2013, pp. 4801–4807, doi:10.1109/ECCE.2013.6647346.

[20] F. Wang, T. Zhu, F. Zhuo, H. Yi, S. Shi, and X. Zhang, "Analysis and optimization of flexible MCPT strategy in submodule PV application," *IEEE Transactions on Sustainable Energy*, vol. 8, no. 1, pp. 249–257, 2017, doi:10.1109/TSTE.2016.2596539.

[21] O. Khan, W. Xiao, and M. S. E. Moursi, "A new PV system configuration based on submodule integrated converters," *IEEE Transactions on Power Electronics*, vol. 32, no. 5, pp. 3278–3284, 2017, doi:10.1109/TPEL.2016.2633564.

[22] T. Duman and M. Boztepe, "Evaluation of zero voltage switching SEPIC converter for module integrated distributed maximum power point tracking applications," in *2017 10th International Conference on Electrical and Electronics Engineering (ELECO)*, 30 Nov.–2 Dec. 2017, pp. 1480–1484.

[23] Y. Nimni and D. Shmilovitz, "A returned energy architecture for improved photovoltaic systems efficiency," in *Proceedings of 2010 IEEE International Symposium on Circuits and Systems*, 30 May–2 June 2010, pp. 2191–2194.

[24] R. Giral, C. A. Ramos-Paja, D. Gonzalez, *et al.*, "Minimizing the effects of shadowing in a PV module by means of active voltage sharing," in *2010 IEEE International Conference on Industrial Technology*, 14–17 March 2010, pp. 943–948.

[25] R. Giral, C. E. Carrejo, M. Vermeersh, A. J. Saavedra-Montes, and C. A. Ramos-Paja, "PV field distributed maximum power point tracking by means of an active bypass converter," in *2011 International Conference on Clean Electrical Power (ICCEP)*, 14–16 June 2011, pp. 94–98.

[26] S. Poshtkouhi, A. Biswas, and O. Trescases, "DC–DC converter for high granularity, sub-string MPPT in photovoltaic applications using a virtual-parallel connection," in *2012 Twenty-Seventh Annual IEEE Applied Power Electronics Conference and Exposition (APEC)*, 5–9 Feb. 2012, pp. 86–92, doi:10.1109/APEC.2012.6165802.

[27] D. Shmilovitz and Y. Levron, "Distributed maximum power point tracking in photovoltaic systems—emerging architectures and control methods,"

Automatika, vol. 53, no. 2, pp. 142–155, 2012, doi: 10.7305/automatika.53-2.185.

[28] S. Qin, C. B. Barth, and R. C. N. Pilawa-Podgurski, "Enhancing micro-inverter energy capture with submodule differential power processing," *IEEE Transactions on Power Electronics*, vol. 31, no. 5, pp. 3575–3585, 2016.

[29] Y. Jeon, H. Lee, K. A. Kim, and J. Park, "Least power point tracking method for photovoltaic differential power processing systems," *IEEE Transactions on Power Electronics*, vol. 32, no. 3, pp. 1941–1951, 2017.

[30] K. A. Kim, P. S. Shenoy, and P. T. Krein, "Converter rating analysis for photovoltaic differential power processing systems," *IEEE Transactions on Power Electronics*, vol. 30, no. 4, pp. 1987–1997, 2015.

[31] P. S. Shenoy, K. A. Kim, B. B. Johnson, and P. T. Krein, "Differential power processing for increased energy production and reliability of photovoltaic systems," *IEEE Transactions on Power Electronics*, vol. 28, no. 6, pp. 2968–2979, 2013.

[32] H. Jeong, H. Lee, Y. Liu, and K. A. Kim, "Review of differential power processing converter techniques for photovoltaic applications," *IEEE Transactions on Energy Conversion*, vol. 34, no. 1, pp. 351–360, 2019.

[33] J. T. Stauth, M. D. Seeman, and K. Kesarwani, "A resonant switched-capacitor IC and embedded system for sub-module photovoltaic power management," *IEEE Journal of Solid-State Circuits*, vol. 47, no. 12, pp. 3043–3054, 2012.

[34] J. T. Stauth, K. Kesarwani, and C. Schaef, "A distributed photovoltaic energy optimization system based on a sub-module resonant switched-capacitor implementation," in *2012 15th International Power Electronics and Motion Control Conference (EPE/PEMC)*, 4–6 Sept. 2012, pp. LS2d.2-1–LS2d.2-6.

[35] P. S. Shenoy and P. T. Krein, "Differential power processing for DC systems," *IEEE Transactions on Power Electronics*, vol. 28, no. 4, pp. 1795–1806, 2013.

[36] S. Qin and R. C. N. Pilawa-Podgurski, "Sub-module differential power processing for photovoltaic applications," in *2013 Twenty-Eighth Annual IEEE Applied Power Electronics Conference and Exposition (APEC)*, 17–21 March 2013, pp. 101–108.

[37] L. F. L. Villa, T. Ho, J. Crebier, and B. Raison, "A power electronics equalizer application for partially shaded photovoltaic modules," *IEEE Transactions on Industrial Electronics*, vol. 60, no. 3, pp. 1179–1190, 2013.

[38] C. Schaef, K. Kesarwani, and J. T. Stauth, "A coupled-inductor multi-level ladder converter for sub-module PV power management," in *2013 Twenty-Eighth Annual IEEE Applied Power Electronics Conference and Exposition (APEC)*, 17–21 March 2013, pp. 732–737.

[39] H. J. Bergveld, D. Büthker, C. Castello, *et al.*, "Module-level DC/DC conversion for photovoltaic systems: the delta-conversion concept," *IEEE Transactions on Power Electronics*, vol. 28, no. 4, pp. 2005–2013, 2013.

[40] J. T. Stauth, M. D. Seeman, and K. Kesarwani, "Resonant switched-capacitor converters for sub-module distributed photovoltaic power management," *IEEE Transactions on Power Electronics*, vol. 28, no. 3, pp. 1189–1198, 2013.

[41] R. Sangwan, K. Kesarwani, and J. T. Stauth, "High-density power converters for sub-module photovoltaic power management," in *2014 IEEE Energy Conversion Congress and Exposition (ECCE)*, 14–18 Sept. 2014, pp. 3279–3286.

[42] C. Schaef and J. T. Stauth, "Multilevel power point tracking for partial power processing photovoltaic converters," *IEEE Journal of Emerging and Selected Topics in Power Electronics*, vol. 2, no. 4, pp. 859–869, 2014.

[43] S. Qin, S. T. Cady, A. D. Domínguez-García, and R. C. N. Pilawa-Podgurski, "A distributed approach to maximum power point tracking for photovoltaic submodule differential power processing," *IEEE Transactions on Power Electronics*, vol. 30, no. 4, pp. 2024–2040, 2015.

[44] M. S. Zaman, Y. Wen, R. Fernandes, *et al.*, "A cell-level differential power processing IC for concentrating-PV systems with bidirectional hysteretic current-mode control and closed-loop frequency regulation," *IEEE Transactions on Power Electronics*, vol. 30, no. 12, pp. 7230–7244, 2015.

[45] F. Wang, T. Zhu, F. Zhuo, and H. Yi, "An improved submodule differential power processing-based PV system with flexible multi-MPPT control," *IEEE Journal of Emerging and Selected Topics in Power Electronics*, vol. 6, no. 1, pp. 94–102, 2018.

[46] K. Kesarwani and J. T. Stauth, "A comparative theoretical analysis of distributed ladder converters for sub-module PV energy optimization," in *2012 IEEE 13th Workshop on Control and Modeling for Power Electronics (COMPEL)*, 10–13 June 2012, pp. 1–6.

[47] A. Blumenfeld, A. Cervera, and M. M. Peretz, "Enhanced differential power processor for PV systems: resonant switched-capacitor gyrator converter with local MPPT," *IEEE Journal of Emerging and Selected Topics in Power Electronics*, vol. 2, no. 4, pp. 883–892, 2014.

[48] H. G. Chiacchiarini, J. G. Ceci, A. R. Oliva, and P. S. Mandolesi, "Individual solar cells balance for maximum power extraction in series arrays," in *2014 IEEE Biennial Congress of Argentina (ARGENCON)*, 11–13 June 2014, pp. 759–764.

[49] P. S. Shenoy, B. Johnson, and P. T. Krein, "Differential power processing architecture for increased energy production and reliability of photovoltaic systems," in *2012 Twenty-Seventh Annual IEEE Applied Power Electronics Conference and Exposition (APEC)*, 5–9 Feb. 2012, pp. 1987–1994.

[50] S. Ben-Yaakov, A. Blumenfeld, A. Cervera, and M. Evzelman, "Design and evaluation of a modular resonant switched capacitors equalizer for PV panels," in *2012 IEEE Energy Conversion Congress and Exposition (ECCE)*, 15–20 Sept. 2012, pp. 4129–4136.

[51] S. Qin, A. J. Morrison, and R. C. N. Pilawa-Podgurski, "Enhancing micro-inverter energy capture with sub-module differential power processing," in

2014 IEEE Applied Power Electronics Conference and Exposition – APEC 2014, 16–20 March 2014, pp. 621–628.

[52] S. W. Moore and P. J. Schneider, "A review of cell equalization methods for lithium ion and lithium polymer battery systems," in *SAE Technical Paper Series*, 2001, pp. 1–5. Available: https://doi.org/10.4271/2001-01-0959.

[53] K. Sano and H. Fujita, "A resonant switched-capacitor converter for voltage balancing of series-connected capacitors," in *2009 International Conference on Power Electronics and Drive Systems (PEDS)*, 2–5 Nov. 2009, pp. 683–688, doi:10.1109/PEDS.2009.5385680.

[54] S. Tan, S. Kiratipongvoot, S. Bronstein, A. Ioinovici, Y. M. Lai, and C. K. Tse, "Adaptive mixed on-time and switching frequency control of a system of interleaved switched-capacitor converters," *IEEE Transactions on Power Electronics*, vol. 26, no. 2, pp. 364–380, 2011.

[55] S. Ben-Yaakov, "On the influence of switch resistances on switched-capacitor converter losses," *IEEE Transactions on Industrial Electronics*, vol. 59, no. 1, pp. 638–640, 2012.

[56] Y. Jeon and J. Park, "Unit-minimum least power point tracking for the optimization of photovoltaic differential power processing systems," *IEEE Transactions on Power Electronics*, vol. 34, no. 1, pp. 311–324, 2019.

[57] C. Olalla, C. Deline, D. Clement, Y. Levron, M. Rodriguez, and D. Maksimovic, "Performance of power-limited differential power processing architectures in mismatched PV systems," *IEEE Transactions on Power Electronics*, vol. 30, no. 2, pp. 618–631, 2015.

[58] T. Shimizu, M. Hirakata, T. Kamezawa, and H. Watanabe, "Generation control circuit for photovoltaic modules," *IEEE Transactions on Power Electronics*, vol. 16, no. 3, pp. 293–300, 2001.

[59] C. Olalla, D. Clement, M. Rodriguez, and D. Maksimovic, "Architectures and control of submodule integrated DC–DC converters for photovoltaic applications," *IEEE Transactions on Power Electronics*, vol. 28, no. 6, pp. 2980–2997, 2013.

[60] J. Du, R. Xu, X. Chen, Y. Li, and J. Wu, "A novel solar panel optimizer with self-compensation for partial shadow condition," in *2013 Twenty-Eighth Annual IEEE Applied Power Electronics Conference and Exposition (APEC)*, 17–21 March 2013, pp. 92–96.

[61] J. Park and K. Kim, "Multi-output differential power processing system using boost-flyback converter for voltage balancing," in *2017 International Conference on Recent Advances in Signal Processing, Telecommunications & Computing (SigTelCom)*, 9–11 Jan. 2017, pp. 139–142.

[62] K. Sun, Z. Qiu, H. Wu, and Y. Xing, "Evaluation on high-efficiency thermoelectric generation systems based on differential power processing," *IEEE Transactions on Industrial Electronics*, vol. 65, no. 1, pp. 699–708, 2018.

[63] P. Sharma and V. Agarwal, "Exact maximum power point tracking of grid-connected partially shaded PV source using current compensation concept,"

IEEE Transactions on Power Electronics, vol. 29, no. 9, pp. 4684–4692, 2014.

[64] A. K. Pati and N. C. Sahoo, "A novel control architecture for maximum power extraction from the photovoltaic system under partially shaded conditions using current equalization approach," in *2015 IEEE International Conference on Signal Processing, Informatics, Communication and Energy Systems (SPICES)*, 19–21 Feb. 2015, pp. 1–5.

[65] J. Biswas, A. M. Kamath, A. K. Gopi, and M. Barai, "Design, architecture, and real-time distributed coordination DMPPT algorithm for PV systems," *IEEE Journal of Emerging and Selected Topics in Power Electronics*, vol. 6, no. 3, pp. 1418–1433, 2018.

[66] M. O. Badawy, A. Elrayyah, F. Cingoz, and Y. Sozer, "Non-isolated individual MPP trackers for series PV strings through partial current processing technique," in *2014 IEEE Applied Power Electronics Conference and Exposition – APEC 2014*, 16–20 March 2014, pp. 3034–3041.

[67] G. Chu and H. Wen, "PV-to-bus DPP architecture based on unidirectional DC–DC converter for photovoltaic application," in *2019 10th International Conference on Power Electronics and ECCE Asia (ICPE 2019 – ECCE Asia)*, 27–30 May 2019, pp. 1–6.

[68] G. Chu, H. Wen, Y. Yang, and Y. Wang, "Elimination of photovoltaic mismatching with improved submodule differential power processing," *IEEE Transactions on Industrial Electronics*, vol. 67, no. 4, pp. 2822–2833, 2020.

[69] G. Chu, H. Wen, Y. Hu, L. Jiang, Y. Yang, and Y. Wang, "Low-complexity power balancing point-based optimization for photovoltaic differential power processing," *IEEE Transactions on Power Electronics*, vol. 35, no. 10, pp. 10306–10322, 2020.

[70] G. Chu, H. Wen, Z. Ye, and X. Li, "Design and optimization of the PV-virtual-bus differential power processing photovoltaic systems," in *2017 IEEE 6th International Conference on Renewable Energy Research and Applications (ICRERA)*, 5–8 Nov. 2017, pp. 674–679.

[71] Y. Levron, D. R. Clement, B. Choi, C. Olalla, and D. Maksimovic, "Control of submodule integrated converters in the isolated-port differential power-processing photovoltaic architecture," *IEEE Journal of Emerging and Selected Topics in Power Electronics*, vol. 2, no. 4, pp. 821–832, 2014.

[72] B. Choi, D. Clement, and D. Maksimovic, "A CMOS controller for submodule integrated converters in photovoltaic systems," in *2014 IEEE 15th Workshop on Control and Modeling for Power Electronics (COMPEL)*, 22–25 June 2014 2014, pp. 1–6.

[73] R. Bell and R. C. N. Pilawa-Podgurski, "Decoupled and distributed maximum power point tracking of series-connected photovoltaic submodules using differential power processing," *IEEE Journal of Emerging and Selected Topics in Power Electronics*, vol. 3, no. 4, pp. 881–891, 2015.

[74] E. Candan, P. S. Shenoy, and R. C. N. Pilawa-Podgurski, "A series-stacked power delivery architecture with isolated differential power conversion for

data centers," *IEEE Transactions on Power Electronics*, vol. 31, no. 5, pp. 3690–3703, 2016.

[75] G. Chu, H. Wen, L. Jiang, Y. Hu, and X. Li, "Bidirectional flyback based isolated-port submodule differential power processing optimizer for photovoltaic applications," *Solar Energy*, vol. 158, pp. 929–940, 2017, doi: https://doi.org/10.1016/j.solener.2017.10.053.

Chapter 7

Dual-axis solar tracking system providing an intelligent step changing range (SCR) approach using real-time FDM with a sensorless design

Gökay Bayrak[1], Alper Yılmaz[1] and Anıl Karadeniz[1]

In this study, an electrical and mechanical design of a sensorless dual-axis solar tracking system (DASTS) is proposed by using a fuzzy logic decision-maker (FDM) method. The proposed DASTS provides maximum efficiency by adaptively changing the step range without using any sensors. DASTS has eliminated sun position detection errors caused by environmental factors in conventional methods. The step ranges determined for the movement of the motors (linear actuators) in sun tracking are given to the system by an FDM developed in real-time. In the intelligent step changing range (SCR) approach, there are two independent FDMs that direct the azimuth angle and tilt (elevation) angle to control the DASTS. The proposed system is implemented in the LabVIEW environment, and 30.5% more electrical energy was obtained compared to the fixed PV system. In addition, the proposed method is 6.64% more efficient than the DASTS using only the mathematical method. Besides, with the FDM with SCR, step times can be determined with an accuracy of 97.24%. As a result, the proposed intelligent sensorless system is especially suitable for rural installations, unlike the systems in the literature, with its cost-effectiveness, low maintenance requirements, and high-efficiency increase.

7.1 Introduction

In the literature, many methods are proposed for photovoltaic (PV) systems to track the sun [1]. These methods are examined under two main headings as active and passive methods. Among these methods, there are more studies on active tracking systems. The primary reason for this is that passive tracking systems are limited and insufficient to achieve the desired efficiency levels. Since active systems generally contain gears, motors, controller cards, and expandable software,

[1]Bursa Technical University, Faculty of Engineering and Natural Sciences Department of Electrical and Electronics Engineering, Turkey

installation and maintenance costs are more expensive than passive systems. In addition, their efficiency is quite high compared to passive systems. Active control methods can be examined under five main headings [2–10]: (1) systems providing sensor control, (2) microprocessor-based control, (3) open-closed loop systems, (4) intelligent systems, and (5) hybrid systems formed by the combination of two or more of these (Figure 7.1).

The general operating principle of passive solar tracking systems (STSs) is to achieve a movement in the tracking system by creating an imbalance in the system thanks to sunlight [9,10]. This movement can be achieved with liquids with thermal expansion or shape alloy materials. The main benefit of passive systems is their ability to track the sun without the use of motors, gears, or controllers. However, these methods have many disadvantages [2]. The biggest disadvantage of passive tracking systems is that they are dependent on weather conditions. Another disadvantage is the difficulties in choosing suitable glass and gas types.

In systems developed using active sun tracking methods, mechanical equipment such as motors and gears and microprocessors that control them are generally used [2]. Here, the motors are controlled by microprocessors using algorithms. The motors are used to enable the system to move and the solar panels perform the tracking process. Active methods can be applied to uniaxial and DASTS [3]. In the study where a DASTS that can move based on real-time solar radiation measurements is designed [4], the signals from the radiation sensors are sent to a high-performance 16-bit digital signal processor. Then, the position of the PV panels is adjusted by comparing the obtained signals with the sun location information determined by mathematical calculations. While the proposed system provides an average of 20% increase in efficiency on sunny days compared to fixed systems, the efficiency is very low on cloudy days. The DASTS proposed in Ref. [5] consists of one ATmega328 microcontroller, two DC motors, four light-dependent resistors (LDRs), and four relays. Single-axis and dual-axis tracking systems are tested from

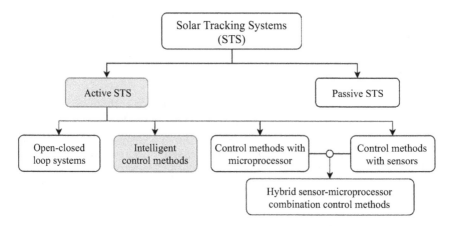

Figure 7.1   Solar tracking systems according to control methods [1]

9 a.m. to 4 p.m. for one month. The single-axis system generated 24.05% more power, while the DASTS generated 26.22% more power. However, the conditions are not adequately evaluated in this study. In a study [11] using the daily elevation angle as input and the torque force of the motors as the output, the open loop tracking system was developed for DASTS using mathematical formulas with the LabVIEW program. An acceptable error rate is determined during the initial and final stages of force action to evaluate the accuracy of the system. It is determined that this error rate is quite low compared to the efficiency obtained from the system.

In recent years, intelligent control methods incorporating techniques like fuzzy logic, neural networks, image processing, and deep learning are proposed for the control of STS in addition to the existing methods [6,12–14]. In the study [6], in which a new approach using image processing and deep learning is developed, a low-cost and intelligent STS is designed. The system is trained using the images obtained during the day with the camera on it. The sun, cloud, shadow, obstacle, and other heliostat systems in these images are taught to the STS. It is determined that the margin of error of the tested system is close to conventional systems. In another study conducted with the machine vision technique [12], unlike traditional methods, STS is developed according to the shadow. An iron bar is placed in front of the camera and the image of the shadow formed by this bar on a plate is processed and the sun is tracked. The movement of the system is provided according to the distance between the center of the shadow and the endpoint. This low-cost system that is developed monitors with 2° accuracy and has increased efficiency by 25%. DASTS with fuzzy logic control is proposed in Ref. [13]. The basis of this system is based on finding the vertical position of the sun on the solar panels. The difference between measured light intensity is used to determine the position and orientation of the solar panels. Here, the Mamdani fuzzy inference system is used to position, rotate the panels, and feed the motor. The rules of fuzzy logic control are chosen based on the differences between horizontal and vertical light intensity, and thanks to the proposed model, the output power increased by 12.45% compared to fixed solar panels. However, the cost of the proposed system is quite high. In Ref. [14], a method using the fuzzy logic controller and METEOSAT (satellite earth images) images is proposed for DASTS. The proposed model used an image processing technique to detect cloud coverage and cloud time in the sky. Cloudiness and its duration are used as the input of the controller, and by evaluating these data, it is ensured that the solar panel came to the most suitable position. Three different conditions are investigated, including clear, cloudy, and partly cloudy skies. Four images are obtained from the satellite at the beginning of each hour and an algorithm based on image processing is used to find the most suitable location. The proposed system is tested for 17 months, proving that the system works better in cloudy skies and has a 23% efficiency increase. Although there is an increase in efficiency, there is no study on the optimization of the times and step intervals determined for the operation of the motors.

Most of the methods in the literature have suggested various control methods that performed based on the information received from the sensors and based on this information, the movement of the motors in the system is provided. Especially in the change of environmental conditions such as cloudy weather, this approach cannot produce accurate results. In addition, the times and step intervals determined for the operation of the motors are defined specifically for the system, and there are deficiencies in determining the step intervals automatically.

In this study, DASTS can detect the position of the sun without a sensor and automatically adjust the running step times of the motors with a fuzzy logic decision maker (FDM) is proposed. In SCR approach, there are two independent fuzzy logic-based decision makers that direct the azimuth angle and elevation angle to control the DASTS. The contribution of the study can be summarized as follows:

- A novel highly accurate intelligent, sensorless DASTS is proposed using the SCR approach with FDM.
- Motor step intervals (angle value) are determined automatically with the proposed FDM.
- Position detection errors caused by environmental factors are eliminated.
- Fabrication cost much less than sensor-based DASTSs.
- The DASTS with FDM using SCR is 6.64% more efficient than the dual-axis solar tracking system using only the mathematical method.
- Suitable for rural installations with its cost-effectiveness, low maintenance requirements, and high-efficiency increase.

The remainder of the chapter is divided into four sections. First, in Section 7.2, conventional sensorless DASTS is explained and information about the mechanical, electrical and control structure of the proposed system is given. Section 7.3 details the proposed FDM-based smart, sensorless and automatic sun-tracking method using SCR. With the proposed method, the results of fixed and conventional DASTS systems are compared in detail in Section 7.4 and conclude our work in Section 7.5.

7.2 DASTS

The system consists of a mechanical structure, electrical hardware, and electronic control systems. While the mechanical structure consists of PV panel carrier assembly and linear actuators providing east–west, and north–south movements, the electronic control system consists of a computer, software, actuator controllers, and microcontrollers. The general block diagram of the developed solar tracking system is shown in Figure 7.2.

7.2.1 *Mechanical system*

The developed STS is designed to carry two panels weighing 17 kg and to be resistant to various weather events. Strong ball casters are placed under the thick floor sheet to make it easier to transport. With the floor plate used, both balances are provided, and space is provided for the control panel and batteries. The mechanical structure of the system is shown in Figure 7.3. The DASTS is intended

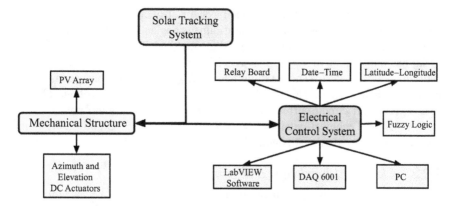

Figure 7.2   DASTS general schematic

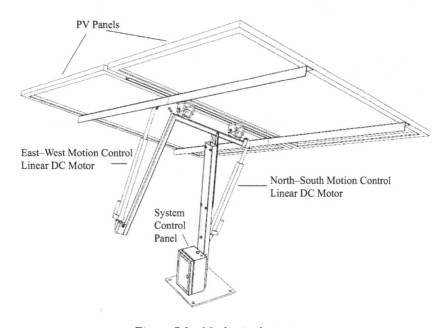

Figure 7.3   Mechanical structure

to move in two axes (north–south, east–west). To provide these movements, two linear DC motors (actuators) of the same size and power are used. Linear actuators are systems that can convert the rotational movement of a low-voltage DC motor into linear (forward–backward) motion via gear wheels. The primary reason for choosing these motors is that they eliminate the need for drivers and encoders. Also, they can move heavy loads with little power. In this study, LFHA12-300 model with 300 mm stroke length at 12 V DC voltage is used. In addition, the nominal speed of the model is 20 mm/s and the torque it produces is 500 N–50 kg.

Selected PV panels in the developed system; it is chosen as the type that an average consumer can use, with standard dimensions and general electrical properties. The PV panel is made of polycrystalline material. The PV panel used weighs 17 kg has dimensions of $1,580 \times 808 \times 35$ mm and can operate at temperatures between $-40\,°C/+85\,°C$. The maximum power (P_{max}) for the panels used is 160 W, the open circuit voltage (V_{oc}) is 43.3 V, the maximum power voltage (V_{pm}) is 35.2 V, the short circuit current (I_{sc}) is 4.98 A, and the maximum power current (I_{pm}) is 4.4 A. Technical details of the PV panels used (25 °C and 1,000 W/m^2) obtained under standard test conditions have tolerance of $\pm5\%$.

7.2.2   Electrical hardware and control system

The electrical elements used in the design of the system can be categorized as hardware and software. As hardware, power supply, relay card, terminals, NI DAQ USB-6001 data acquisition card, and PC are used. In terms of software, the LabVIEW program and related tools are used. The general block diagram of the electrical control system of the designed DASTS is shown in Figure 7.4.

In the designed system, USB-6001 DAQ is used to control the motors that provide the movement of the panels and to monitor the power data generated by the panels in real-time. Besides, the voltage required to drive the linear actuators can be supplied from the USB-6001 DAQ card outputs. For this reason, four relays and 12 V DC power supply are used to feed the motors and to make the north—south/east—west movements. All equipments are brought together on a panel designed as shown in Figure 7.5.

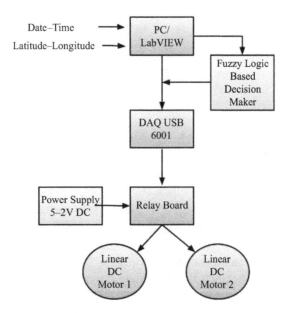

Figure 7.4   Electrical control structure of the developed DASTS

To measure, store, and use the energy obtained from the system, a distribution board is designed as shown in Figure 7.6. In the figure, the inverter shown with number 1 is a device containing a charge controller with a rated power of 3,000 VA/2,400 W and the battery input is 24 V. The device shown in number 2 is a 600 W inverter. The device shown in number 3 is a charge controller working with the inverter with 600 W power. With the package type switch located at number 4 just to its left, it is possible to select which inverter is desired to be operated. The devices shown with 5 are batteries with 12 V 120 Ah voltage and current values placed in series. The materials in number 6 are the fuse and terminal groups. Terminals are placed to facilitate the connection of cables coming from outside and fuses are placed to protect the system from overload.

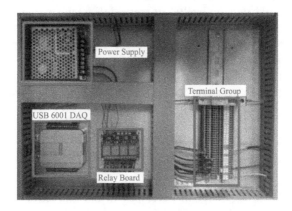

Figure 7.5   System control board

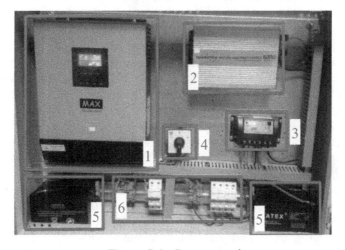

Figure 7.6   Power panel

7.3   Proposed FDM-based method using SCR

In the proposed method, the sun is tracked by calculating the position of the sun according to the mathematical model, without using any sensor. The main difference between the proposed method and the literature is the SCR-based FDM method, which is developed by utilizing the error between the calculated sun position values and the values obtained from the United States Naval Observatory (USNO) database. There are two FDMs in the method. The inputs of the first FDM are the azimuth angular error and the error rate of change, respectively. The inputs of the second FDM used are in elevation angular error and error change rate. By defining the fuzzy rule table with $7 \times 7 = 49$ states and related rules created for these input values, it is automatically decided what step interval (angle) linear actuators will move with the output of the fuzzy controller. As a result, the proposed method works with an intelligent method that can automatically decide and adjust the motor step intervals with FDM. In this section, first, the LabVIEW program, which is used to determine the sun's position, is introduced and the conventional sensorless DASTS algorithm is discussed. Then, the determination of the relevant state variables and membership functions of the FDM used in the proposed method, the creation of the rule base, and the determination of the decision-maker engine parameters are explained in detail.

7.3.1   Determining the position of the sun

To determine the sun's position, parameters such as the sundial, zenith angle, inclination angle, declination angle, elevation angle, and azimuth angle must be calculated [8,15]. In the program developed in the LabVIEW environment, the latitude–longitude information of the location of the system is entered into the system. In addition to this information, the system clock is automatically included in the calculation in the program and the position of the sun is determined and real-time tracking is performed. Figure 7.7 shows the software developed to calculate the position of the sun. In the specified number 1 part, the date and time information for which the position of the sun is requested is entered. In part 2, there is the option to automatically obtain the date and time information from the computer clock or enter it manually. In section 3, the latitude and longitude location information of the point where the PV panels are located is entered. After these settings are made, the program is run. The results are shown in 4, 5, and 6. The 4 is the graphic that shows the sun's movement, path, and position in real-time in 3D. Part 5 shows the azimuth angle, elevation angle, and tilt angle information of the sun with the specified position. The 6 shows the time data when the sun will be overhead on the specified date and the elevation angle data.

The developed software allows remote monitoring and control. In this direction, various settings must be made to monitor and control the LabVIEW program interface from any point on the network server with an Internet connection. After the network server settings are made, the name and uniform resource locator (URL) address of the document to be created via the network publishing tool are

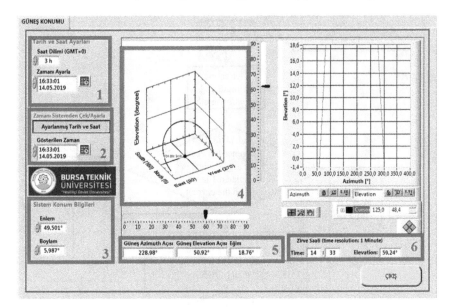

Figure 7.7   LabVIEW program that calculates sun position

determined. With the settings made, sensorless FDM-based DASTS can be remotely controlled or monitored.

7.3.2   Conventional sensorless DASTS algorithm

Figure 7.8 shows the flowchart of the performing principle of the sensorless automatic STS. The first step of the DASTS with mathematical method is completed by calculating the position of the sun in the sky and the route that the sun will follow in the sky during the day according to the location of the system and the date–time information. The next step, there is the process of moving the system by starting the motors. The developed dual-axis position and time-based (sensorless) solar tracking system uses azimuth angle and elevation angle calculations to track the sun. During the period from sunrise to sunset, linear DC motors are operated according to the angle difference between the angle calculated at the previous hour and the current time in 60-min periods, and follow-up is carried out. With the information that the motors travel 4.75° in 1 s, it is calculated how many seconds they will operate according to the angle difference value calculated.

7.3.3   Proposed FDM-based method using SCR

The proposed FL module consists of two independent FDM that direct the azimuth angle and elevation angle to control the DASTS. Module controls two linear DC motors to orient the PV panels at right angles to the sun's rays. During sun tracking, the step ranges of the motors that provide movement in the east–west and north–south axes are calculated automatically with a fuzzy logic-based intelligent

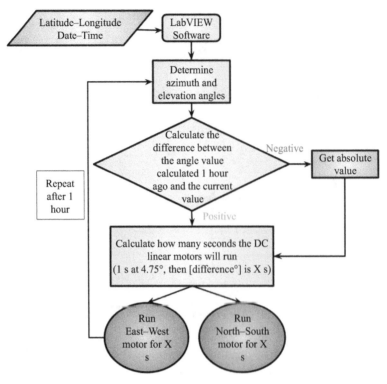

Figure 7.8   Sensorless automatic solar tracking system flow diagram

method, and higher efficiency is obtained than traditional sensorless solar tracking systems.

7.3.3.1   Determining state variables

In the proposed FDM, the differences in sun angles are defined as state variable inputs. Here, $H_{Azimuth}$ is the change in azimuth angle and $H_{Elevation}$ is the change in elevation angle. Error change rates are also expressed as $dH_{Azimuth}$ and $dH_{Elevation}$. In the literature, the sensor information is generally used to obtain angular differences. In this study, the differences between the azimuth and elevation angles [8,15] are calculated according to the location–date–time information of the system with mathematical calculations and the sun angles obtained from the USNO database are used as inputs. Evaluation of angular errors and rates of change is given in (7.1)–(7.4).

- A_{calcA} = Calculated azimuth angle value
- A_{dataA} = Azimuth angle value from USNO database
- A_{calcE} = Calculated elevation angle value
- A_{dataE} = Elevation angle value from USNO database

$$H_{Azimuth} = A_{calcA}(k) - A_{calcA}(k-1) \tag{7.1}$$

$$H_{Elevation} = A_{calcE}(k) - A_{dataE}(k-1) \tag{7.2}$$

$$dH_{Azimuth} = \Delta(H_{Azimuth}(k)) = A_{calcA}(k) - A_{dataA}(k) \tag{7.3}$$

$$dH_{Elevation} = \Delta(H_{Elevation}(k)) = A_{calcE}(k) - A_{dataE}(k) \tag{7.4}$$

The values calculated here are the data created at 1-hour intervals from sunrise to sunset.

7.3.3.2   Developing FDM for azimuth angle

The FDM state variable inputs for the azimuth angle are the angular error $H_{Azimuth}$ and the rate of change of error $dH_{Azimuth}$. Linguistic variables for azimuth angular error: it is defined as the maximum deviation of the solar panel from its optimal position relative to the sun's rays. As seen in Figure 7.9, when the sun rays are normal to the surface of the panel, the angle of incidence (θ) is zero. This angle can vary from $-90°$ to $+90°$ from sunrise to sunset.

Determination of I/O membership functions
The azimuth angular error ($H_{Azimuth}$) and the *error rate of change ($dH_{Azimuth}$)* are defined as seven triangular membership functions for the range $H_{Azimuth}$ $[0°, 45°]$ and $dH_{Azimuth}$ $[-0.6°, +0.6°]$, respectively. The angle changes between sunrise and sunset on June 21, the longest day, are taken as a basis. It was observed that the highest difference between the azimuth angle changes calculated with a 1-hour interval is $45°$. For this reason, the variables are positioned at equal intervals of $5.625°$ between $0°$ and $45°$. Due to its sensitivity, seven levels are used to provide sufficient resolution for the azimuth angle. Membership functions representing two input variables are shown in Figure 7.10. Because of simplicity and computation efficiency, the triangular membership function is preferred in this study.

Input variables are specified as fuzzy sets defined in linguistic terms: Very Small (VS), Slightly Small (SS), Small (S), Medium (M), Big (B), Slightly Big (SB), Very Big (VB); Negative Big (NB), Negative Medium (NM), Negative Small (NS), Near Zero (NZ), Positive Small (PS), Positive Medium (PM), and Positive Big (PB).

The control output signal of the FDM is on time with the signal applied to the DC motor relays, and the output values are calculated by the fuzzy inference

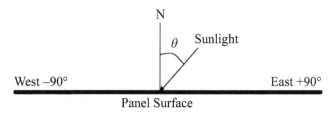

Figure 7.9   The angle of incidence of the sun's rays on the panel surface

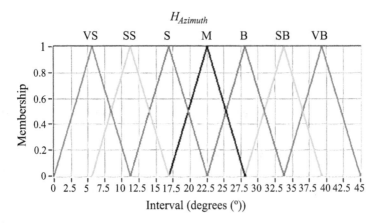

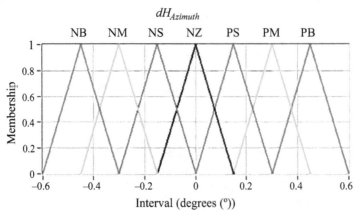

Figure 7.10   Membership functions for azimuth angle error and rate of change of azimuth angle error

system. It is defined by seven triangular membership functions in the interval [0, +10] s. Linear DC motor travels 4.75° in 1 s. Accordingly, the step interval times is determined equally at 1.2 s intervals. As shown in Figure 7.11, the seven triangular membership functions by seven linguistic terms. It is defined as very slow motion (VS), slow motion (S), little slow motion (LS), medium motion (M), little fast motion (LF), fast motion (F), and very fast motion (VF).

Establishing the rule base
The fuzzy rules are derived in such a way that the deviation from the desired posture (azimuth angular error) is minimized to achieve the control object. FDM consists of rules of the following form:

$$R^i : IF \ H_{Azimuth} = A_1^i \quad AND \quad dH_{Azimuth} = A_2^i \quad THEN \quad u = B^i i = 1 \dots.$$

$$(7.5)$$

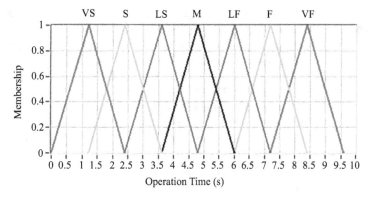

Figure 7.11   Membership functions for the azimuth angle output (u)

Table 7.1   Fuzzy logic-based 7 × 7 rule matrix created for azimuth angle

u		$H_{Azimuth}$: **azimuth angular error**						
		VS	**SS**	**S**	**M**	**B**	**SB**	**VB**
$dH_{Azimuth}$: error rate of change	NB	M	LS	S	VS	VS	VS	VS
	NM	LS	M	LS	S	VS	VS	VS
	NS	F	LS	M	LS	S	VS	VS
	NZ	VF	F	LS	M	LS	S	VS
	PS	VF	VF	F	LS	M	LS	S
	PM	VF	VF	VF	F	LS	M	LS
	PB	VF	VF	VF	VF	S	LS	M

The A_1^i and A_2^i values are linguistic terms that represent previous pairs. B^i is the fuzzy set representing the result for the i value. Table 7.1 shows the fuzzy logic-based 7 × 7 rule matrix for the azimuth angle. Control rules are visualized as a 2-dimensional matrix structure in which the leftmost column and top row contain fuzzy sets.

Here are some examples of how the rules work:

Rule 7: IF $H_{Azimuth}$ = VB AND $dH_{Azimuth}$ = NB THEN u = VS

"The angular error is very big (VB)" indicates that the angle value from the data is greater than the calculated angle value. The expression "error change rate negative big (NB)" indicates that the tracking system is moving eastward. For this reason, to follow the sun, the controller must apply the "very slow (VS) motion" action.

Rule 25: IF $H_{Azimuth}$ = M AND $dH_{Azimuth}$ = NZ THEN u = M

This rule means that where the bias error is within the fuzzy M region and there is also a rate of error variation within the NZ region, the controller applies its current state as moving at medium (M) speed.

The generated rule base matrix has a diagonal symmetric feature. That is, the slow-motion terms of the control action (VS, S, and SS) are placed on the diagonal. Another feature of this matrix is that the step time to be applied for linear actuators in both directions increases with the distance from the diagonal.

7.3.3.3   Developing FDM for elevation angle

The second FDM is used to adjust the tilt angle of the tracking system. It operates together with the azimuth angle controller. FDM inputs are elevation angle error ($H_{Elevation}$) and error change rate value ($dH_{Elevation}$), respectively. Its output is the control process that drives the second linear actuator.

The tilt angle is the angle made by the line connecting the equatorial plane and the centers of the sun and the earth, and it varies between $-23.45°$ and $+23.45°$. Fewer membership functions are used as the variation for the inclination angle changes only on a seasonal basis. Input variables are specified as fuzzy sets defined in linguistic terms: Negative big (NB), Negative small (NS), Near zero (NZ), Positive small (PS), and Positive big (PB). The duration of driving the motors was defined as very slow motion (VS), slow motion (S), medium speed motion (M), fast motion (F), and very fast motion (VF). Figure 7.12 shows the graphical representation of the membership function values of the elevation angle error and the error change rate. Figure 7.13 shows the number of steps of the single membership functions of the output variable "u."

Table 7.2 shows the FDM rule base in matrix form. Similar to the rule base matrix of the fuzzy controller controlling the azimuth angle, the elevation angle rule base matrix also has a diagonally symmetric property. The important feature of this matrix is that the number of steps to be applied for the actuators in both directions increases with the distance from the diagonal.

7.4   Results

Sensorless automatic solar tracking system results: With the developed software, between 05:00 in the morning and 21:00 in the evening (UTC +3) for the location of 40.19° N and 29.12° E (BTÜ Mimar Sinan, Bursa, TR) on 29.06.2019, in 60-min periods. The data of the calculated solar motion were taken. The calculated sun angles are shown in Table 7.3.

As a result of the movement of the sun, it is observed that the azimuth angle values between noon hours (12.00–14.00) are quite high compared to other times of the day. The graph of the change between the hourly differences of these angles is shown in Figure 7.14(a). The length of time the linear actuator would operate is calculated mathematically based on the magnitude of the change in the angles. It is determined that the panels move with an angle value of 4.75° with the linear actuator moving for 1 s. The developed software in the LabVIEW environment has managed how long the linear actuator will work according to this angle value and how it will follow the hourly angle difference. The data of the operating times of the linear actuator according to the angle changes during the day are shown in Figure 7.14(b).

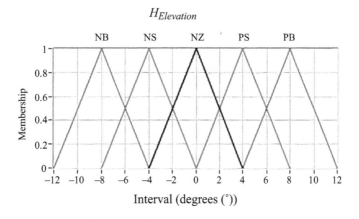

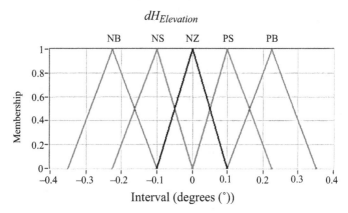

Figure 7.12   Membership functions for elevation angle error and rate of change of elevation angle error

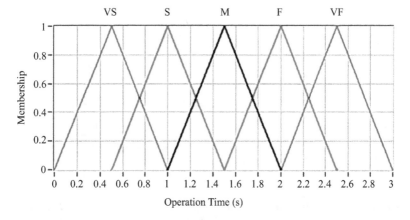

Figure 7.13   Single membership functions step time of output variable "u"

Table 7.2   Fuzzy logic-based 5 × 5 rule matrix created for elevation angle

u		$dH_{Elevation}$				
		PB	**PS**	**NZ**	**NS**	**NB**
$H_{Elevation}$	NB	M	S	VS	VS	VS
	NS	F	M	S	VS	VS
	NZ	VF	F	M	S	VS
	PS	VF	VF	F	M	S
	PB	VF	VF	VF	F	M

Table 7.3   Calculated sun angles

Time	Elevation (angle)	Azimuth (angle)
05:00	−6,09	52,48
06:00	3,56	62,33
07:00	14,09	71,34
08:00	25,18	80,03
09:00	36,58	89,09
10:00	47,99	99,59
11:00	58,96	113,74
12:00	68,38	136,89
13:00	73,02	177,15
14:00	69,29	219,25
15:00	60,23	244,11
16:00	49,36	258,98
17:00	37,97	269,76
18:00	26,56	278,93
19:00	15,42	287,63
20:00	4,82	296,57
21:00	−4,96	306,29

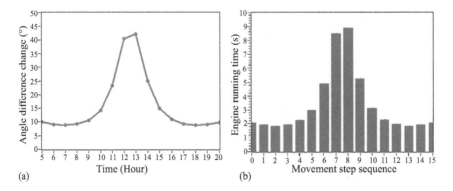

(a)

(b)

Figure 7.14   (a) Azimuth angle hourly change graph. (b) East–west engine operating times during the day

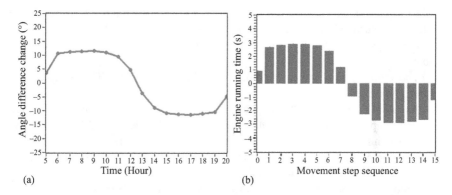

Figure 7.15 *(a) Elevation angle hourly change graph. (b) North–south engine running times during the day*

As a result of the movement of the sun, it is determined that the elevation angle values decreased in the afternoon and the sun is positioned toward its old position in the north–south movement. In the graph shown in Figure 7.15(a), it is shown that the hourly elevation angle differences are below zero. With the calculations made according to the amount of change between the angles, it is determined how long the motor that provides the north–south movement would operate. It is determined that the panels change position with an angle of 4° with the motor moving for 1 s. The graph shown in Figure 7.15(b) shows the operating times of the engine during the day. The times indicated by the (−) sign below zero in this graph indicate that the motor is moving in the reverse direction.

7.4.1 Measurements and comparison

The amount of energy produced by the fixed panel and the panel placed on the movable system during the day are measured hourly. For June 30, 2019, fixed system, sensorless DASTS, and FDM-based DASTS data are collected at Bursa Technical University Mimar Sinan Campus between 09:00 and 15:00. The fixed panel is positioned in the southeast direction with an inclination of 28.5°. Figure 7.16 shows the fixed panel and the movable system.

It is measured that the motors on the two axes of the proposed FDM-based DASTS consume a minimum of 2.1 W and a maximum of 5.25 W during movement during the day. The amount of power consumed by the motors during solar tracking corresponds to 7% of the power generated by the sensorless DASTS system. It is observed that the power consumed by the developed DASTS is quite low compared to the amount of power generated by the PV system. The current drawn by the two 12 V 120 Ah batteries in the power panel while charging and the current drawn by the inverter while operating is calculated as the load of the system. While ensuring the operation of the system, measurements are taken on how much voltage and current are generated by the panels.

Comparison of FDM-based sensorless DASTS and fixed PV system: The voltage and current values generated by the FDM-based DASTS system and the fixed PV system in the measurements made on June 30, 2021 are shown in Table 7.4.

The output power values obtained from FDM-based DASTS and fixed PV systems are given in Table 7.5.

According to the results obtained, there is a 30.5% efficiency difference between the power generated by the panel placed on the developed FDM with the SCR algorithm and the PV panel placed permanently. The developed system has succeeded in producing the highest power values it can achieve by tracking hourly, even if the sun is covered with clouds.

Comparison of conventional sensorless automatic DASTS and FDM-based DASTS using SCR: In Table 7.6, the obtained power values and the total efficiency ratio of the proposed FDM-based DASTS system and the traditional DASTS

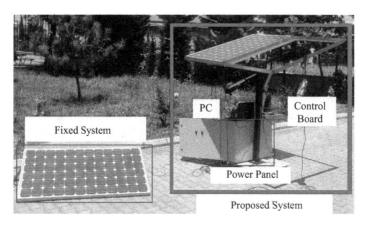

Figure 7.16   Experimental setup

Table 7.4   Current and voltage values generated by FMD-based DASTS with FDM and fixed system

Time	FDM-based DASTS		Fixed system	
	Current (A)	Voltage (V)	Current (A)	Voltage (V)
9:00	2.51	35	2.32	33
10:00	2.37	36	1.95	34
11:00	1.84	36	1.55	36
12:00	1.56	37	1.23	36
13:00	1.04	26	0.94	25
14:00	2.15	35	1.87	33
15:00	1.56	37	0.85	26

Table 7.5 The output power values obtained from FDM-based DASTS and fixed PV

Time	FDM-based DASTS power (W)	Fixed PV system power (W)	Increase in efficiency (%)
9:00	87.85	76.56	14.7
10:00	85.32	66.30	28.7
11:00	66.24	55.80	18.7
12:00	57.72	44.28	30.4
13:00	27.04	23.50	15.1
14:00	75.25	61.71	21.9
15:00	57.72	22.10	161.2
All	457.14	350.25	30.5

Table 7.6 The output power values obtained from FDM-based DASTS and conventional sensorless DASTS

Time	FDM-based DASTS power (W)	Conventional DASTS power (W)	Increase in efficiency (%)
9:00	87.85	78.54	11.85
10:00	85.32	67.90	25.65
11:00	66.24	63.00	5.14
12:00	57.72	30.00	92.40
13:00	27.04	87.72	−69.17
14:00	75.25	34.32	119.25
15:00	57.72	67.20	−14.10
All	457.14	428.68	6.64

using the mathematical model are given. The ratio of the total power values produced during the day was taken as efficiency.

As a result of this comparison, it has been observed that the proposed FMD-based DASTS is more efficient by producing 6.64% more power than the sensorless conventional DASTS, with the total data discussed at the end of the day.

7.5 Conclusion

In this study, an electrical and mechanical design of a DASTS is implemented and an intelligent control method using a mathematical model based on FDM is proposed. The proposed FDM-based DASTS provides maximum efficiency by adaptively changing the step range without using any sensors according to the sun's azimuth/elevation angles and the position of the region. FDM-based DASTS has eliminated sun position detection errors caused by environmental factors in conventional methods. Besides, fabrication cost much less than sensor-based DASTSs. The control of the motors and the entire system is ensured by real-time software developed in the LabVIEW environment. With the proposed DASTS with the SCR method, 28.5% more electrical energy was obtained compared to the fixed PV system. In addition, the proposed method is 6.64% more efficient than the dual-axis solar tracking system using only the mathematical method. Besides, with the FDM with SCR, step times can be determined with an accuracy of 97.24%. The proposed intelligent sensorless system is especially suitable for rural installations with its cost-effectiveness, low maintenance requirements, and high-efficiency increase.

Future studies will focus on the proposed system to track the sun more precisely and increase system efficiency. In the proposed method, these measurement intervals can be reduced with more advanced controllers to be used while determining the engine operating times according to hourly movements. The proposed intelligent FDM, which is developed according to the angle differences and the error between the angle differences, can be developed to determine the motion time intervals of the system according to the angle differences. In this way, during the hours when the sun emits more radiation, more power can be obtained by tracking the sun at shorter intervals.

References

[1]   Nadia, A. R., Isa, N. A. M., and Desa, M. K. M. (2018). Advances in solar photovoltaic tracking systems: a review. *Renewable and Sustainable Energy Reviews*, *82*, 2548–2569.

[2]   Nsengiyumva, W., Chen, S. G., Hu, L., and Chen, X. (2018). Recent advancements and challenges in Solar Tracking Systems (STS): a review. *Renewable and Sustainable Energy Reviews*, *81*, 250–279.

[3]   Garg, A. (2015). Solar tracking: an efficient method of improving solar plant efficiency. *International Journal of Electrical and Electronics Engineers*, *7*(1), 199–203.

[4]   Robles Algarín, C., Ospino Castro, A., and Casas Naranjo, J. (2017). Dual-axis solar tracker for using in photovoltaic systems, *International Journal of Renewable Energy Research*, *7*(1), 137–145.

[5]   Akbar, H. S., Siddiq, A. I., and Aziz, M. W. (2017). Microcontroller based dual axis sun tracking system for maximum solar energy generation, *American Journal of Energy Research*, *5*(1), 23–27.

[6]   Carballo, J. A., Bonilla, J., Berenguel, M., Fernández-Reche, J., and García, G. (2019). New approach for solar tracking systems based on computer vision, low-cost hardware and deep learning. *Renewable Energy*, *133*, 1158–1166.

[7]   Almonacid, F., Fernandez, E. F., Mellit, A., and Kalogirou, S. (2017). Review of techniques based on artificial neural networks for the electrical characterization of concentrator photovoltaic technology. *Renewable and Sustainable Energy Reviews*, *75*, 938–953.

[8]   Hafez, A. Z., Soliman, A., El-Metwally, K. A., and Ismail, I. M. (2017). Tilt and azimuth angles in solar energy applications – a review. *Renewable and Sustainable Energy Reviews*, *77*, 147–168.

[9] Clifford, M. J. and Eastwood, D. (2004). Design of a novel passive solar tracker. *Solar Energy*, *77*(3), 269–280.

[10] Parmar, N. J., Parmar, A. N., and Gautam, V. S. (2015). Passive solar tracking system. *International Journal of Emerging Technology and Advanced Engineering*, *5*(1), 138–145.

[11] Alexandru, C. (2013). A novel open-loop tracking strategy for photovoltaic systems. *The Scientific World Journal*, 205396.

[12] Abdollahpour, M., Golzarian, M. R., Rohani, A., and Zarchi, H. A. (2018). Development of a machine vision dual-axis solar tracking system, *Solar Energy*, *169*, 136–143.

[13] Diego F. Sendoya-Losada (2017). Design and implementation of a photo-voltaic solar tracking using fuzzy control for Surcolombiana University, *ARPN Journal of Engineering and Applied Sciences*, *12*(7), 2272–2276.

[14] Zaher, A., N'goran, Y., Thiery, F., Grieu, S., and Traoré, A. (2017). Fuzzy rule-based model for optimum orientation of solar panels using satellite image processing, *Journal of Physics: Conference Series*, *783*(1), 012058.

[15] Goswami, D. Y., Kreith, F., and Kreider, J. F. (2015). *Principles of Solar Engineering*, Boca Raton, FL: CRC Press.

Chapter 8

Design and realization of a solar remote tracker system in a rural area

S. Rahal[1] and A. Abbassi[2]

This chapter in this book will focus on the study of a remote photovoltaic (PV) installation using a tracker system in a rural area in Morocco. We start our work by designing a general concept of a solar tracker system in an off-grid site to respond to the need for energy needs for a small farm which is taken as an example in this study. This system will be studied based on different phases of the study life: design, installation, operation, and maintenance.

We are mainly based on the Sun position data's tracking with a photocell sensor and the field configuration that we will describe in more detail in this chapter. By following the Sun throughout the day, this system maximizes power generation compared with traditional fixed inclination installation. The generating capacity can be increased by 15%–20% with a limited increase in investment.

The system is controlled by the remoting used tracker which gets updates every 10 s of the Sun's position from a sensor. And with the simplest design structure, two actuators move the solar panel on the horizontal axis (East/West) and the vertical axis (North/South). This double-axis system has shown flexible response, good reliability, and fault identification ability. This work may contribute to the development of a solar digital tracker system based on astronomical data by the exploitation of the exact Sun's positions throughout a cloudy day in which a photocell sensor can provide inaccurate measurements.

Acronyms

ADC: analogue to digital converter
ASTS: active solar tracking system
CT: clock time
EVA: ethylene-vinyl acetate
I2C: inter-integrated circuit

[1]Socy BENOMAR OPTIC, Laayoune, Morocco
[2]Research Laboratory in Physics and Sciences for Engineers, Polydisciplinary faculty of Beni-Mellal, Moulay Sultan University, Morocco

I_{mp}:	current at P_{max}
I_{sc}:	short-circuit current
LDR:	light-dependent resistor
NOCT:	nominal operating cell temp
PV:	photovoltaic
PWM:	pulse width modulation
SPI:	serial peripheral interface
ST:	solar time
TE:	equation of time
USART:	Universal Synchronous Asynchronous Receiver Transmitter
UV:	ultraviolet
V_{mp}:	voltage at P_{max}
V_{oc}:	open-circuit voltage
Wp:	peak power

8.1 Introduction

In the last decades, energy consumption is increasing dramatically which influenced the environment and then our survival. Indeed, fossil energy sources (coal, oil, and natural gas) are used in continually abusive ways [1–3]. They are extremely polluting when they undergo a combustion process, with a limited reserve over time. Thus, the scientific community now recognizes the responsibility of this overconsumption of global warming [4,5], which risks having dramatic effects on the physical, political, and social-economic balances on our planet. The use of clean energies to ensure energy security and the development of populations is becoming unavoidable, that is why now research in the field of alternative energy resources, therefore, becomes a crucial issue.

Renewable energies, and especially solar energy, are nowadays the solution to being independent of fossil sources and the problems they cause. This choice, motivated by the fact that the quantity of solar energy received each day by the planet, is five times greater than all human daily consumption (nuclear, coal, wood, fuel oil, gas, hydraulics, etc.), which seems to be the best solution, especially in southern countries which are extremely sunny. Much scientific research has been carried out not only in the field of solar energy conversion but also in every specific field like materials and others [6]. The design, optimization, and realization of systems based on new technologies of materials are current issues since they can lead to better exploitation of solar energy if they are properly done. One of the ways taken is to increase the solar radiation captured by solar panels, maximum efficiency is achieved when the incident light is perpendicular to the cells of the photovoltaic panels: this is now the concept of "solar tracking."

Therefore, the energy efficiency of these systems depends on the degree of use and conversion of solar radiation [7]. There are two ways to maximize the useful energy rate: by optimizing the conversion and degree of absorption and by increasing the incident radiation rate by employing mechanical steering systems with the use of solar trackers controlled by embedded systems.

Solar tracking is a system that tracks the apparent radiations of the Sun in real-time. Solar trackers are designed and built to optimize the power yield of photovoltaic modules by ensuring that they are always perpendicular to the Sun during the day. A few research reported interest in new technologies such as Embedded systems, big data, cloud, and artificial intelligence in the development of PV applications [8–12]. The work presented in this chapter focused on the study, design of a controlled tracker, and implementation of a PV system: we are first interested in the fundamental PV technologies especially to establish various parameters that can be used to describe the trajectory of the Sun. We will discuss solar tracking techniques and their different technical types. The last part of this chapter is devoted to the realization of our embedded solar concept of the tracking system.

After having seen the fundamental concepts of the solar tracking techniques, the results of the simulations carried out that the use of a mobile system could significantly increase the quantity of PV energy produced, particularly when the solar tracking is done according to the four directions: North, South, East, and West.

8.2 PV and solar tracking

PV energy is the direct conversion of light into electricity using solar cells. The photovoltaic effect, which means the production of electricity directly from light, was first observed in 1839 by Edmond Becquerel. However, until the 1950s, that researchers from the Bell Telephone company in the United States succeeded in manufacturing the first solar cell. More than 90% of solar cells in the world are mainly composed of silicon [13]. This semiconductor material has the advantage of being able to be produced from an almost inexhaustible natural resource: quartz, a component of granites, and sands. For the composition of solar panels, it is possible to choose between several types of voltaic cells: monocrystalline silicon, poly-crystalline silicon, and amorphous silicon. In addition, the choice of cells is also based on durability and price. The choice can also vary depending on sunshine, geographical location, and orientation. A PV solar module (or PV solar panel) is a panel made by a set of PV cells electrically interconnected, and therefore a group of these solar panels represents an array as shown in Figure 8.1.

Figure 8.2 shows the different layers of a typical solar panel, and they consist of the following:

- A support frame and a seal (1 and 2).
- An extra clear glass plate (promoting light transmission) (3).
- Two layers of ethylene-vinyl acetate (EVA) coat the cells ensuring their protection against bad weather and humidity (4).
- Different strings of cells (5).
- A sheet of Tedlar.
- This polymer with high resistance to UV and high temperature ensures the module's mechanical resistance to external shocks (wind, transport, etc.) (6).

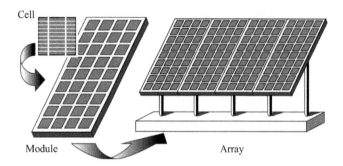

Figure 8.1   A solar cell, module, and array [14]

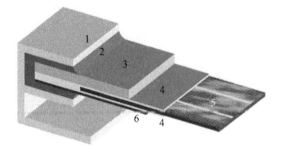

Figure 8.2   Components of a typical module [15]

Each cell can produce a certain voltage and a certain current, especially with special electrodes [16,17]. To obtain the voltage and current required by our application, we connect the cells electrically, and form a photovoltaic module (also called a sensor or panel), in parallel to increase the current and in series to increase the voltage. This association must be carried out while respecting specific criteria, due to the imbalances existing in a network of solar cells. Although the numerous cells present different characteristics due to the inevitable construction dispersions, the non-uniform illumination, and the temperature over the entire network. However, it is important to take some precautions because the existence of less efficient cells or the occlusion of one or more cells (due to shading, dust, etc.) can permanently damage the cells.

Before treating the technical aspect of this study, it was necessary to give some notions relating to the movements of the Sun (and the Earth). Latitude is a geographic coordinate represented by an angular value, an expression of the position of a point on Earth (or on another planet), north or south of the equator which is the reference plane. Longitude is geographical data represented by an angular value of the East–West positioning of a point on Earth (or on another planet). The longitude of reference on Earth is the meridian of Greenwich as illustrated in Figure 8.3.

In Figure 8.4, is the angle between the horizontal and the line to the Sun, which is called the solar altitude angle. It is the complement of the zenith angle θ_z. It is a quantity that expresses a difference between a given point and a reference level; by convention, on Earth, this level is most often sea level (or "zero level"). We also use the term elevation. The solar azimuth angle γ_s is the angular displacement from

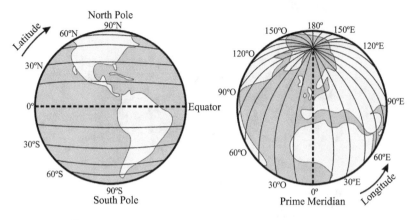

Figure 8.3   Illustration of geographic latitude and longitude of the Earth [18]

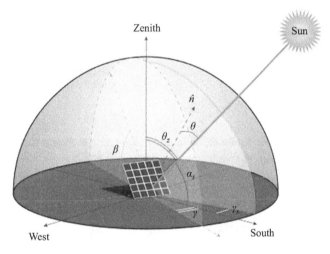

Figure 8.4   Daily path of the Sun across the sky from sunrise to sunset [19]

the south of the beam radiation projection on the horizontal plane. The displacements east from the south are negative and the west of the south are positive. In Figure 8.5, the two angles related to the Sun's position and the angles β and γ define the photovoltaic module position. β is the tilt angle and γ is the surface azimuth angle. The angle θ is between the normal to the PV surface and the incident beam radiation is indicated.

The angle θ is given by the mathematical expression:

$$\cos(\theta) = \cos(\theta_z)\cos(\beta) + \sin(\theta_z)\sin(\beta)\cos(\gamma_s - \gamma) \tag{8.1}$$

To determine the solar ray angles, we need further information about the declination angle δ which is the angular position of the Sun at noon concerning the

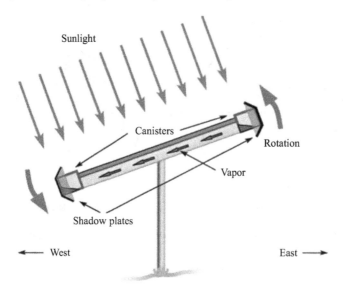

Figure 8.5   A typical passive solar tracking system [20]

plane of the equator (north positive) and given by the following expression:

$$\delta = 23.45 \times \sin\left[\frac{2\pi \times (284 + n)}{365}\right] \tag{8.2}$$

With the declination angle δ, the latitude angle, and the hour angle ω, the solar zenith angle θ can then be found from the following equation:

$$\sin(a_s) = \sin(al)\sin(\delta) + \cos(\Phi)\cos(\delta)\cos(\omega) \tag{8.3}$$

And:

$$\sin(\gamma_s) = \left[\frac{\cos(\delta) \cdot \sin(\omega)}{\cos(a_s)}\right] \tag{8.4}$$

The hour angle is the number of degrees that the earth must rotate before the Sun will be directly over the local meridian. Considering the earth rotates $360°$ in 24 hours, the hour angle can be described as follows:

$$\omega = \frac{15°}{hour} \tag{8.5}$$

The solar time is different from the local time also called clock time (CT). It is defined by the expression:

$$ST - CT = 4(L_{ST} - L_{loc}) + TE \tag{8.6}$$

where ST and CT are expressed in minutes, L_{loc} is the longitude of the location, and L_{ST} is the standard meridian for the local time zone. Both are given in degrees and

TE is the equation of time:

$$TE = 9.87 \cdot \sin \left[\frac{4\pi \cdot (n - 81)}{365} \right] + 7.67 \cdot \sin \left[\frac{2\pi \cdot (n - 1)}{365} \right] \tag{8.7}$$

where n is the number of days.

8.3   Solar tracking techniques

We note that a solar tracker is a device allowing a heliographic telescope (to observe the Sun or some of its effects on the atmosphere). This system directs the solar panels to increase their productivity [21]. The system aims to orient the sensors in real-time toward the Sun, to place the panel in an optimal position concerning the incidence of solar radiation (perpendicular to the radiation if possible), because throughout the day and the year (according to the seasons), the position of the Sun varies constantly and in a different way according to the latitude. This adaptation in real-time has the effect of substantially increasing the capture and production of energy. It can be done on two axes: in azimuth (from east to west, as the day progresses) and in height (depending on the season and, the progressed Sun in the day). The ideal is to use a tracker with two axes, but there is also one-axis systems (typically with tracking only in azimuth, the angle relative to the ground being fixed according to the local optimum, which depends on the latitude).

This study overviews the parameters, types, and drive system techniques covering different use applications. Two main solar tracking system types depend on their movement degrees of freedom which are single-axis solar tracking systems and dual-axis solar tracking systems. The solar tracker drive systems encompassed five categories based on the tracking technologies, namely, passive tracking, active tracking, semi-passive tracking, manual tracking, and chronological tracking. In this study, we focus on active solar tracking systems and their technologies, as in recent research studies, this type of system is the most common solar tracker drive type with 76.42% of exploitation.

The passive solar tracking systems are generally composed of two copper tubes mounted on the East and West sides of the PV panel. The tube of copper material (Figure 8.5) is filled with chemical fluids capable of vaporizing at low temperatures. When exposed to solar radiation, the temperature increases on one side of the panel, and the compound in the copper tube vaporizes. The gaseous part of the compound occupies a larger internal volume, and its liquid part is shifted to the shaded side. This mass transfer operation adjusts the balance of the PV panel by rotating it towards the source of the solar rays. This type of follower does not consume any energy for the repositioning of the panel.

However, active solar trackers use the principle of light detection, following the solar trajectory while seeking to optimize the angle of incidence of solar radiation on their surface as much as possible to increase the uptake of energy. ASTS have many types and advancements. Some solutions use various logic or artificial intelligence (AI) to track the Sun with high efficiency, but most of these systems are still in the phase of trial and development. Many approaches are used to

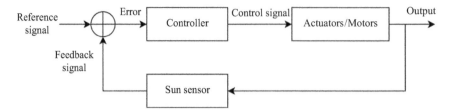

Figure 8.6   Closed-loop control strategy for solar tracking

solve the problem of precise solar tracking in some form. They cannot handle environmental and weather conditions perfectly [22,23]. A greater part of these solutions adopts sensors, GPS, or other software solutions. The prevailing dis-advantage of ASTS is the imprecise control procedures (e.g., cloud management and sensors with high sensitivity thresholds) [24].

The control algorithms adapted for active solar tracking systems command and employ the electrical signals of actuators, or electric motors, to achieve an accurate and precise angle of incidence. The required precision depends fundamentally on the acceptance angle of the system, which is generally about a tenth degree. These algorithms are classified according to three strategies: open-loop, closed-loop, and combined (called hybrid-loop). Research has shown that the most widely used solar tracking control strategy is closed-loop, which is illustrated in Figure 8.6, repre-senting 54.39% of all the publications consulted, on the other hand, open-loop represents 28.95%, and 16.67% for hybrid-loop strategy [23].

Control algorithms such as on–off, proportional-integral-derivative (PID), and proportional-integral (PI) were implemented by 67.55% and the on–off control is the most used of them representing 57.02%.

A study of the existing active solar tracking systems reviewed patents and technologies available based on their technical features. These are presented to compare the solar tracking solutions that use sensor driver systems entirely or partially, as well as control mechanisms. The following market products and patents were reviewed in this research [25]:

- FUSIONSEEKER DS-50D6W and FUSIONSEEKER DS-100D10 [26];
- ECO-WORTHY, dual-axis solar tracker controller [27];
- WST03-2 [28];
- Luoyang Longda Bearing Co., Ltd., solar-tracking controller [29];
- SunTura solar tracker [30];
- STA2000-HW [31];
- MLD sensor [32,33];
- WO 2020/185271 A1 patent description [34];
- EP 2 593 759 B1 patent description [35];
- WO 2013/074805 A1 patent description [36].

The characteristics of control mechanisms of these solutions reviewed above are summarized in Table 8.1. The ●/○/– marks mean yes/partially or another device is required for operation / no.

Table 8.1 Summary of the characteristics of the control mechanisms of the solar tracking solutions using active sensor driver systems reviewed above [25]

Solar tracking solutions	Seeking the brightest point in the sky		Protection against clouding	Protection against the unnecessary swaying of motors	Protection against wind	Assuming position after sunset
	Continuous	Periodic				
FUSIONSEEKER DS-50D6W and DS-100D10 [26]	●	–	○	○	○	●
ECO-WORTHY, dual-axis solar tracker controller [27]	●	–	○	○	○	●
WST03-2 [28]	●	–	–	–	○	–
Luoyang Longda Bearing Co., Ltd., solar-tracking controller [29]	●	–	–	–	○	–
SunTura solar tracker [30]	●	–	–	–	–	–
STA2000-HW [31]	●	●	○	○	●	○
MLD sensor [32,33]	●	–	–	○	–	●
WO 2020/185271 A1 patent description [34]	●	–	○	○	●	●
EP 2 593 759 B1 patent description [35]	●	–	○	○	–	●
WO 2013/074805 A1 [36]	●	–	–	–	–	–

These show that currently there is no control mechanism available for these solutions that can solve the following issues:

- Periodic searching for the brightest point in the sky;
- Preventing excessive motor activity resulting from the periodic search due to cloudiness;
- Setting the position of the system after sunset to an ideal position at sunrise the following day;
- Reducing the effects of storms on the mechanism without an anemometer;
- Absence of significant protection for the electronics from the elements.

During the elaboration of our solar tracker presented, we used the ECO-WORTHY dual-axis solar tracker controller [27] in a rural area. The main emphasis was placed on examining this solution in terms of these five detected problems above in rural conditions.

8.4   The experimental installation used in the study

8.4.1   *Needs analysis*

The first step of this work is the definition of the need. The objective is to define the elements to be addressed, as well as the issues and constraints of the realization. The solar tracker allows the structure to follow the path of the Sun, and thus maximize the yield of production. The FAST diagram (Figure 8.7) highlights the following main functions and the constrained functions:

We have established the function characterization in Table 8.2.

- FP: Main function
- HR: Constraint function

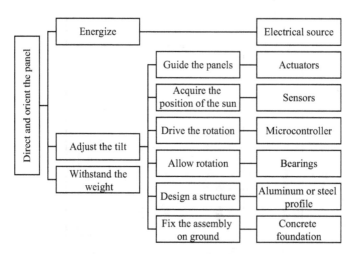

Figure 8.7   FAST diagram of the solar tracking system used in this study

Table 8.2   FAST diagram functions table

	Service function	Criteria	Requirement	Flexibility
FP	Direct and orient the panel	Rotation	Electric actuators	F1
		Autonomy	No intervention of the user	F0
		Steering	Microcontroller	F1
FC1	Withstand outdoor conditions	Metallic structure	No traces of rust and wear	F0
FC2	Position of the system	Sun exposure	Ground	F1
FC3	Facilitate the maintenance	Electric equipment	Accessibility	F0
FC4	Power the system	Cylinder supply	Socket	F0
FC5	Resist the weight	Metallic structure	No deformations	F0

- Flexibility F0: Zero flexibility
- Flexibility F1: Low flexibility
- Flexibility F2: Good flexibility
- Flexibility F3: Strong flexibility

8.4.2   Design

To carry out the design for any solar tracker, we started by schematizing the general configuration of the concerned tracker by dividing it into four large blocks (electronic control block, mechanical structure, movement block, and a power supply and electrical power block). The system will therefore consist of motorized photovoltaic panels and the equipment necessary for energy storage and energy regulation. This part essentially consists of electrical and electronic components used, which can command and control the various maneuvers and movements that the system can perform during the solar tracking operation. In this study, we used the ECO-WORTHY dual-axis solar tracking controller [27] (Figure 8.8).

The control unit uses a light-sensing sensor (photosensitive sensors) to search for the brightest spot in the sky. This solution performs continuous Sun tracking from sunrise to sunset. In physics, photosensitivity is described as the ability of certain materials to react to light. From there, we understand that there is a close link between this type of material and the field of solar energy. A light sensor, or photosensitive sensor, can generate an electrical output signal indicating the intensity of the light it receives. This operation is done by measuring the radiant energy that exists in a range of frequencies ranging from infrared to ultraviolet.

The main objective of the use of a solar tracking system is to continuously keep the PV panels facing the Sun. This maximizes the amount of radiation received and thus, the electrical energy produced will be greater. To achieve this, our sensor must have a very precise structure.

The photosensitive sensor provided by ECO-WORTHY is made up of four solar cells sitting at a 45-degree angle (Figure 8.9). We noticed the direct exposure

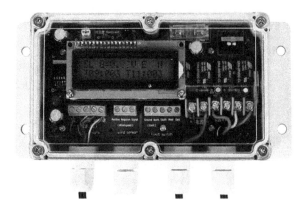

Figure 8.8   ECO-WORTHY dual-axis solar tracking controller [27]

Figure 8.9   The photosensitive sensor is provided by the ECO-WORTHY dual-axis tracking system [27]

of wiring to the elements and concluded that the ideal would be to protect our cells against the possible infiltration of water and humidity by covering the front face with a transparent cover.

Based on the constructor's manual, this sensor must be fixed on the bottom part of the PV panel for correct functioning. In addition to supporting the photovoltaic panel, thanks to a meticulously schematized metal structure, this part is responsible for moving and rotating the entire system in four directions (East, West, North, and South) with two electric actuators.

Mechanical energy, or work, is usually done by converting electrical energy. In a Sun tracking system, the mechanical movements are ensured by motors that work with direct current to facilitate their power supply directly, thanks to the PV panels, and by browsing the various works and their multitudes, research made in this field, we used linear actuators. A linear actuator is essentially an asynchronous motor whose rotor has been "unwound" so that instead of producing a rotational force by a rotating electromagnetic field, it produces a linear force along its length by installing a displacement electromagnetic field. These engines are divided into two main groups:

- Long-stroke motors.
- Limited stroke motors, called linear actuators, and are what we will use in our study.

In our experiment, we chose to use the linear actuators provided by the ECO-WORTHY dual-axis tracking system (Figure 8.10) to be able to steer it in the four planned directions, after having received the appropriate instructions from the control part. For the North -South movement (less frequent than that going from East to West) we have opted for the model (1) in Figure 8.10. For the East -West movement, we opted for the model (2) in Figure 8.10. The advantage of having a longer arm is greater freedom of movement, which is why we used it for the East–West maneuver.

We sketched a model on the 3D drawing software AUTOCAD for the metal structure (Figure 8.11).

To operate, the actuators used in our study as well as the daily needs of 1,079 Wh/day for lighting and pumping water from an underground water tank, a voltage of 12 V is needed. We have chosen to power these motors with seven lead batteries

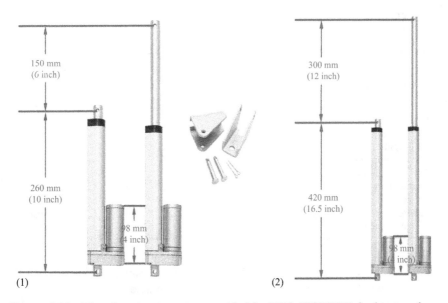

Figure 8.10   The electric actuators provided by ECO-WORTHY dual-axis solar tracking system: (1) is for the North/South axis and (2) is for the East/West axis [27]

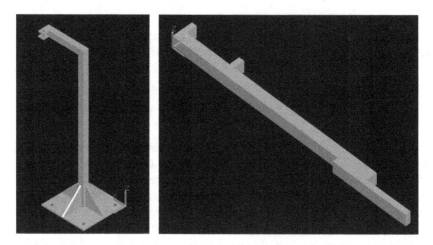

Figure 8.11 Modeling of the pole and the East/West pivot arm

Table 8.3 Technical specifications of the linear actuators [27]

Characteristics	North/South axis		East/West axis
Voltage		12 V	
Load capacity		1,500 N	
Stroke length	150 mm		300 mm
Speed		5.7 mm/s	
Max current		3 A	
Min current		0.8 A	
Waterproofing		IP65	

Table 8.4 Technical specifications of the solar charger controller [37]

Charge current	30 A
Rated voltage	12 V/24 V DC
USB output	5 V/2.5 A
Working temperature	$-35\,°C$ to $+60\,°C$
Self-consumption	<10 mA
Floating charge voltage	13.7 V (adjustable)
Discharge up to voltage	10.7 V (adjustable)
Discharge recovery voltage	12.6 V (adjustable)
Max solar panels	3×100 W

connected in parallel (12 V/45Ah). This type of battery can be recharged by a PV panel that will be installed on the metal structure of our solar tracker.

The charge/discharge regulator is associated with our photovoltaic generator, its role, among other things, is to control the charge of the battery and limit its discharge. Its function is essential because it has a direct impact on the life of the battery. We have chosen to implement a pulse width modulation (PWM) type charge regulator. Our regulator is inserted between the PV field and the battery. It is composed of an electronic switch operating in PWM and an anti-return device (diode). The opening and closing of the electronic switch take place at a certain frequency, which makes it possible to regulate the charging current according to the state of charge precisely. The switch is closed when the battery voltage is lower than the regulator limiting voltage. The battery is then charged with the current corresponding to the sunshine. We are in the "Bulk" phase. When the battery voltage reaches a predetermined regulation threshold, the switch opens and closes at a fixed frequency to maintain an average current injected into the battery. The battery is charged, and we are in the "Floating" phase.

Note that this PWM regulator can only work exclusively with PV modules ranging from 36 or 72 cells (12 V or 24 V). And to power, this system and mainly generate electricity, we used a 12 V and 150 W PV module with the characteristics as shown in Table 8.5.

The solar inverter converts a 12 V, 24 V, or 48 V DC battery voltage (in our case the pack voltage is 12 V) into a 220–240 V AC voltage identical to that of the domestic electrical network. We estimated 1,079 W of electrical energy needed on the farm per day. These needs consist of one pump for pumping water during the

Table 8.5 Technical specifications of the solar panel

Rated maximum power	150 W
Open-circuit voltage (V_{oc})	22.32 V
Short-circuit current (I_{sc})	9.00 A
The voltage at P_{max} (V_{mp})	18.0 V
Current at P_{max} (I_{mp})	8.33 A
Nominal operating cell temp (NOCT)	48±2 °C
Maximum system voltage	1,000 V DC
Maximum series fuse rating	18 A
Operating temperature	−40 °C ~+ 85 °C
Cell technology	Polycrystalline
Manufacturer	SOLON SOLAR

Table 8.6 Technical specifications of the inverter (COTEK SP-1000) [38]

AC output voltage	200/220/230/240 V AC
Rated power	1,000 W
Surge power	<1,750 W
Output waveform	100 V series
Frequency	50/60 Hz ±0.5%
DC input voltage	12 V; 24 V; 45 V
DC input voltage range	10.5–16.5 V DC; 21–33 V DC; 42–66 V DC
Efficiency	92%; 93%; 93%; 92%; 94%; 94%
Operating temperature	−20~40 °C

day and flood lights for lighting during the night. For our 12 V solar installation, we used the inverter COTEK SP-1000 with the following characteristics.

8.4.3   Realization

Our study was carried out within a farm located in DOUAR TIGUEMI IZDERN, TAMANAR rural province, where this site is semi-connected to the national electricity grid in addition to using the electrical energy produced mainly for water pumping and lighting. In this context, we deemed it appropriate to install our solar tracking system within this location (Figures 8.12–8.14).

Figure 8.12   Satellite image showing the location of the solar tracker (30.993165, −9.674119) [39]

Figure 8.13   Path of the Sun during the day (January 2, 2021) [40]

Figure 8.14   Sunsets in summer (image 1) and in winter (image 2)

We have chosen these exact coordinates in Figure 8.13 because it is the highest point on the property (altitude of 336 m) and due to the absence of shade throughout the year (Figure 8.14).

Figure 8.14 shows two different sunsets in the same location of our installation, and this remark was the main reason we carried out this study in the first place.

Once all the elements and blocks necessary for the proper functioning of our solar tracking system were ready, we began to install it. The first step was to prepare and pour the concrete platform needed to fix the main post. We can see that a concrete base is necessary to fix the Sun tracker system. This platform, which is above the ground, is mainly used as a counterweight to support the heavy weight of the metal structure (Figure 8.16). Furthermore, Figure 8.15 represents the schematic of this installation.

After the installation of our system, we ran some tests regarding the mechanical movements of the East–West arm also both actuators, the sensitivity level of the sensor, the resistance against the elements (wind, water, rain, and dust), and the production of electricity generated by the solar panel.

We noticed that in cloudy conditions, the search for the brightest point in the sky becomes continuous, and this contributes to the decay and swaying of the actuators as we have seen only after four months of use. In other words, in the case of weak light intensity, the linear actuators performing the four directions maneuvers due to constant seeking of the brightest point of the sky will be subjected to increased degradation and more deterioration due to unnecessary operation. We also noticed that the system takes a horizontal position at night. Furthermore, we learned that if an anemometer was integrated into the solar tracking controller, the system will take this horizontal position also in rainy conditions and in the case of strong winds to secure the structural integrity of the system. In summary, this study shows adequate results in terms of generating electricity in a remote area with the usage of an active solar tracking system from the global market, and the energy needs of the farm were satisfied.

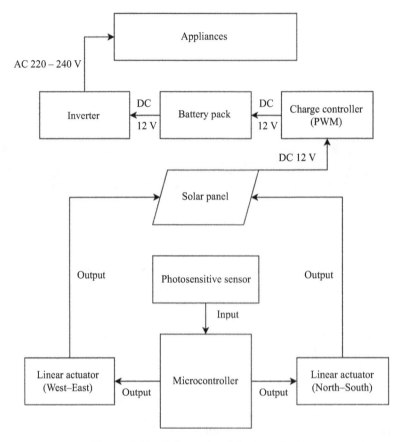

Figure 8.15   Schematic of the installation

Figure 8.16   View of the solar tracking system

Table 8.7   *Comparison of results between the solar tracking system and the fixed system*

	Fixed system	Solar tracking system
Available energy	258.6 kWh/year	345.5 kWh/year
Used energy	350.6 kWh/year	337.2 kWh/year
Specific production	1,724 Wh/kWp/year	2,304 Wh/kWp/year
Performance ratio	68.26%	67.41%
Solar fraction	63.57%	85.53%

In comparison with using the same components in a traditional fixed system, we recorded an increase of 22.17% in terms of energy production including the energy needs to run the solar tracking system (actuators, controller, etc.). More in-depth results are summarized in Table 8.7.

8.5   Conclusion

This work is part of the design of a connected and controlled PV system. It is maintained in a rural area to encourage the population to ensure stability and energy independence. We have given a general overview of renewable energies, in particular solar energy, and the type of trackers that can be useful. The experimental test carried out gave us the feasibility of these systems being an alternative with maximum energy production. The different elements of our solar tracker were exposed, microprocessor, photosensitive sensor as well as linear actuators give a configuration that showed better sensitivity in terms of tracking the trajectory of the Sun, acquiring a conversion of energy produced from DC to AC also, the test of the instantaneous production of energy by the single PV panel showed that following the Sun would be very beneficial for the future photovoltaic installation that the Sun tracking system will ensure.

To enhance our research, we provide in Appendix A an in-depth study of a PIC-based solar tracking controller which is an embedded system used in this case to perform the essential functions of a dual-axis solar tracking system and to solve the problem of continuous search of the brightest point of the sky that we noticed in our solar tracker. This controller can be in a future study for other iterations such as data collection of the performances of the system and monitoring.

Appendix A

A.1   PIC-based solar tracking controller

When choosing a microcontroller chip for the tracking system, it is essential to consider the functions it would need to perform. The functions include the conversion of analog voltage from the light sensors into digital data that can be later used for comparisons. The microcontroller also needs the capacity to handle inputs from the sensors and the outputs to the actuator's circuit. After researching an appropriate microcontroller chip for this type of application, we found the PIC18F452 microcontroller (Table A.1) to be the preferred choice. This chip is an 8-bit with 100 nanosecond instruction execution FLASH-based microcontroller that has 34 input/output (I/O) pins out of 40 pin packages. With 16-bit timers, 8 channels, 10-bit analog to digital converters, and I2C, SPI, and USART peripherals, this powerful microcontroller can perform the instructions much more easily and yet consumes less than 0.2 μA standby current and 1.6 mA normal current during 5 V and 4 MHz operations.

In this part, we detail the electronic circuits needed to run the solar tracker system and its program. This includes:

- Oscillator and reset circuit
- Voltage regulation circuit
- Light sensors circuit
- Actuators circuit

A microcontroller (MCU) is a compact integrated circuit designed to command a specific operation in an embedded system. Typical MCUs contain one or multiple processors, memory, and input/output peripherals. These microcontrollers are found in almost all types of electronics such as vehicles, medical devices, vending machines, and household appliances, among a long list of devices. They are essentially miniature computers designed to control features of a much larger component without a complex operations system.

In a PIC microcontroller, which is a part of the mid-range MCU family, an internal circuit is used that is called an oscillator for generating a stable and accurate

Table A.1   Features and specification of the PIC18F452 microcontroller [41]

CPU	8-bit
Number of pins	40
Operating voltage (V)	2 ... 5.5
Number of I/O pins	34
ADC module	1 (8-Channels, 10-bits)
Timer module	8-bit (1), 16-bit (3)
Comparators	0
DAC module	0
Communication peripherals	SPI, I2C, UART
External oscillator	Yes
Internal oscillator	No
Program memory (KB)	32
CPU speed (MIPS)	10
RAM bytes	1,536
Data EEPROM	256

periodic clock signal to the device. The frequency of this clock can range from a few kilohertz (kHz) to dozens of megahertz (MHz). This clock is essential and required for executing the instructions and peripherals of the program.

Besides the oscillator, another additional circuit added to the microcontroller is the master reset circuit (Figure A.1). This circuit is connected to the Master Clear Pin External Reset (MCLR)/VPP pin which is an optional external reset that is activated by pulling the pin low. The MCLR pin, when enabled, will hold the microcontroller in reset mode if the pin is pulled low. When the external MCLR reset occurs, the program counter will be reset to the top of the program execution or memory location which is the start of all program memory.

Since there are different power supplies in our tracking system, 12 V for the actuators circuit and 5 V for the microcontroller circuit, two power supplies are needed to accommodate both with enough current. Especially the actuators because they demand high currents when they are initialized.

In this study, two voltage regulator circuits are needed (Figures A.2 and A.3), one 5 V is supplied from the 12 V lead–acid battery system, and the other 12 V signals are sent to the actuators from our 5 V microcontroller.

The other voltage regulator circuit needed is for the voltage supply from the solar panel to the microcontroller, from 12 V to 5 V. This circuit can be implemented with an LM7805CT linear voltage converter.

In this application, we implement the usage of a combination of four small photovoltaic cells (Figure 8.14) to program the system to track Sun radiation. Each cell acts as a sensor used to sense the light source in all four directions (North, South, East, West). These sensors are connected to the ADC input on the micro-controller as described in Table 8.8. The pins PA0, PA1, Pa2, and PA3 are chosen (Figure A.4).

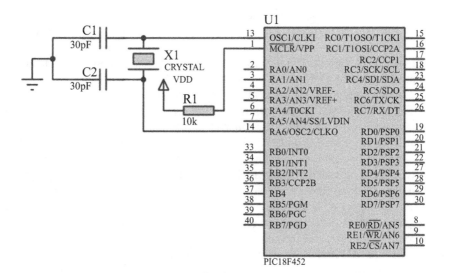

Figure A.1   Schematic of oscillator and reset circuit for PIC18F452

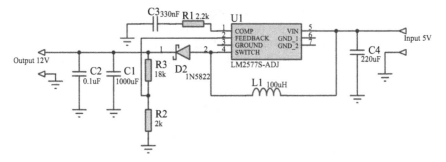

Figure A.2   Voltage regulator circuit with LM2577S-ADJ

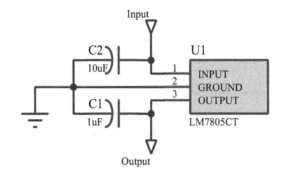

Figure A.3   Voltage regulator circuit with LM7805CT

Two linear actuators are used in this system, one for the horizontal axis movement (North–South) and the other for the vertical axis movement (East–West). Due to the unavailability of actuators in the library of Proteus, we represent them with servo motors. The pins chosen are pin PD4 for the north–south actuator and pin PD7 for the east–west actuator (Figure A.5).

The assembly of all these circuits is shown in Figure A.6.

The program used for the tracking controller has been written in C language code for the PIC18F452 microcontroller by using mikroC PRO for PIC software which is an embedded C compiler that allows us to convert a code written in C programming language to machine language using analog to digital converters (ADC). Many embedded systems need to read sensors that produce analog signals, in our case the light sensors, and since processors are built to interpret digital data (1s and 0s), they are not capable of processing the analog signals sent by sensors. The purpose of ADCs is to convert incoming data into a form that the microcontroller's processor can recognize.

From the ADC, the processor accesses the voltages sent and performs the comparisons then uses a step algorithm to command the actuators in the directions determined by the program written.

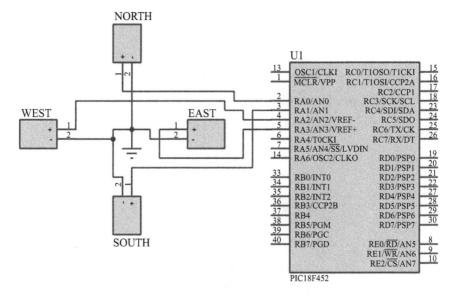

Figure A.4   Light sensors circuit on PIC18F452

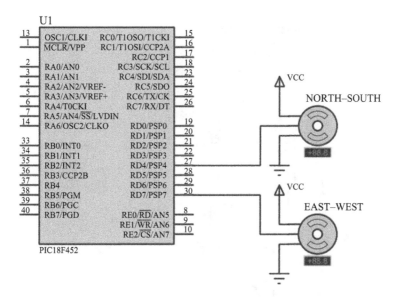

Figure A.5   Actuators circuit on PIC18F452

The code written is exported to the microcontroller to be tested and modified when it is needed. The overall advantage of this type of solar tracking system is the flexibility in modifying the code at any time which can give us room for more improvements in the future.

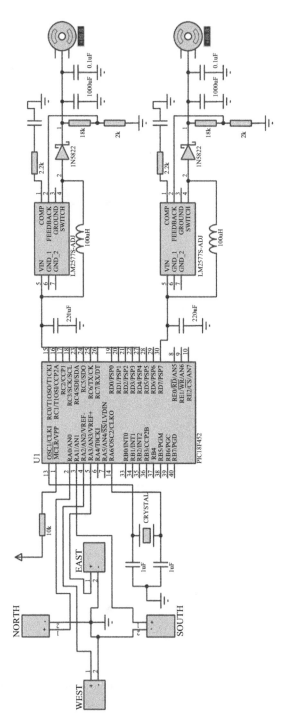

Figure A.6   Schematic of the whole solar tracker circuit

References

[1] Syvitski J., Waters C.N., Day J., *et al.* Extraordinary human energy consumption and resultant geological impacts beginning around 1950 CE initiated the proposed Anthropocene Epoch, Commun Earth Environ. 1, 32, 2020. https://doi.org/10.1038/s43247-020-00029-y.

[2] Florinda, M., Carlos, F., and Miroslava, S., Fossil fuel energy consumption in European countries, Energy Procedia 153, 107–111, 2018, https://doi.org/10.1016/j.egypro.2018.10.050.

[3] Tanveer A. and Dong dong Z., A critical review of comparative global historical energy consumption and future demand: the story told so far, Energy Rep. 6, 1973–1991, 2020, https://doi.org/10.1016/j.egyr.2020.07.020.

[4] Rafael M.S. and Reza B., Climate change/global warming/climate emergency versus general climate research: comparative bibliometric trends of publications, Heliyon 7(11), e08219, 2021, https://doi.org/10.1016/j.heliyon.2021.e08219.

[5] Gabriele C.H., Brönnimann S., Cowan T., *et al.*, Causes of climate change over the historical record, Environ. Res. Lett. 14, 123006, 2019, https://doi.org/10.1088/1748-9326/ab4557.

[6] Abbassi A., Nainaa F., Arejdal M., *et al.*, Structural and optical properties of $Zn1-x-yAlx\ SiyO$ wurtzite heterostructure thin film for photovoltaic applications, Mater. Sci. Eng.: B, 260, 114614, 2020, ISSN 0921-5107, https://doi.org/10.1016/j.mseb.2020.114614. https://sciencedirect.com/science/article/pii/S0921510720301215.

[7] Tawalare P.K., Optimizing photovoltaic conversion of solar energy, AIP Adv. 11, 100701, 2021; https://doi.org/10.1063/5.0064202.

[8] Motahhir S., El Hammoumi A., and El Ghzizal A., The most used MPPT algorithms: review and the suitable low-cost embedded board for each algorithm, J Cleaner Product. 246, 118983, 2020, ISSN 0959-6526, https://doi.org/10.1016/j.jclepro.2019.118983.

[9] Motahhir S., El Ghzizal A., Sebti S., and Derouich A., Modeling of photovoltaic system with modified incremental conductance algorithm for fast changes of irradiance, Int. J. Photoenergy, 2018, 13 pp., https://doi.org/10.1155/2018/3286479.

[10] Dana-Alexandra C., Linda B., and Gheorghe L., Real-time stochastic power management strategies in hybrid renewable energy systems: a review of key applications and perspectives, Electric Power Syst. Res. 187, 106497, 2020, ISSN 0378-7796, https://doi.org/10.1016/j.epsr.2020.106497.

[11] Kaile Z., Chao F., and Shanlin Y., Big data driven smart energy management: from big data to big insights, Renew. Sustain. Energy Rev. 56, 215–225, 2016, ISSN 1364-0321, https://doi.org/10.1016/j.rser.2015.11.050.

[12] Liu Q. and Zhang Q.-J., Accuracy improvement of energy prediction for solar-energy-powered embedded systems, IEEE Trans Very Large Scale Integration (VLSI) Syst. 24(6), 2062–2074, 2016, doi:10.1109/TVLSI.2015.2497147.

[13] Segal M., Material history: learning from silicon, Nature 483, S43–S44, 2012, https://doi.org/10.1038/483S43a.

[14] Martin A.G., Silicon solar cells: state of the art, Phil. Trans. R. Soc. 371, 1996, A.3712011041320110413 [online], Available from https://royalso-cietypublishing.org/doi/full/10.1098/rsta.2011.0413.

[15] Jimmy R., Thomas D., Eric S., and Bocar S., Le pompage photovoltaïque: Manuel de cours à l'intention des ingénieurs et des techniciens. Québec, 1998, p. 22.

[16] Wolfe K.D., Dervishogullari D., Passantino J.M., Stachurski C.D., Jennings G.K., and Cliffel D.E., Improving the stability of photosystem I-based bioelectrodes for solar energy conversion, Curr. Opin. Electrochem. 19, 27–34, 2020. ISSN 2451-9103, https://doi.org/10.1016/j.coelec.2019.09.009.

[17] Yolina H. and Mario M., Conversion of solar energy into electricity by using duckweed in direct photosynthetic plant fuel cell, Bioelectrochemistry 87, 185–191, 2012. ISSN 1567-5394, https://doi.org/10.1016/j.bioelechem.2012.02.008.

[18] Djexplo. Illustration of geographic latitude and longitude of the earth [online] 2011. Available from: https://commons.wikimedia.org/wiki/File:Latitude_and_Longitude_of_the_Earth.svg.

[19] Marco R. and Giuseppe T., Submerged and Floating Photovoltaic Systems, London: Academic Press, 2018, pp. 13–32.

[20] Ahmed F.S., Single-Axis and Dual-Axis Solar Tracker, Figure 1, Available from: https://electricalacademia.com/renewable-energy/single-axis-dual-axis-solar-trackers/#

[21] Hafez A.Z., Youssef A.M., and Harag N.M., Renew. Sustain. Energy Rev. 91, 754–782, 2018.

[22] AL-Rousan N., Isa N.A.M., and Desa M.K.M., Advances in solar photo-voltaic tracking systems: a review, Renew. Sustain. Energy Rev. 82, 2548–2569, 2018.

[23] Fuentes-Morales R.F., Diaz-Ponce A., Peña-Cruz M.I., *et al.*, Control algo-rithms applied to active solar tracking systems: a review, Sol. Energy, 212, 203–219, 2020.

[24] PANNON Green Power Ltd. Interview on the Practical Experiences of the Existing Solar Tracking Systems and About Their Own Solar Tracking Developments. Available from: https://pannongreenpower.hu/en/home/ (accessed on 16 October 2021).

[25] Zsiborács H., Pintér G., Vincze A., and Hegedűsné Baranyai N., A control process for active solar-tracking systems for photovoltaic technology and the circuit layout necessary for the implementation of the method, Sensors 22, 2564, 2022. https://doi.org/10.3390/s22072564.

[26] FUSIONSEEKER. FUSIONSEEKER Solar Tracker Controllers, FUSIONSEEKER DS-50D6W, FUSIONSEEKER DS-100D10. Available from: http://fusionseeker.com/ (accessed on 22 September 2021).

[27] ECO-WORTHY. Dual Axis Solar Tracker Controller with Remote Control. Available from: https://eco-worthy.com/products/dual-axis-solar-tracker-controller-with-remote-control (accessed on 22 September 2021).

[28] Aliexpress.com. WST03-2 Dual Axis Solar Tracker Controller. Available from: https://aliexpress.com/item/32555639188.html (accessed on 22 September 2021).

[29] Luoyang Longda Bearing Co. Ltd. Solar Tracking Controller. Available from: http://ldb-slewdrive.com/other-slewingdrive-spare-part/solar-tracking-controller.html (accessed on 22 September 2021).

[30] WindyNation. *SunTura Solar Tracker*; Windy Nation: Ventura, CA, USA, 2012, pp. 1–11.

[31] SolarStalker. STA2000-HW. Available from: http://solarstalker.com/solar-trackercontroller.aspx (accessed on 22 September 2021).

[32] DEGERenergie GmbH & Co. KG MLD SENSOR. Available from: https://degerenergie.de/mld-sensor/ (accessed on 22 September 2021).

[33] DEGERenergie GmbH & Co. KG Advantages of Deger Tracking Systems. Available from: https://degerenergie.de/advantage-degertracker/ (accessed on 22 September 2021).

[34] Palmer, D., Palmer, J., and Palmer, L., WO2020/185271A1. Available from: https://patents.google.com/patent/WO2020185271A1/en (accessed on 22 September 2021).

[35] Moser M.K., EP 2 593 759 B1. Available online: https://patents.google.com/patent/EP2593759B1/en?oq=EP2593759B1 (accessed on 22 September 2021).

[36] Walsh J.P., Novak P.J., Fenwick-Smith R., and Mccluney W.R., WO2013/074805A1. Available from: https://patents.google.com/patent/WO2013074805A1/en?oq=WO2013074805A1 (accessed on 22 September 2021).

[37] PANTHER RV PRODUCTS, PWM 30A Solar Power Charge Controller / Voltage Regulator. Available from https://pantherrvproducts.com/pwm30a/.

[38] COTEK, SP-1000 (1000W) Pure Sine Wave Inverter. Available from https://cotek.com.tw/product/Pure-Sine-Wave-Inverter-SP1000/.

[39] Google Earth. Available from https://earth.google.com/web/.

[40] Sun Locator Pro [application]. Available from http://sunlocator.com/

[41] Features and Specification of the PIC18F452 Microcontroller. Available from: https://components101.com/microcontrollers/pic18f452-8-bit-pic-microcontroller (accessed on 15 February 2023).

Chapter 9

Comprehensive literature review on the modeling and prediction of soiling effects on solar energy power plants

Bouchra Laarabi[1,2], Gabriel Jean-philippe Tevi[3], Wim C. Sinke[4], Amadou Seidou Maiga[3], Natarajan Rajasekar[2] and Abdelfettah Barhdadi[1]

The investigation of soiling effects on solar installations' performance is a topic of fast-increasing interest and importance. All over the world, works are conducted on measurement, analysis, modeling, and mitigation of these effects to perform accurate yield predictions and develop optimal maintenance strategies. This is important for uncertainty and risk reduction in large-scale investments. For this reason, the review work reported here is fully focused on the modeling and prediction of the soiling effect on solar power plants' performance. A summary of the main characteristics of each model has been described. The papers summarized have been analyzed and gathered into groups to better present the trend that research on this phenomenon follows. This review will help researchers working on this topic gain an overview of the models applied. It will also help to better choose the right direction in modeling and predicting the soiling phenomenon. The aim is to serve as a reference for upcoming works and for a better consideration in the future monitoring systems.

Nomenclature

ACF	autocorrelation function
$adjR^2$	adjusted coefficient of determination
AFWA	US Air Force Weather Agency
AIC	Akaike Information Criterion
AM	air mass
ANN	artificial neural network

[1]Physics of Semiconductors and Solar Energy Research Team (PSES), Energy Research Center (CRE), High College for Education and Research (ENS), Mohammed V University in Rabat (UM5R), Morocco
[2]Solar Energy Research Cell, School of Electrical Engineering, VIT University, Vellore, India
[3]Gaston Berger University, Saint Louis, Senegal
[4]Institute of Physics, University of Amsterdam, Netherlands

AOD	aerosol optical depth
ARIMA	auto-regressive integrated moving average
ASHRAE	American Society of Heating, Refrigeration and Air Conditioning
BIC	Bayesian Information Criterion
BNN	Bayesian Neural Networks
CF	cleanliness factor
CI	Cleanness Index
CMAQ	community multiscale air quality
CPV	concentrator photovoltaic
CSD	classical seasonal decomposition
CSP	concentrated solar power
DFA	dynamic factor analysis
DLM	dynamic linear models
DNI	direct normal irradiance
ELM	extreme learning machine
Ew/Ewo	energy production with snow/energy production without snow
FQSD	fused air quality surface using downscaling
FRP	fixed rate precipitation
G	irradiance
GIS	Geographic Information System
GOCART	Georgia Institute of Technology – Goddard Global Ozone Chemistry Aerosol Radiation and Transport
HCPV TJ	high concentrating photovoltaic triple junction
RH	relative humidity
HTAP	hemispheric transport of air pollution emissions
HYSPLIT	hybrid single-particle Lagrangian integrated trajectory
IEA PVPS	international energy agency photovoltaic power systems
JB	Jarque-Bera normality test
Jsc	short-circuit current density
LB	Ljung-Box test
LCA	life-cycle assessment
LMA	Levenberg–Marquardt Algorithm
LR	linear regression
MAD	mean absolute deviation
MAE	mean absolute error
MAPE	mean absolute percentage error
MBE	mean bias error
ME	modeling efficiency

MENA	Middle East and North Africa
MISR	multi-angle imaging spectro-radiometer
MLP	multi-layer perceptron
MLPN	multi-layer perceptron network
MLR	multiple linear regression
MLRIT	multiple linear regression of interaction terms
MS	Markov regime switching
MSE	mean square error
NARX	nonlinear autoregressive with eXogenous inputs
NASA	National Aeronautics and Space Administration
PCD	Pollution Control Department
PID	potential-induced degradation
PM_n	PM is particulate matter and n is the diameter in micron (e.g. PM_{10} is particulate matter with diameter of 10 μm or smaller)
PRM	polynomial regression model
PS	partial shading
PTC	parabolic-trough collectors
PV	photovoltaic
Q–Q	quantile–quantile plot
R	correlation coefficient
R&D	research & development
R^2	coefficient of determination
RACM	regional atmospheric chemistry mechanism
RBFN	radial basis function network
RH	relative humidity
RMSE	root mean square error
RMSEn	normalized root-mean-square error
ROC	receiver operation characteristic
RS	series-resistance
S	span of the local regression
Sd	snow depth
SMARTS	simple model of the atmospheric radiative transfer of sunshine
SR	spectral response
SRR	stochastic rate and recovery
T_a	ambient temperature
US EPA	United States Environmental Protection Agency
VWS	vertical wind speed
WRF	Weather Research and Forecasting

WRF-Chem	Weather Research Forecasting with Chemistry
WRF-DuMO	Weather Research and Forecasting-Dust Model
WS	Wind Speed

9.1   Introduction

Evolving renewable energy-based solutions to mitigate today's most critical environmental issues as greenhouse gas emission is of high priority in the global arena. For this reason, the renewable energy market grows faster through the years and many energetic strategies are upfront to implement ambitious projects in the short, medium, and long terms.

9.1.1   *Solar PV plants over the world*

Because of advantages including abundant availability, easy installation, minimal maintenance and simple interfacing, solar PV technologies have become the preferred source of renewable energies. Moreover, solar PV is the top source of new power capacity in several major markets [1]. Worldwide, in 2017, around 98 GW_p of PV capacities are installed (on- and off-grid) increasing its total capacity by nearly one-third with a cumulative total of approximately 402 GW_p [1]. Among the largest PV plants, we find Bhadla Solar Park which is located in Bhadla village. Bhadla is located about 200 km north of Jodhpur and about 320 km west of the state capital, Jaipur, Rajasthan in India. The area has been described as "almost unlivable" due to its climate. Normal temperatures in Bhadla hover between 46 °C and 48 °C, with hot winds and sandstorms occurring frequently. The power plant was commissioned in March 2020 with an installed capacity of 2.25 GW and spans 14,000 acres [2]. There is also the Huanghe Hydropower Hainan Solar Park which is located in the remote province of Qinghai in China. The solar photovoltaic plant has an installed capacity of 2.2 GW, and was built in five phases and includes 202.8 megawatts (MW)/ megawatt hours (MWh) of storage capacity [3]. We find as well Benban solar park which is located about 650 km south of the Egyptian capital Cairo in Benban. It has an installed capacity of 1.65 GW corresponding to an annual production of approximately 3.8 terawatt hours (TWh) [4].

In relationship to the nature of their locations, the above-mentioned installations have a common point. This manifests in being installed in remote areas that have the advantage of being large areas, very sunny, and with no obstacles. However, these locations present the inconvenience of being characterized by harsh conditions, experiencing high temperature, being hard to monitor and being very dusty in most cases. One of the solutions used to control, manage, and monitor these installations to facilitate the generation and utilization of their produced energy is the usage of digital technologies [5].

9.1.2 Digital technologies applied in the monitoring of remote PV plants

Digital technologies are defined as the digital tools, sets of technologies, systems, devices and resources that generate, store, or process data. Their main characteristic is the possibility of solving numerous problems in a relatively short time [6]. These devices are embedded in PV plants to optimize their energy generation [5].

In this context, many studies have been conducted. These studies include prototypes and conceptual systems to monitor the state of a photovoltaic system through an Internet of Things (IoT)-based network to control it remotely [7]. In addition, they are designed to centralize the monitoring of environmental parameters that impact the solar power plant operation efficiency and its operating characteristics (voltage, electric current, power, etc.) [8]. As per the change in the atmospheric or weather conditions, the user can control the solar panels also [9]. Predictive maintenance which includes localization and definition of related faults and failures in a PV system is very important [7]. These systems should present the advantage of being cost-effective and adaptable to large PV plants which can be controlled during their entire lifetime [10].

One of the phenomena that is prevailing in these remote areas and which has a considerable effect on the production of solar PV plants is soiling. Nevertheless, this phenomenon was sometimes overlooked and not taken into consideration in the developed monitoring systems [11]. This can be due to its complex nature and the lack of precise information on its effect. For these reasons, many studies have been conducted to model its effect and predict its impact. These could be included in the monitoring systems to present useful information, mainly on the need for cleaning or not.

9.1.3 Soiling phenomenon and its modeling

Soiling, *also named dusting, is usually a mixture of small amounts of organic particles and/or minerals from geomorphic fallout such as sand, clay, or eroded limestone. Atmospheric airborne dust (aerosols) is attributed to various sources, such as soil elements lifted by the wind (Aeolian dust), volcanic eruptions, vehicle movement, and pollution* [12]. Studies regarding the soiling phenomenon started in 1942 and since then numerous contributions have been made in analyzing soiling effects and research on this topic is increasing over the years [12].

Works on the above subject investigate it following a specific direction and focusing on a precise aspect. The different aspects that are generally analyzed are categorized in Figure 9.1 'Assessment' category includes the evaluation of soiling impact or the discussion of the factors influencing soiling [13–18]. 'Particle properties' include the determination of soiling particles' composition, morphology, or size [19–21]. 'Mitigation/cleaning' takes into account the different techniques, systems, or methods, whatever preventive or corrective, developed to eliminate, reduce, or avoid the soiling effect [22–24]. 'Instrumentation' includes the effect of soiling not on solar energy systems but instruments such as pyranometers or pyrheliometers [16, 25]. 'Dust simulators' are mainly instruments that have been developed to simulate the soiling effect in indoor conditions [26–28]. 'Modeling'

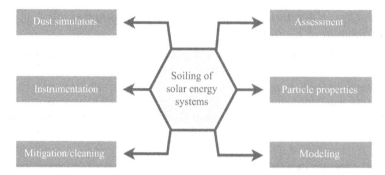

Figure 9.1   The different aspects of the soiling phenomenon

considers all models established to model the soiling phenomenon or to predict its effect over time [29]. This latter refers to the establishment of a physical, conceptual, or mathematical representation of real phenomenon that is not easy to understand directly which will help to explain its behavior [30]. It is an aspect of high importance since it will help to estimate the losses and to draw suitable and optimized strategies for soiling mitigation. Even though modeling is very important, the generated models remain only an approximation of reality. This leads researchers to improve the developed models continuously. For that, an important step is to provide researchers with a comprehensive overview of existing knowledge and gaps in modeling and prediction of soiling, which can be used to define and prioritize further research on this issue. The review work proposed here is part of this step and it is fully dedicated to the modeling and prediction of soiling effects on solar power plants.

For a better understanding of the work in this review, articles on the topic from 1998 until the beginning of 2020 are exhaustively summarized with specific points highlighted. The second section is an overview of soiling and the modeling of soiling. The third section is showing the different paths, presented as categories, which can be followed to model the soiling phenomenon.

The fourth section is dedicated to the description and summary of the models applied per each category. These have been discussed and analyzed in the fourth section. While the last section is the conclusions and recommendations of the work.

9.2   Overview of soiling and modeling of soiling

Soiling-induced losses are influenced by many factors which make their impact vary from one location to another. According to a study presented in [31], in 2018, soiling has been revealed to reduce global power production by around 3%–4%. This is equivalent to 3–4 billion euro losses on global revenue. Unfortunately, dust concentration is mainly high in desert areas where the use of solar energy is very important and profitable (Figure 9.2).

Various reviews regarding the soiling of solar energy systems can be seen and its consolidated list is shown in Table 9.1. Some reviews include almost all aspects

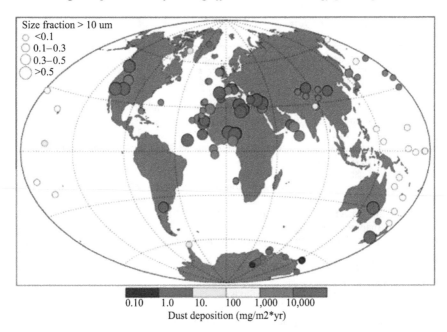

Figure 9.2   Size and density distribution of dust aerosols [32]

of the soiling phenomenon as in the case of [12,33–37] with a difference in the covered period. Other reviews concern a specific aspect of the soiling phenomenon. We found, cleaning methods, self-cleaning systems, preventive techniques, and anti-soiling methods [38], [39], [40], and [41] respectively. In the same way, the review detailed in [42] focuses on the pollen effect, while in [43], authors were interested more in discussing the interrelation between soiling and the spectral response of modules. Regarding the adhesion mechanisms, there is [44–47]. For the techno-economic evaluation, there are the works referenced in [31,48]. Other reviews followed a different direction based on a specific geographic area such as the studies Ref. [49] for the MENA region, Ref. [50] related to Iraq, Ref. [51] for the Nigerian context, and Ref. [52] regarding Morocco and India countries.

All aspects signify that the review is more general and takes into account almost all sides (assessment, modeling, cleaning/mitigation techniques, particle properties, etc.).

Through all these reviews aforementioned, it is obvious that although modeling of soiling is an important aspect, it is rare when it is reviewed. Nevertheless, studies on this subject are continuously published. Based on the literature survey conducted in this review, these studies are summarized and presented in the following map (Figure 9.3) according to their geographic location of measurements. A high number is attributed to lab work. This latter and 'Unknown' which refers to unknown locations have been added to the legend to show the other publications to where they belong. From Figure 9.3, it can be seen that the studies are mainly concentrated in the middle built known by high insolation and high dust concentration as has been observed in Figure 9.2.

Table 9.1   Important reviews published on the subject of the effect of soiling

Reference	The focus of the review and the aspects discussed
[33]	All aspects
[12]	All aspectsPeriod covered: from 1942 to 2012
[34]	Summary of parameters affecting the soiling phenomenon
[35]	All aspects/ flat surfaces/ focus on cleaning mechanismsPeriod covered: the past 70 years
[36]	All aspects/ identification of representative scientific contributionsPeriod covered: from 1942 to 2013
[50]	All aspectsFocus on Iraq country
[53]	Discussion of soiling impact, influencing factors on soiling and cleaning methods
[54]	Discussion of soiling impact
[55]	All aspectsPeriod covered: from 2012 to 2015
[56]	Shading by soiling
[57]	Discussion of soiling impact, influencing factors, and measurement methods
[44]	Adhesion mechanisms
[38]	Discussion of soiling impact and influencing factors on soiling but focus on cleaning methods
[58]	Focus on measurement methods
[37]	All aspects
[59]	Discussion of influencing factors, mitigation and cleaning techniques as well as measurement methods
[45]	Modeling of adhesion mechanisms
[46]	Discussion of soiling impact, mitigation and cleaning methods and adhesion mechanisms
[60]	Focus on mitigation and cleaning methods
[49]	Discussion of soiling impact (focus on MENA region), mitigation techniques and modeling
[47]	Summary of macroscopic and microscopic factors influencing adhesion mechanisms
[39]	Self-cleaning methods
[40]	Preventive cleaning solutions
[61]	Discussion of soiling impact, influencing factors and mitigation techniques
[62]	Discussion of soiling impact and mitigation techniques
[41]	Analysis of the so-far developed anti-soiling systems
[31]	Techno-economic evaluation of the strategies for soiling mitigation
[51]	Discussion of soiling impact, influencing factors, mitigation methods and soiling in the Nigerian context
[63]	Discussion of soiling impact and mitigation techniques
[42]	In-depth analysis of the interrelationship between pollen and soiling of solar photovoltaic systems
[43]	Analysis of interrelation soiling and spectral response of modules
[48]	Photovoltaic systems monitoring
[52]	Discussion of soling process, the influencing factors, the mitigation techniques as well as Morocco and India's contributions

9.3   Modeling of soiling, which path to follow?

According to the recommendation in [64], the modeling of soiling should be done in two parts; the first one should concern the model which will predict the amount

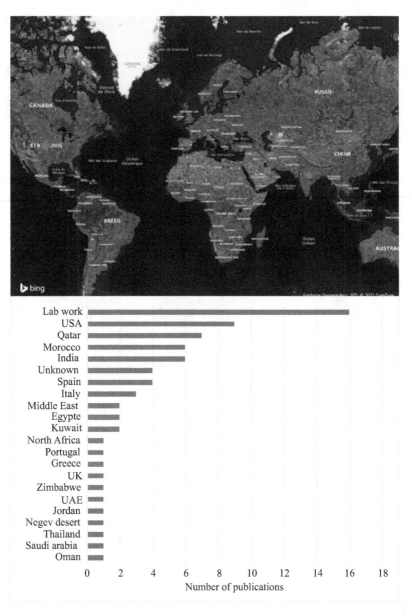

Figure 9.3   Statistics of publications on modeling and prediction of soiling and their geographic location. 'Unknown' refers to an unknown location

of dust mass deposited according to the climatic conditions while the second part should focus on the model which will correlate the amount of dust mass deposited and PV modules losses due to soiling. The classification presented in Figure 9.4 has been done based on that statement. From what is shown, the models are divided into six groups to which different letters (A, B, C, AB, AC, BC) and colors have been assigned to highlight parts. 'A' refers to the models that deal with the prediction of soiling potentials and dust concentration, 'B' the models which study the interaction of the deposited pollutants and the surface while 'C' corresponds to the models which determine the patterns of losses due to soiling. 'AB' groups the models that use weather conditions' variables to predict the deposited pollutants, 'BC' the correlations between the deposited pollutants and the soiling losses, whereas 'AC' is related to the correlations between weather conditions and the soiling losses (Table 9.2).

Figure 9.5 shows the share of groups in the gathered papers. Based on this figure, it can be seen that the highest percentage belongs to the group BC followed respectively by AC, C, and A and AB. The two models suggested in [64] are AB and BC and these two types of modeling are complementary to each other. For part

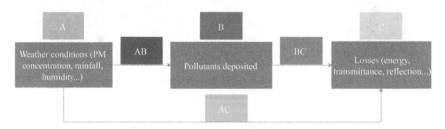

Figure 9.4. Classification of publications according to the desired objective of modeling

Table 9.2   Classification of publications according to the desired objective of modeling

A	Modeling and prediction of soiling potentials/dust concentration	[65–68]
C	Determining the patterns of losses due to soiling	[69–74]
AB	Use of weather conditions to predict deposited pollutants	[75–78]
BC	Correlation between losses due to soiling and deposited pollutants (mass, composition, etc.)	[26, 29, 64, 75, 76, 78–96]
AC	Correlation between weather conditions and losses due to soiling	[15, 77, 97, 98–116]

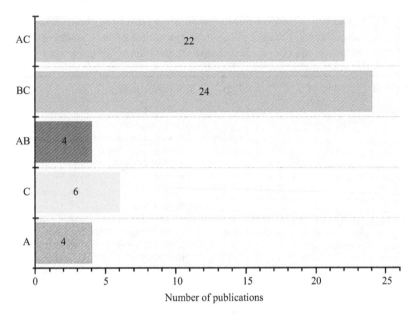

Figure 9.5   Share of each group in the published papers

B which is about deposited pollutants, many published papers describe and model the interaction between particles and the surface as can be seen in [47]. This part is also an important one but it is far beyond the scope of the present review.

In the following section, a summary of the different applied models per category is presented. Tables 9.3–9.7 present the different and miscellaneous studies conducted in the framework of modeling and prediction of soiling phenomenon from 1998 (where, to our knowledge, the first paper on modeling of soiling has been published) until 2019 (including the beginning of 2020) for each category. They give a summary related to the performance of the model based on the conclusions drawn by its authors. They highlight the applied technology as well as the experimental place of these studies when specified by authors, just to help the interested to have a quick look at the geographical location or to know if it is a laboratory simulation.

9.4   Applied models per each category

9.4.1   Modeling and prediction of soiling potential/dust concentration (category A)

In Table 9.3, authors of the work [65] performed a dust risk map for solar energy systems in the arid area based on the correlation between Aerosol Optical Depth (AOD), provided by the Multi-angle Imaging Spectro-Radiometer (MISR) (NASA satellite data), wind regime, and air temperature. While in the study [66], a simulation of dust

Table 9.3   Summary of articles published on the modeling of the soiling
phenomenon (category A). The performance of each model discussed is
based on the conclusions drawn by the authors of each work

Reference	Location of measurements	Technology	Main conclusions
[65]	Oman	All solar energy systems	No comment about the performance of the model. The rate of dust deposition could be estimated from the open-source HYSPLIT-Hybrid Single Particle Lagrangian Integrated Trajectory Model has been presented as a perspective.
[66]	Middle East	Useful for all technologies	The dust model implemented was performant enough; it has been improved by applying a salt effect correction and the use of the salt map.
[67]	The Middle East and North Africa	PV glazing (useful for other technologies)	The results presented are preliminary and more studies should be conducted to validate the used model.
[68]	Western Saudi Arabia	PV-modules (useful for other technologies)	Field tests should be done to evaluate the spatial assessment of the real soiling risk.

events based on the model WRF-DuMO which has been already applied in [117]. Following the same path, authors of [67] used the Geographic Information System (GIS) as a base for modeling the distribution of soiling potential in the Middle East and North Africa (MENA). Similarly, in the article [68], soiling risk maps have been created using GIS to expect the area or regions where dust emissions are highest. In this model, soiling has been defined as a function of the following meteorological and environmental factors: land cover, vegetation, land use, soil type and soil texture, precipitation, wind speed and wind direction, and humidity.

9.4.2   Determining the patterns of losses due to soiling (category C)

Based on the theoretical analysis proposed in [118], Ref. [70] proposed a model which calculates the reflection losses in PV modules. For CSP aluminum mirrors, there are the works [69, 71–73]. The modeling of the cleanliness factor (CF) of CSP aluminum mirrors has been studied in [71] using locally weighted regression (loess). CF factor refers to the reflectance of a soiled mirror divided by the reflectance of a clean one. The CF data have been treated as time series with the time index as a regressor. Two cases have been studied based on the span s of the local regression (the number of data points considered for each local regression); $s = 0.2$ and $s = 0.75$. Two tests have been used to examine the normality of the

Table 9.4 *Summary of articles published modeling of soiling phenomenon (category C). The performance of each model discussed is based on the conclusions drawn by the authors of each work*

Reference	Location of measurements	Technology	Main conclusions
[70]	South of Navarre (Spain)	PV plant	Losses were correctly estimated for tracking and also horizontal surfaces, each one with its performance but there was no specific comments about the performance of the model.
[71]	Southwest Morocco	CSP aluminum mirrors	The results showed that $s = 0.75$ is considered proper for defining the studied data and their principal pattern. The examination of (ACF) function and (Q–Q) plot allowed the confirmation of the result obtained.
[69]	Agadir, Morocco	CSP solar reflectors	The results showed that the model that can be considered the best-fitted to describe the cleanliness long-term change is the local linear trend. When an optimal discount factor of 0.95 is taken into consideration, the model performs even better. The model could be improved by including different explanatory variables (like weather conditions) and taking into account the effect of seasonality.
[73]	Tantan, Agadir Morocco	CSP	The Dynamic Factor Analysis (DFA) discovers that the best model has two shared trend patterns. This has been found based on the Akaike Information Criterion (AIC).
[72]	Almeria, Spain	CSP mirrors	The approach proposed in the paper presents a promising method to model the soiling of CSP reflectors as it takes into consideration the seasonal soiling and the washing cycle frequency. Nevertheless, the model should be tested on a real CSP plant.
[74]	Golden in Colorado and San José in California (USA), Chennai and Tezpur (India), Jaén (Spain), Penryn (UK) and El Shourouk City (Egypt)	PV	The main findings are: the hemispherical transmittance can be described by the Ångström turbidity equation, the distribution of particle size found following the IEST-STD-CC 1246E cleanliness standard and the PV panels' optical losses generated under the effect of soiling can be quantified from the cleanness value and the distribution of particle size.

Table 9.5   Summary of articles published modeling of soiling phenomenon (category AB). The performance of each model discussed is based on the conclusions drawn by the authors of each work

Reference	Location of measurements	Technology	Main conclusions
[75]	Measurements at Doha, Qatar and theoretical modeling	PV modules	There are no comments on the performance of the model as it has been validated, only on a typical day.
[77]	Doha, Qatar	PV modules	The model showed a good correlation between the change in the PV energy yield and the PM dry deposition against PM10 and PM20 concentrations. However, the test of the model in other environments and for longer periods is still needed to validate it.
[78]	Arizona and Central/ southern California states, USA	PV modules	The model was judged performant. However, to be more accurate, the model should include other impacting factors such as wind, dew formation and particulate matter larger than 10 μm, also the performance of the model should be checked in more geographic areas.
[76]	Simulation and field measurements in three locations (two in Colorado and one in Texas) USA	Solar PV glass	The particulate matter (PM) dry deposition estimates are more minor than the field measurements because of the underestimation of the ambient total particle concentration and the underestimation of the deposition rates (due to the underestimation of surface wind speed input and more frequent precipitation input). However, it shows a strong correlation between the transmittance loss and the total particle deposition.

residuals and their independence, the Quantile–Quantile (Q–Q) plot and the auto-correlation function (ACF). Likewise, the work [69] concerns the analysis of the cleanliness of the solar reflectors exposed outdoors. This has been performed using the free R software for statistical computing and graphics together with dynamic linear models (DLM) package, the zoo package [119] for indexing the time series, and Amelia's package for multiple imputations [120]. A comparison between three models has been conducted: the local level model (i.e., a first-order polynomial model), the local linear trend model (that is, a second-order polynomial model), and the local level with intervention model by including an intervention variable. The

Table 9.6 Summary of articles published modeling of soiling phenomenon (statistical modeling). The performance of each model discussed is based on the conclusions drawn by the authors of each work

Reference	Location of measurements	Technology	Main conclusions
[26]	The theoretical model has been compared with the results of an indoor simulation	PV module (glazing surface)	After almost 50% decrease in light transmittance a disagreement between the calculated and measured values. The mathematical prediction could be improved if the real shape and the transparency to incident light of the particles have been better analyzed.
[94]	Minia, Egypt	Glass plates	The model showed a good performance. However, an application on other climate types would be valuable.
[79]	Theoretical model and field measurements in Shuwaikh, Kuwait	PV modules	The results showed a good correlation between the model and the measurements. However, the model is limited to 1.5 g/m^2 of the amount of sand dust particles.
[80]	Theoretical study	PV panels	The shape and the size distribution of particles should be analyzed to better estimate the impact of particle deposition on transmittance.
[81]	Simulation and field measurements in Athens, Greece	PV modules	The model is judged to be quite performant. However, the test of the model on different applications and different pollutants is still needed. The test of the long-term prediction capability of the model is still required as well.
[83]	Laboratory simulation	PV modules	Good agreement between the models and the results of the indoor simulation. Other variables should be taken into consideration such as moisture, wind speed and tilt angle to approach the models to the real behavior of dust settlement.
[82]	Kuwait	PV modules	No comments about the performance of the model.
[86]	Thailand	PV module	No comment on the performance but the results obtained allowed the development of a mathematical relationship between dust on PV modules and the reduction in the electrical energy output.
[84]	Lab Simulation	PV module	No comment on the model's performance.
[85]	Colorado, USA	Glass samples	The model was found acceptable.

(Continues)

Table 9.6   (Continued)

Reference	Location of measurements	Technology	Main conclusions
[90]	Soil samples collected from Shekhawati region in India were used in an artificial soiling experiment	PV panels	The validation of the empirical model is achieved experimentally and its efficiency is shown in the obtained results.
[29]	Shekhawati region, India	PV modules	From the results obtained it has been concluded that regression models are useful in determining the relationship between the particle size compositions, and irradiance with power output but it is the neural networks that are efficient in power prediction based on the particle size composition.
[88]	Experiments conducted in India	PV module	The model was found robust, especially in stochastic and dynamic conditions.
[89]	Experiments in India	PV panel	The model was performant enough.
[87]	Theoretic method tested based on field measurement in an unknown location	PV systems	This model can practically predict module performances under different weather conditions. Further investigations on particle size, particle material and PV module type factors are needed.
[75]	Measurements at Doha, Qatar and theoretical modeling	PV modules	There are no comments on the performance of the model as it has been validated, only on a typical day.
[91]	Numerical simulation	PV panel	The model shows good accuracy.
[64]	Theoretical modeling	PV modules	The model proposed, has been judged to be applicable in the majority of cases met in urban and non-urban polluted areas. The authors recommend that the modeling of soiling should be done in two parts: The first part: concerns the model which will predict the amount the dust mass deposited according to the climatic conditions. The second part: concerns the model which will correlate the amount of dust mass deposited and PV modules losses (to which their proposed model belongs)
[92]	Laboratory simulation	Glass slides	The theoretical models show a good agreement with the measurements. However, to be completed, models to forecast the mass density of particles deposited per unit area are still needed.
[93]	Theoretical modeling and simulation	PV modules	Based on the results obtained the model shows satisfactory accuracy to predict efficiency losses due to soiling.

(Continues)

Table 9.6 (*Continued*)

Reference	Location of measurements	Technology	Main conclusions
			However, the effect of the solar PV module type, the particles deposited composition and the solar density need to be included considering their important impact. Furthermore, in areas with highly soiling potential, the linear relation between the dust deposition density and efficiency loss has to be reviewed.
[78]	Arizona,and Central/ southern California states, USA	PV modules	The model was judged performant. However, to be more accurate, the model should include other impacting factors such as wind, dew formation and particulate matter larger than 10 μm, also the performance of the model should be checked in more geographic areas.
[95]	Negev desert	CSP glass reflectors	The model presents a new way to assess the losses due to soiling in reflectors. The model can be improved by the application on a large plant with different compositions of dust. The application of the model can also be extended to predict the optical losses of PV panels.
[76]	Simulation and field measurements in three locations (two in Colorado and one in Texas) USA	Solar PV glass	The particulate matter (PM) dry deposition estimates are more minor than the field measurements because of the underestimation of the ambient total particle concentration and the underestimation of the deposition rates (due to the underestimation of surface wind speed input and more frequent precipitation input). However, it shows a strong correlation between the transmittance loss and the total particle deposition.
[96]	Southern Spain	CSP and PV	The model showed promising results. However, the authors thought that its performance is less because the following assumptions are made: spherical particles, low soiling level (<2 g/m^2), no agglomerations, constant size distribution and the optical properties of the dust considered

Table 9.7   Summary of articles published modeling of soiling phenomenon (category AC). The performance of each model discussed is based on the conclusions drawn by the authors of each work

Reference	Location of measurements	Technology	Main conclusions
[97]	The countryside of southern Italy	PV plants	The coefficient of correlation is more than 99% which shows the effectiveness of the model to estimate the power.
[98]	Southern Italy	PV plant	The accuracy of the BNN model is higher than the polynomial one, which can be influenced by the size of the database. A large database is required to adjust better the polynomial coefficients, while in the case of BNN; a 4,000 samples dataset is largely sufficient to adapt the weights and bias of the network. Concerning the comparison between BNNs, MLPN and RDFN, the three models have practically the same MAPE (1.1%), but the BNNs present the most sample structure.
[99]	Lab simulation and real condition measurements, Italy	PV module	No indication of the performance of the model in estimating the effect of soiling.
[101]	Doha, Qatar	PV modules	More work is needed to better understand and to better predict the effect of dust and weather conditions on PV performance losses.
[100]	Lab simulation and measurements	HCPV TJ cells	Good results were obtained for the predictions but the variation noted for spectral response represents a challenge in performance modeling and monitoring.
[112]	Tantan, Agadir, Morocco	CSP	According to mean square error, the best model found is the one with 30 neurons and 2 tapped delay. However, the model would give better results if a large amount of data is used and by joining multiple architectures of NARX.
[102]	Field measurements (but no indication of the location)	PV systems	The model used was judged as consistent and robust. Further on its comparison with least-squares linear regression. The Theil-Sen estimator makes the proposed method more easily scalable to large numbers of sites without the need for an anomaly filtering.

(Continues)

Table 9.7   (Continued)

Reference	Location of measurements	Technology	Main conclusions
[111]	Field measurements (but no indication of the location)	PV systems	The model is judged robust and low cost since the need for new hardware is not required.
[107]	Gandhinagar in Ahmedabad, India	PV modules	No comments about the accuracy of the model. This latter is limited as it considers a linear relationship between the change in transmittance and PM mass flux. The authors believe as well that further studies which include more influencing parameters are still needed.
[15]	Computational simulations based on field measurements in Doha (Qatar)	PV modules	The models were found sufficiently accurate and performant.
[103]	Doha, Qatar	PV modules	The ANN model has been recognized as a suitable tool to model the complex relationship between soiling losses and environmental conditions. The ANN model has shown, significantly, a better prediction of soiling loss against the multivariable linear regression model. The correlation coefficient between the measured and the ANN predicted value was not highly significant which could be explained by the presence of other parameters affecting the soiling such as the probability of dew formation and condensation, the physicochemical properties of dust, the residence time of dust particles on the critical surfaces and the prevailing effect of dust loading.
[104]	Rabat, Morocco	Solar PV glass	The results obtained show a good correlation between the soiling rate and the weather conditions. It also shows that rainfall and wind direction impact more the soiling rate.
[106]	Rabat, Morocco	PV modules	The results have shown a good correlation between the soiling rate and the weather conditions. The sensitivity analysis indicates that the relative humidity followed by wind direction is the most influential parameter among the inputs used on the soiling rate.

(Continues)

Table 9.7   (Continued)

Reference	Location of measurements	Technology	Main conclusions
[105]	Data were recorded at 20 sites around the United States of America USA	PV systems	No comments about the performance of the model but the results have shown that the annual average of the mean daily particulate matter values recorded are the best predictors of soiling losses.
[108]	Field measurements (United States) and simulation	PV systems	The method outperforms the commonly used fixed rate precipitation (FRP) method.
[77]	Doha, Qatar	PV modules	The model showed a good correlation between the change in the PV energy yield and the PM dry deposition against PM10 and PM20 concentrations. However, the test of the model in other environments and for longer periods is still needed to validate it.
[110]	Evora, Portugal	CSP mirror	Based on the results obtained, the comparison between the three models has shown that ANN is a promising tool to predict soiling losses. However, better results would be achieved if more data, variables and different time scales have been taken into consideration. The MLRIT model is also important as it helps to understand more how the variables interlinked with each other and how they affect the output.
[109]	USA (California, Oceanic Islands (Hawaii and Virgin), East Coast and the Southeast)	PV modules	The results obtained have confirmed that environmental parameters such as particulate matter and rainfall can be used to estimate soiling losses in the USA. Further studies and a larger number of data points are still needed to improve the model and make it more general.
[116]	Zarqa, Jordan	PV modules	The results revealed that both the optimized models of ANN and ELM managed to predict the efficiency of the PV panels. However, the optimized ELM was more accurate.
[113]	Harare, Zimbabwe	PV modules	The two models ANN and MLR have shown good performance in modeling of soiling of PV modules.
[115]	Arizona and California states, USA	PV modules	The improvement applied to the method [108] has led to a better estimation of the optimal cleaning day and

(Continues)

Table 9.7 (*Continued*)

Reference	Location of measurements	Technology	Main conclusions
			the energy gains. The other model judged well in predicting the week when the cleaning should be applied. However, to reach better results, the authors believe that their methods should be applied on a large number of sites and data and also the use of more robust weather generation algorithms.
[114]	Sharjah, UAE	PV modules	The results have shown that no difference in accuracy was shown between the two models. Both models showed good accuracy.

selection criteria used are: mean square error (MSE), mean absolute deviation (MAD), mean absolute percentage error (MAPE), Akaike Information Criterion (AIC), and Bayesian Information Criterion (BIC). To detect the most important exposed mirror material trends and the impact of weather factors when taken into consideration as explanatory variables, the dynamic factor analysis (DFA) has been applied to the cumulative soiling of CSP mirrors [73]. Authors of Ref. [72] proposed a set of seven non-linear time series models based on Markov regime Switching (MS) to model the soiling of solar reflectors regularly cleaned. The models have been evaluated using the AIC and the BIC. Moreover, a diagnostic of the model's residuals has been done based on the Jarque-Bera (JB) normality test [121] to check the normality assumption and the ACF and the Ljung-Box (LB) test [122] to check the independence of the residuals. After the evaluation of the models, a simulation of 30 MWe parabolic-trough collectors (PTC) plant without thermal storage has been conducted and different scenarios based on the chosen models have been developed.

In 2020, an investigation of the spectral and optical characteristics of various types of soiling has been conducted [74]. Four correlations or comparisons have been studied. First, the hemispherical transmittance vs. wavelength correlation is compared to the Ångström turbidity equation (based on the Mie-scattering theory). Second and third, the size distribution and the fractional coverage area have been compared to IEST-STD-CC 1246E cleanliness standard. Fourth, the transmittance has been correlated to the fractional coverage area.

9.4.3 Use of weather conditions to predict deposited pollutants (category AB)

Authors of the work [75] presented in their study a model to predict soil mass accumulated from its morphology. A correlation between particulate matter (starting with PM10 concentration, then PM20 concentration, then PM dry deposition) and the variation in the PV energy yield taking into consideration the effect of humidity, ambient temperature, wind speed, and wind direction is suggested in [77]. The particulate matter has been modeled based on the Weather Research Forecasting with Chemistry (WRF-Chem), the GOCART aerosol scheme (Georgia Institute of Technology – Goddard Global Ozone Chemistry Aerosol Radiation and Transport) and the RACM chemistry (Regional Atmospheric Chemistry Mechanism), the HTAP (Hemispheric Transport of Air Pollution emissions), and the AFWA emission scheme (US Air Force Weather Agency).

Another model to predict soiling losses has been established in [78]. This model is based on the evaluation of the deposited dust based on the particulate matter concentration (PM10 and PM2.5). This latter is adjusted using rainfall data and the soiling ratio or transmittance loss is estimated based on the correlation carried out in [123]. The modeled data were compared with the experimental data from [124].

In 2019, authors of [76] presented a correlation between soiling losses and particulate matter (PM) dry deposition estimates from the Community Multiscale Air Quality (CMAQ) model [125]. The initial and boundary conditions have been provided from [126] while the meteorological input data were provided from the Weather Research and Forecast (WRF) model [127]. The deposition measurements and transmittance losses were provided from the study of Boyle *et al.* [128,129].

9.4.4 *Correlation between losses due to soiling and deposited pollutants (category BC)*

As for our knowledge, the first model established has been published in [26]. This evaluates the transmittance coefficient for beam light based on the estimation of the amount of sand dust particles per unit area of glazing surface, the size of the particles, beam light incidence angle and wavelength. Continuity of this work has been conducted in [79] where they validated their model [26] using field measurements and the equation which links the decrease in the PV panels' efficiency and the increase in the sand dust particles amount has been presented.

In 2001, authors of the study [94] performed a correlation between the reduction factor of transmittance and dust accumulation taking into account the tilt angle and number of days of exposure since the last cleaning. The accuracy of the multi-variable nonlinear regression model has been evaluated based on the MAE and the RMSE. While in [80], authors estimated the dirt effect on the transmittance, which depends on the shape and size distribution of particles and incident angle, based on the calculation of the probability that a single particle will cover a given area on a surface (overlapping model proposed by NASA). In the same year, authors of [81] correlate the energy yield and dust deposited based on artificial soiling composed of three pollutant types. The results obtained from the artificial soiling have been compared with a natural one. From these results, a model (exponential regression) has been developed.

Authors of [83] correlate between the free fractional area A (area not covered by particles) of the glass and the quantities of dust deposited using two models: first model: An analytical model based on [26] and use a grain threshold algorithm in the software package Gwyddion, Second model: Simulation using Monte Carlo method. While in [82], authors correlate between spectral transmittance and dust density based on 'Simple Model of the Atmospheric Radiative Transfer of Sunshine (SMARTS)' code to account for the tilt angle of the surface plane [130] combined with a MATLAB® code (MathWorks, Natick, MA, USA).

Based on the International Energy Agency Photovoltaic Power Systems TASK 2 – Performance, Reliability and Analysis of Photovoltaic Systems (IEA PVPS Task 2) [131] and the dust accumulated on PV modules using Dust-fall jar methods from Pollution Control Department (PCD) and standard for detecting dust by US EPA Method 5 of the United States Environmental Protection Agency (US EPA: US Environmental Protection Agency), a mathematical model for the analysis of the efficiency is performed in [86]. Optical modeling of the impact of dust on PV panels' performances by simulating the electromagnetic wave propagation in a three-layer surface (air-dust-glass) using COMSOL Multiphysics software has been proposed in [84]. The model is based on the analysis of the transmission and reflection coefficients at the limits of each layer of the structure (air-dust-glass). A linear regression model has been adopted in [85] to describe the relationship between dust mass accumulation and transmission loss for different angles and different locations.

In 2016, based on the Multiple variable regression model, the power output generated from soiled photovoltaic modules, incident irradiance and the particle size composition of soil have been correlated in [90]. In the work [29], they tried to predict the power output of soiled PV panels. To do this, two models have been chosen. The first model is artificial neural networks based on multi-layer perceptron (MLP) model: the ANN has been implemented using MATLAB with a feed-forward network that uses Levenberg–Marquardt algorithm (LMA) for training. MSE and gradient have been used to test the performance of the model. The percentage of different particle sizes present in the different types of soil collected and the horizontal irradiance value has been used as inputs for training the network while the power generated has been used as output. Seventy percent of the data collected were used for training, 15% for validation and the rest 15% for testing. The second model is a multiple linear regression model and the analysis was performed using a regression add-on in Microsoft Excel. The percentage of different particle sizes present in the different types of soil collected and the horizontal irradiance value have been used as independent variables while the power generated has been used as a dependent variable. The analysis has been tested using the coefficient of determination R^2. In the same year, a hybrid data clustering and a data division method are studied in [88] and have been adapted for the prediction of PV power in dusty environments. The same authors suggested a bi-harmonic interpolation that was first applied to develop an analysis of contours to define connections between the tilt angle of a soiled module and irradiance losses. Second, those contours are compared to power contours that follow a theoretical power expression [89]. In the study [87], authors analyzed a tilted PV panel with dust depositions on its front surface based on two models: first model: optical modeling which presents a combination of three studies: [26], [118], and [123]; to find a new correlation based on different incident angles of a tilted PV module, and second model: thermal modeling based on the general form of the energy conservation equation to estimate the absorbed energy from dust and other layers like glass.

Authors of the work [75] presented in their study a model to predict soil mass accumulated from its morphology. A numerical simulation has been performed in [91] where an accurate model to predict the output power of a panel is proposed. Thanks to this model and also by assuming that dust accumulation follows a Poisson distribution with a constant arrival rate which depends on the location considered. Two relative transmittance models were developed to assess soiling: one considers the radius of a single dust particle while the other considers the whole particle size distribution. Based on the works [79,81,132], Ref. [64] tried, somehow, to overpass some limits of those works. The model developed is a mass-dependent model for low and moderate naturally deposited dust concentration has been validated using data from the literature. Two other models have been proposed in [92] and have been compared to experimental measurements: first model: multiple scattering (the two-stream approximation) [133], and second model: the Monte Carlo method [134]. The models and measurements concern a correlation between transmittance and mass density of dust under different monochromatic light and for two types of dust (absorbing and non-absorbing). A model to predict the soiling of PV modules based on the theoretical modeling of particle deposition is presented in [93].

Another model to predict soiling losses has been established in [78]. This model is based on the evaluation of the deposited dust based on the particulate matter concentration (PM10 and PM2.5). This latter is adjusted using rainfall data. The soiling ratio or transmittance loss is estimated based on the correlation carried out in [123]. The modeled data were compared with the experimental data from [124]. Continuity of the work [135] is presented in [95]. The model suggested, highlights the relationship between the reflectance and cleanness variation with the incidence angle. The model is based on the attenuation of radiation taking into consideration the reflector cleanness, the type of dust, the spatial distribution, and the homogeneity of the dust sample. In [76], a correlation between soiling losses and particulate matter (PM) dry deposition estimates from the Community Multiscale Air Quality (CMAQ) model [125] is presented. The initial and boundary conditions have been provided from [126] while the meteorological input data were provided from the Weather Research and Forecast (WRF) model [127]. The deposition measurements and transmittance losses were provided from the studies [128, 129]. In their paper, the authors present a Mie-scattering theory-based model that evaluates the soiling optical losses from the particle mass density [96].

9.4.5   Correlation between weather conditions and losses due to soiling (category AC)

The work published in [97] proposed a polynomial regression model to estimate the power generated by dirty and cleaned modules based on ambient temperature and global irradiance. This has been compared with Bayesian Neural Networks (BNNs) in [98]. This last model has three layers: an input layer, a single hidden layer, and an output layer. The first layer has two inputs (the solar irradiance and the cell temperature), while the output layer has one node which is the power produced by the PV plant. About 4,800 samples have been divided into three subdivisions: 70% for the training, 20% for testing, and 10% for validation. The effectiveness of the model has been evaluated using the mean correlation coefficient R, the RMSE, the MAE, and the MAPE. In the same paper, another comparison has been made also between the BNNs, the multi-layer perceptron network (MLPN), and the radial basis function network (RBFN). In the paper [99], the power produced by a PV panel is estimated based on solar irradiance and cell temperature [136]. The difference between the estimated power and the measured power presents the losses due to the effect of aging and dust.

Guo *et al.* tried in their paper [101] to correlate the daily change of the cleanness index and the daily ambient environmental conditions (dust concentration, wind speed, and relative humidity) has been examined using a multi-variable linear regression model. This has used these parameters as independent variables to predict the dependent variable (cleanness index). Based on this, in [15], the accuracy of the linear multivariate regression model and a semi-physical model for the prediction of daily change in the cleanness index is evaluated.

The soiling effect has been studied as well on HCPV TJ cells in [100]. While in the work [112], the reflectance has been modeled using Nonlinear AutoRegressive with exogenous inputs (NARX) (a special class of ANN) and Bayesian regularization. The inputs were humidity, temperature, rainfall, wind velocity, and direction. The accuracy of the model has been tested using the correlation coefficient (R) and MSE.

Based on PV production data, ambient temperature, plane-of-array irradiance, and daily precipitation totals, the median daily soiling rates are extracted [102]. The method is built considering and using the robust Theil-Sen estimator for slope extraction. A continuity of this is presented in [111]. In this study, the quantification of energy losses due to soiling is determined based on the interplay between soiling, precipitation, and solar resource patterns. Based on this, authors in [108] proposed a field-data-based soiling assessment algorithm that they named stochastic rate and recovery (SRR). An improvement of this is suggested by authors of [115]. This improvement is manifested by taking into consideration the seasonal change in the soiling deposition rate.

A correlation between meteorological parameters (irradiance, ambient temperature, humidity, rainfall, wind speed and wind direction) and soiling rate of solar PV glass using artificial neural networks has been proposed in [104,106]. The model developed is based on the algorithm of Levenberg–Marquardt backpropagation, and the active functions Tansig, and Purline. The accuracy of the model has been estimated based on the RMSE, the MAPE, and the coefficient of determination (R^2). In [104], a sensitivity analysis based on Garson algorithm [137] has been also applied to quantify the effect of each variable on the soiling rate while in [106] the sensitivity analysis is based on PAWN index [138].

Data recorded at 20 sites around the United States of America have been used in [105] in a linear regression between two environmental and meteorological parameters (particulate matter and precipitation) and soiling losses occurring on PV panels. This study has been extended in [109]. The authors tried to correlate between soiling ratio and parameters such as PM and rainfall based on the measurements of different stations and different databases. Two models have been used: single-variable regression and two-variable regression. The quality of the correlations has been tested using the adjusted coefficient of determination (adjR^2), the RMSE and a normalized Root-Mean-Square Error (RMSEn). The PM data have been extracted from the US Environment Protection Agency (US EPA) database [139], the Fused Air Quality Surface Using Downscaling (FQSD) [140], and from the Dalhousie model [141]. The rainfall data have been recorded from Oregon State University's PRISM database [142].

Authors of [107] present in their study a model to estimate the influence of particulate matter (estimated based on NASA GISS ModelE2 [143]) and dust on the transmittance based on their work [144]. In the same year, a correlation between environmental conditions (PM10, relative humidity (RH) and wind speed (WS)) and PV soiling in terms of the Cleanness Index (CI; a measure of PV performance losses due to soiling) has been studied in [103]. Based on this correlation, an ANN model has been developed to predict the daily cumulative PV soiling losses as a function of environmental parameters. The model has been compared to the multivariable linear regression model to evaluate the prediction accuracy. Another correlation between particulate matter (starting with PM10 concentration, then PM20 concentration, and then PM dry deposition) and the variation in the PV energy yield taking into consideration the effect of humidity, ambient temperature, wind speed and wind direction is suggested in [77]. The PM has been modeled based on the WRF-Chem (Weather Research Forecasting with Chemistry), the GOCART aerosol scheme (Georgia Institute of Technology – Goddard Global Ozone Chemistry Aerosol Radiation and Transport) and the RACM chemistry (Regional Atmospheric Chemistry Mechanism), the HTAP (Hemispheric Transport of Air Pollution emissions), and the AFWA emission scheme (US Air Force Weather Agency).

A comparison between three models to predict losses due to soiling based on weather conditions (PM$_{0.5-2.5}$, PM$_{0.5}$, vertical wind speed (VWS), temperature T, and relative humidity (RH)) has been studied in [110]. First model: multiple linear regression (MLR); second model: multiple linear regression of interaction terms (MLRIT), and third model: ANN. The models have been tested using the coefficient of determination (R^2), the adjusted coefficient of determination (adjR^2), the RMSE, and the MSE. Similarly, the authors in [114] predict PV output based on the irradiance of clean and soiled panels using two models: multiple linear regression and neural networks. The accuracy of the models has been evaluated based on the coefficient of determination (R^2) and the RMSE. Almost the same models have been used in [113] to correlate environmental parameters and efficiency losses due to the soiling of PV modules. The two based techniques are ANN and MLR. The selection of the environmental parameters to take into consideration has been done using the Burota algorithm [145]. The performance of the models has been evaluated based on the criteria: R^2_{adj}, MAPE, modeling efficiency (ME), mean bias error (MBE), and RMSE. Following the same path [116] proposed a model to predict soiling loss profiles based on weather generation algorithms. In this work, the authors tried to predict the PV panels' efficiency based on the soiling and temperature effects. The effect of soiling is considered as the number of exposure days. Five models have been applied: MLR, ANN, extreme learning machine (ELM), and the optimized versions of ANN and ELM. The performance of the models has been tested based on the coefficient of determination (R^2) the adjusted coefficient of determination (adjR^2), the accuracy, and the MSE.

9.5   Analysis and discussion of the applied models

From the summary of models applied so far, it can be deduced that many models have been used to study the soiling phenomenon. The application of these models has been done on various technologies as can be seen in Figure 9.6. From this, it can be seen that PV technology is ranked first and reaches 79% of the share of models conducted on it. This result is almost similar with a slight decrease to what has been found recently by Costa *et al.* [37], where they said that 85% of publications on soiling concern PV technology. The slight decrease is probably because the effect of soiling on other technologies, particularly CSP is getting more attention. The last part of the graph which is 'All technologies' includes models that we expect their usefulness for all technologies as they concern the simulation of soiling potential.

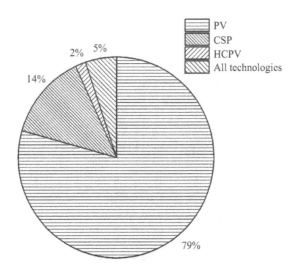

Figure 9.6   Share of models performed according to the technologies on which they have been applied

9.5.1   *Theoretical modeling/statistical modeling*

To be able to better analyze the state of the art, the articles gathered and summarized above have been divided into two groups:

- *Theoretical modeling*: this category presents all models based on physics laws.
- *Statistical modeling*: this category focuses on models based on a statistical analysis of data.

The different applied models have been gathered in Table 9.8. Among the theoretical models, there is the work proposed in [26,79]. Their research focused on a mathematical correlation experiment whom results should have improved the predictive capability of simulation models taking into account the sand dust

Table 9.8   Analysis of articles published from 1998 to 2019 (including the beginning of 2020) about modeling and prediction of the soiling phenomenon

	Based techniques	References
Theoretical modeling	Laws of solar radiation attenuation through the atmosphere (absorption, emission, reflection, transmission, Rayleigh scattering, Mie scattering...) [26,70,75,79,84,87,92,95,96] ASHRAE incidence angle modifier [75,80] Simple Model of the Atmospheric Radiative Transfer of Sunshine (SMARTS) [82,100] Grain threshold algorithm in the software package Gwyddion [83] The general form of the energy conservation equation [87] PV power generation models [91,99] Van Der Waals adhesion, particle–particle collisions, particle–surface collisions, electrodynamics, the fluid (air), and gravity forces [93,78] Ångström turbidity equation [74]	[26,70,74,75,78–80, 82,84,87,91,95, 96,99,100,146]
Statistical modeling	Polynomial regression [97,98] Linear regression [79,85,102,105,108,109] Multi-variable/multiple linear regression [29,78,90,94,101,103,109,110,113,114] Multivariate linear regression [15,116] Exponential regression [64,81] BNNs [98] ANNs (Levenberg Marquardt back propagation) [29, 103,104,106,110,113,114,116] Monte Carlo approach [83,92,111] ANNs (hybrid data clustering and data division techniques) [88] ANN (nonlinear auto-regressive with exogenous inputs NARX) [112]	[15,29,64,66,67–69, 71–74,76–79, 81,83,85,86,88,89, 90,92,94,97,98, 101–106,111–116]

(Continues)

Table 9.8   (Continued)

Based techniques	References
ANN (ELM) [116]	
Simulation of dust event and soiling potentials (dust-fall jar methods from Pollution Control Department (PCD) [86], United States Environmental Protection Agency (US EPA) database [86,109], WRF-DuMO and GIS [66–68], NASA GISS ModelE2 [107], Community Multiscale Air Quality (CMAQ) model [76], Fused Air Quality Surface Using Downscaling (FQSD) [109], Dalhousie model [109], Weather Research Forecasting with Chemistry (WRF-Chem) [77], Georgia Institute of Technology–Goddard Global Ozone Chemistry Aerosol Radiation and Transport (GOCART) aerosol scheme [77], Regional Atmospheric Chemistry Mechanism (RACM) [77], Hemispheric Transport of Air Pollution emissions (HTAP) [77], US Air Force Weather Agency (AFWA) emission scheme [77])	
DFA [73]	
Dynamic Linear Models [69,115]	
Contours analysis [89]	
Sensitivity analysis (Garson algorithm [104], PAWN algorithm [106])	
Theil-Sen estimator [102,111]	
MS [72]	
IEST-STD-CC 1246E cleanliness standard [74]	

accumulation on PV surfaces. Authors of [70] have also proposed a method based on a theoretical model to calculate the optical losses resulting from soiling. In the work [84], the weather conditions were considered to estimate by simulation the attenuation of the light reaching a glass surface. The simulation model proposed took into account the 'air-dust-glass' structure; different values were put for the thickness of the dust to assess the impact on modules' performance. In the paper of [87] an optical-thermal model of a dusty tilted panel has been performed. The authors used a six-layer module in the modeling. This later allowed us to reasonably predict the performance of the module in different weather conditions. In Ref. [80], the computational model that is given evaluates how particles deposition on surfaces leads to a decrease in performance. Recently, authors of [74] found that the transmittance is following the Ångström turbidity equation. This relevant result can be useful to characterize soiling losses in different PV panels.

Concerning statistical modeling, the soiling phenomenon has been treated in different ways. Some studies were aiming at the correlation between the power generated by a soiled PV module/soiling rate or losses and parameters such as solar irradiance, temperature, rainfall, wind speed and direction, humidity and dust characteristics (size and composition) [29,88,90,98,102,105,106,108,111].

While in other studies, the authors were interested more in the correlation between the dust accumulated and the transmission losses or the output energy of a PV module [85,86]. In other works based both on CSP and PV technologies, the main considered factor was the cleanness index/factor which was correlated to parameters like dust concentration, wind speed, and relative humidity [15,69,71,101,103]. Apart from this, other studies were interested in generating, representing, or simulating the dust events, the soiling potential and soiling risk maps which will help to have an idea about the amount of settlement that a module could receive [65–68].

The research papers gathered here show that approximately 73% of studies conducted concern statistical modeling based on the analysis of data, while just 27% were about theoretical modeling based on physics laws (Figure 9.7). Figure 9.7 presents as well the share of publications per theoretical modeling/statistical modeling per year. It shows, on the one hand, that research regarding the soiling phenomenon increases through the years. On the other hand, it confirms as well that the focus on statistical modeling has increased in the last years against the theoretical one. This last part shows a variety in terms of used models with high use of solar radiation attenuation through the atmosphere laws (Table 9.8). The analysis of this revealed that the construction of this type of modeling is based on some hypotheses that are in most cases far from reality. This explains the low percentage of use of this kind of modeling. For instance, in the study [26] , which was the basis of many conducted studies, they assumed that:

- The particles are spherical and have the same size.
- The particles distribution is uniform.
- The second layer of particles is deposited or formed just after the full formation of the first layer.

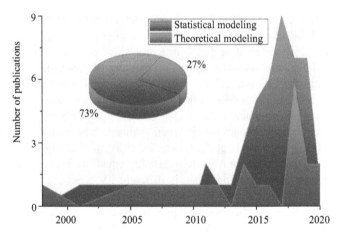

Figure 9.7   Repartition of the different studies conducted on the subject of modeling and prediction of soiling effect

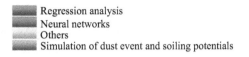

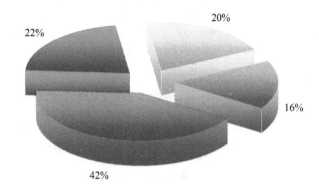

Figure 9.8 Repartition of the different studies performed on statistical modeling of soiling effect topic

Those assumptions might not be true in all cases in real operating conditions, hence results obtained in [26].

For the statistical analysis, the repartition of the applied models is presented in Figure 9.8. From this, it can be seen that most of the applied models were based on the regression analysis with a percentage of 42%. However, all comparisons conducted between neural networks and regression analysis have shown that neural networks were more accurate and reached better results as can be seen in the studies [29,98,103,110].

In some works such [15,101,103,106,110], the authors were interested to predict the soiling effect based on weather conditions. This can be very valuable work if all the parameters affecting the soiling phenomenon are taken into consideration. This can help to forecast the losses due to soiling everywhere a station with the necessary sensors with the same inputs is available. In other works [102,105,108,109,111], the authors were more interested to predict soiling losses based on PV plant production. Though they made field measurements, the models and methods that they proposed can be very useful everywhere a PV plant is available. The simulation of dust events and soiling potential also have a non-negligible share in the publications about soiling as can be seen in Figure 9.8. Furthermore, it has been reported in [47,105] that the dust concentration in the air seems to be a useful parameter for predicting soiling rate changes between different locations. From this and from what we have found, the simulation of dust events and soiling potential is expected to have more attention in future studies. Finding the model linking these potentials to the losses due to soiling will maybe end the need to do on-site measurements.

9.5.2   *Modeling-based soiling influencing factors*

From the summary conducted in [52], the factors that affect the soiling phenomenon can be categorized and summarized as follows: module characteristics (nature of the surface, technology), surrounding environment (land cover, soil texture, bird dropping), climatic conditions (temperature, relative humidity, rainfall, wind speed, wind direction, dew formation, airborne dust), exposure conditions (tilt angle, orientation, the height of installation, tracking, season, exposure time), and soil properties (chemical composition, particle size, particle shape). Based on this summary and based on Table 9.2, from which we selected the papers of 'category AB' and 'category AC', we determined the factors that have been used widely to predict soiling and those that are rarely taken into consideration and present them in Figure 9.9. From this, we can clearly see that the factors that are mostly used are the ones that belong to the category of climatic conditions especially temperature, relative humidity, rainfall, wind speed, and airborne dust. This can be explained by the fact that with almost every solar installation, a weather station will be available which makes it easy to consider the weather conditions. Exposure conditions, module characteristics, and the surrounding environment are mostly fixed for every model due to the difficulty to obtain data for all conditions. However, this represents the lack in almost all the models and many papers; this has been considered as one of the causes to obtain low performance of their models.

9.5.3   *Evaluation metrics*

Varieties of criteria have been applied to evaluate the accuracy of the models developed. The choice between one criterion and another depends on the nature of

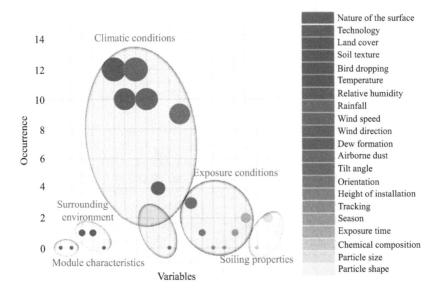

Figure 9.9   Factors influencing soiling and their occurrence in the models investigated

Table 9.9   Evaluation metrics applied for accuracy and performance assessment

Criterion	Symbol	Equation	
The coefficient of determination	R^2	$$R^2 = 1 - \frac{\sum_{i=1}^{N}(\hat{x}_i - x_i)^2}{\sum_{i=1}^{N}(x_i - \bar{x}_i)^2}$$	(1)
Mean absolute error	MAE	$$MAE = \frac{1}{N}\sum_{i=1}^{N}|\hat{x}_i - x_i|$$	(2)
Root mean square error	RMSE	$$RMSE = \left[\frac{1}{N}\sum_{i=1}^{N}(\hat{x}_i - x_i)^2\right]^{1/2}$$	(3)
Mean absolute percentage error	MAPE	$$MAPE = 100 \times \frac{1}{N}\sum_{i=1}^{N}\left|\frac{\hat{x}_i - x_i}{\hat{x}_i}\right|$$	(4)
Mean square error	MSE	$$MSE = \frac{1}{N}\sum_{i=1}^{N}(\hat{x}_i - x_i)^2$$	(5)
Akaike Information Criterion	AIC	$AIC = 2k - 2\ln(\hat{L})$ k: the number of model parameters $\hat{L}$: the maximum value of likelihood function of the model	(6)
Bayesian Information Criterion	BIC	$BIC = 2\ln(\hat{L}) + k.\ln(n)$ n: the number of observations k: the number of regressors $\hat{L}$: the maximum value of the likelihood function of the model	(7)
Modeling efficiency	ME	$$ME = 1 - \left(\frac{\sum_{i=1}^{N}(\hat{x}_i - x_i)^2}{\sum_{i=1}^{N}(x_i - \bar{x}_i)^2}\right)$$	(8)
Mean bias error	MBE	$$MBE = \frac{1}{N}\sum_{i=1}^{N}(\hat{x}_i - x_i)$$	(9)
The receiver operation characteristic analysis	ROC	The ROC is a method for visualizing, organizing, and selecting classifiers based on their performance [147]	

the models. These criteria are gathered in Table 9.9 along with their corresponding equations.

9.6   Conclusions and recommendations

Soiling is acknowledged as the third parameter determining the productivity of solar energy systems after irradiance and temperature. Its considerable effect has made it a hot topic during the last few years. Mastering this phenomenon or predicting it will significantly help to save many investments from losses, especially since it has a big impact in areas known to be the most investment-attractive regions of solar energy exploitation.

In the present chapter, a review of the different studies conducted to model the soiling phenomenon has been presented. A summary of the most important information about each model has been described. The papers summarized have been analyzed and gathered into groups to better present the trend that research on this phenomenon follows.

The review shows a variety of applied models and methods. Each model has its strengths and weaknesses. According to the objective desired from the modeling, five groups were identified. These, describe the different parts of the cycle of soiling, from weather conditions to end up in losses. The parts are complementary to each other and they will be useful for researchers to identify on which they will focus.

The results obtained from the analysis of the applied models have shown that statistical modeling has been applied more than theoretical modeling. The analysis of the latter has revealed that studies conducted have taken into consideration hypotheses that were, in most cases, far from reality. The results have shown as well that in the statistical modeling itself, although regression analysis has been used more frequently, neural network approaches performed better in the prediction of the soiling phenomenon.

From the analysis conducted, it has been also found that two kinds of modeling have taken part in the works performed. The first one concerns the prediction of soiling effects based on factors affecting it and the second one is concluding soiling losses from the data of existing solar plants. These models can be very valuable respectively when a station with the necessary sensors is available and only when a solar plant exists.

However, to approach the complex behavior of the soiling phenomenon and to predict it over time accurately, more data for long periods are still needed. Furthermore, all parameters affecting soiling should be taken into consideration. Additionally, it is highly recommended that the same model should be tested in different climates and on different technologies to validate its usefulness. Furthermore, as it has been proved that the dust concentration ranks first in predicting soiling losses, the use of existing databases of soiling potential adjusted using weather conditions from existing databases or meteorological stations seems to be a low-cost and valuable method to predict soiling losses. This will be achieved, however, only after finding the accurate relationships between the parameters and validating them on the site. Finally, it is highly recommended as well that instead of suggesting new models, it will be wiser to test the published models that have proved good results in different sites and technologies to overcome their limits and to achieve an accurate and well-tested model.

In conclusion, the review presented here summarizes almost all models applied in the framework of studies of soiling effects on solar systems from 1998 until 2019 (including the beginning of 2020). This review will help researchers working on this topic gain an overview of the models applied. It will also help to better choose the right direction in modeling and predicting the soiling phenomenon. Understanding and mastering this phenomenon will not only prevent investments from losses but also save the precious natural resource 'water' by determining the mitigation techniques wisely.

Acknowledgments

This work is supported by the Moroccan Research Institute in Solar Energy and New Energies (IRESEN) in the framework of SOLEIL Inno-PV project and by the National Center for Scientific and Technical Research (CNRST) in the framework of the Research Excellence Scholarship Program and Priority Research Project PPR1 Nr. 14/216. It is also supported by the Department of Science & Technology (DST), Ministry of Science and Technology, Government of India with Federation of Indian Chambers of Commerce and Industry (FICCI) in the framework of the Research Training Fellowship-Developing Countries Scientist (RTFDCS) 2019-2020.

The authors thank Dr. M. Azeem Arshad from Faculty of Sciences, Mohammed V University of Rabat-Morocco, Mr. S. Pulipaka from Birla Institute of Technology and Science of Pilani-India and Dr. Nikhil P. G. from the National Institute of Solar Energy NISE of India for their considerable improvement support of this work.

References

[1] *Renewable Energy Policy Network for the 21st Century. Renewables2018 global status report*, 2018.

[2] "Profiling the top five largest solar power plants in the world." Available: https://nsenergybusiness.com/features/largest-solar-power-plants/. [Accessed: 06-Aug-2022].

[3] "Largest solar PV power plants worldwide 2021 | Statista." Available: https://statista.com/statistics/217265/largest-solar-pv-power-plants-in-opera-tion-worldwide/. [Accessed: 06-Aug-2022].

[4] "Glint Solar | Blog." Available: https://glintsolar.ai/blog/glimpse-at-top-5-largest-parks-in-the-world. [Accessed: 06-Aug-2022].

[5] M. Pagliaro and F. Meneguzzo, "Digital management of solar energy en route to energy self-sufficiency," *Glob. Challenges*, vol. 1800105, p. 1800105, 2019.

[6] "What is Digital Technologies | IGI Global." Available: https://igi-global. com/dictionary/the-uptake-and-use-of-digital-technologies-and-profes-sional-development/51457. [Accessed: 06-Aug-2022].

[7] S. Adhya, D. Saha, A. Das, J. Jana, and H. Saha, "An IoT based smart solar photovoltaic remote monitoring and control unit," in *2016 2nd Int. Conf. Control. Instrumentation, Energy Commun. CIEC 2016*, January, pp. 432–436, 2016.

[8] Ž. Kavaliauskas, I. Šajev, G. Blažiūnas, G. Gecevičius, V. Čapas, and D. Adomaitis, "Electronic system for the remote monitoring of solar power plant parameters and environmental conditions," *Electron.*, vol. 11, no. 9, p. 1431, 2022.

[9] G. V. Karbhari and P. Nema, "Digital control system for solar power plant using iot," *Int. J. Recent Technol. Eng.*, vol. 8, no. 2, pp. 3394–3396, 2019.

[10] M. Aghaei, F. Grimaccia, C. A. Gonano, and S. Leva, "Innovative automated control system for PV fields inspection and remote control," *IEEE Trans. Ind. Electron.*, vol. 62, no. 11, pp. 7287–7296, 2015.

[11] F. Bandou, A. Hadj Arab, M. S. Belkaid, P. O. Logerais, O. Riou, and A. Charki, "Evaluation performance of photovoltaic modules after a long time operation in Saharan environment," *Int. J. Hydrogen Energy*, vol. 40, no. 39, pp. 13839–13848, 2015.

[12] T. Sarver, A. Al-Qaraghuli, and L. L. Kazmerski, "A comprehensive review of the impact of dust on the use of solar energy: history, investigations, results, literature, and mitigation approaches," *Renew. Sustain. Energy Rev.*, vol. 22, pp. 698–733, 2013.

[13] S. A. M. Said and H. M. Walwil, "Fundamental studies on dust fouling effects on PV module performance," *Sol. Energy*, vol. 107, pp. 328–337, 2014.

[14] W. Anana, F. Chaouki, B. Laarabi, *et al.*, "Soiling impact on energy generation of high concentration photovoltaic power plant in Morocco," in *2016 International Renewable and Sustainable Energy Conference (IRSEC)*, 2016, pp. 1–5.

[15] B. Guo and B. Figgis, "Output degradation based on environmental variables in Doha," in *Proceedings of the ASME 2016 10th International Conference on Energy Sustainability*, 2017, pp. 1–8.

[16] E. Zell, S. Gasim, S. Wilcox, *et al.*, "Assessment of solar radiation resources in Saudi Arabia," *Sol. Energy*, vol. 119, pp. 422–438, 2015.

[17] H. G. Silva, L. Fialho, F. M. Lopes, R. Conceiç, and M. Collares-pereira, "PV system design with the effect of soiling on the optimum tilt angle," *Renew. Energy*, vol. 133, pp. 787–796, 2019.

[18] K. Dastoori, G. Al-Shabaan, M. Kolhe, D. Thompson, and B. Makin, "Impact of accumulated dust particles' charge on the photovoltaic module performance," *J. Electrostat.*, vol. 79, pp. 20–24, 2016.

[19] G. Singh, D. Saini, N. Yadav, R. Sarma, L. Chandra, and R. Shekhar, "Dust deposition mechanism and cleaning strategy for open volumetric air receiver based solar tower sub-systems," *Energy Procedia*, vol. 69, pp. 2081–2089, 2015.

[20] K. K. Ilse, J. Rabanal, L. Schönleber, *et al.*, "Comparing indoor and outdoor soiling experiments for different glass coatings and microstructural analysis of particle caking processes," *IEEE J. Photovolt.*, vol. 8, pp. 1–7, 2017.

[21] K. Ilse, M. Werner, V. Naumann, B. W. Figgis, C. Hagendorf, and J. Bagdahn, "Microstructural analysis of the cementation process during soiling on glass surfaces in arid and semi-arid climates," *Phys. Status Solidi – Rapid Res. Lett.*, vol. 10, no. 7, pp. 525–529, 2016.

[22] M. K. Mazumder, M. N. Horenstein, N. R. Joglekar, *et al.*, "Mitigation of dust impact on solar collectors by water-free cleaning with transparent electrodynamic films: progress and challenges," *IEEE J. Photovoltaics*, pp. 2052–2057, 2017, doi: 10.1109/PVSC.2016.7749990.

[23] M. Piliougine, C. Cañete, R. Moreno, *et al.*, "Comparative analysis of energy produced by photovoltaic modules with anti-soiling coated surface in arid climates," *Appl. Energy*, vol. 112, pp. 626–634, 2013.

[24] D. Deb and N. L. Brahmbhatt, "Review of yield increase of solar panels through soiling prevention, and a proposed water-free automated cleaning solution," *Renew. Sustain. Energy Rev.*, vol. 82, pp. 3306–3313, 2018.

[25] R. Mudike, I. Barbate, A. K. Tripathi, Y. B. K. Reddy, and C. Banerjee, "Soiling effect on the performance of solar radiometers in composite climatic zone of India," in *Proc. – 2019 Int. Conf. Electr. Electron. Comput. Eng. UPCON 2019*, pp. 1–6, 2019.

[26] A. Y. Al-Hasan, "A new correlation for direct beam solar radiation received by photovoltaic panel with sand dust accumulated on its surface," *Sol. Energy*, vol. 63, no. 5, pp. 323–333, 1998.

[27] B. Laarabi, M. Rhourri, D. Dahlioui, and A. Barhdadi, "Experimental simulation of the effect of soiling on a solar PV glass," in *2018 6th International Renewable and Sustainable Energy Conference (IRSEC)*, 2018, pp. 1–3.

[28] P. D. Burton and B. H. King, "Artificial soiling of photovoltaic module surfaces using traceable soil components," in *Conf. Rec. IEEE Photovolt. Spec. Conf.*, pp. 1542–1545, 2013.

[29] S. Pulipaka, F. Mani, and R. Kumar, "Modeling of soiled PV module with neural networks and regression using particle size composition," *Sol. Energy*, vol. 123, pp. 116–126, 2016.

[30] "Scientific modeling | science | Britannica." Available: https:/britannica. com/science/scientific-modeling. [Accessed: 04-Jun-2021].

[31] K. Ilse, L. Micheli, B. W. Figgis, *et al.*, "Techno-economic assessment of soiling losses and mitigation strategies for solar power generation," *Joule*, vol. 3, no. 10, pp. 2303–2321, 2019.

[32] N. Mahowald, S. Albani, J. F. Kok, *et al.*, "The size distribution of desert dust aerosols and its impact on the Earth system," *Aeolian Res.*, vol. 15, pp. 53–71, 2014.

[33] M. Mani and R. Pillai, "Impact of dust on solar photovoltaic (PV) performance: research status, challenges and recommendations," *Renew. Sustain. Energy Rev.*, vol. 14, no. 9, pp. 3124–3131, 2010.

[34] Z. Ahmed, H. a Kazem, and K. Sopian, "Effect of dust on photovoltaic performance: review and research status," in *Latest Trends Renew. Energy Environ. Informatics*, pp. 193–199, 2013.

[35] S. Ghazi, A. Sayigh, and K. Ip, "Dust effect on flat surfaces – a review paper," *Renew. Sustain. Energy Rev.*, vol. 33, pp. 742–751, 2014.

[36] A. Butuza, "Reference selection procedure for studying the soiling effect on pv performance," *Appl. Math. Mech.*, vol. 57, no. I, pp. 63–65, 2014.

[37] S. C. S. Costa, A. Sonia, A. C. Diniz, and L. L. Kazmerski, "Solar energy dust and soiling R & D progress: literature review update for 2016," *Renew. Sustain. Energy Rev.*, vol. 82, pp. 2504–2536, 2018.

[38] W. J. Jamil, H. Abdul Rahman, S. Shaari, and Z. Salam, "Performance degradation of photovoltaic power system: review on mitigation methods," *Renew. Sustain. Energy Rev.*, vol. 67, pp. 876–891, 2017.

[39]   A. Syafiq, A. K. Pandey, N. N. Adzman, and N. A. Rahim, "Advances in approaches and methods for self-cleaning of solar photovoltaic panels," *Sol. Energy*, vol. 162, pp. 597–619, 2018.

[40]   D. Deb and N. L. Brahmbhatt, "Review of yield increase of solar panels through soiling prevention, and a proposed water-free automated cleaning solution," *Renew. Sustain. Energy Rev.*, vol. 82, pp. 3306–3313, 2018.

[41]   P. P. Brito, A. S. A. C. Diniz, and L. L. Kazmerski, "Materials design and discovery: potential for application to soiling mitigation in photovoltaic systems," *Sol. Energy*, vol. 183, pp. 791–804, 2019.

[42]   C. S. Saiz, J. P. Martínez, and N. M. Chivelet, "Influence of pollen on solar photovoltaic energy: literature review and experimental testing with pollen," *Appl. Sci.*, vol. 10, no. 14, p. 4733, 2020.

[43]   Á. Fernández-Solas, L. Micheli, F. Almonacid, and E. F. Fernández, "Optical degradation impact on the spectral performance of photovoltaic technology," *Renew. Sustain. Energy Rev.*, vol. 141, p. 110782, 2021.

[44]   B. Figgis, A. Ennaoui, S. Ahzi, and Y. Rémond, "Review of PV soiling particle mechanics in desert environments," *Renew. Sustain. Energy Rev.*, vol. 76, pp. 872–881, 2017.

[45]   G. Picotti, P. Borghesani, M. E. Cholette, and G. Manzolini, "Soiling of solar collectors – modelling approaches for airborne dust and its interactions with surfaces," *Renew. Sustain. Energy Rev.*, vol. 81, pp. 2343–2357, 2018.

[46]   S. A. M. Said, G. Hassan, H. M. Walwil, and N. Al-Aqeeli, "The effect of environmental factors and dust accumulation on photovoltaic modules and dust-accumulation mitigation strategies," *Renew. Sustain. Energy Rev.*, vol. 82, pp. 743–760, 2018.

[47]   K. K. Ilse, B. W. Figgis, V. Naumann, C. Hagendorf, S. Csp, and H. Saale, "Fundamentals of soiling processes on photovoltaic modules," *Renew. Sustain. Energy Rev.*, vol. 98, pp. 239–254, 2018.

[48]   J. G. Bessa, L. Micheli, F. Almonacid, and E. F. Fernández, "Monitoring photovoltaic soiling: assessment, challenges, and perspectives of current and potential strategies," *iScience*, vol. 24, no. 3, p. 102165, 2021.

[49]   B. Hammad, M. Al-Abed, A. Al-Ghandoor, A. Al-Sardeah, and A. Al-Bashir, "Modeling and analysis of dust and temperature effects on photovoltaic systems' performance and optimal cleaning frequency: Jordan case study," *Renew. Sustain. Energy Rev.*, vol. 82, pp. 2218–2234, 2018.

[50]   A. A. Kazem, M. T. Chaichan, and H. A. Kazem, "Dust effect on photovoltaic utilization in Iraq: review article," *Renew. Sustain. Energy Rev.*, vol. 37, pp. 734–749, 2014.

[51]   Y. N. Chanchangi, A. Ghosh, S. Sundaram, and T. K. Mallick, "Dust and PV performance in Nigeria: a review," *Renew. Sustain. Energy Rev.*, vol. 121, p. 109704, 2020.

[52]   B. Laarabi, Y. El Baqqal, N. Rajasekar, and A. Barhdadi, "Updated review on soiling of solar photovoltaicsystems Morocco and India contributions," *J. Clean. Prod.*, vol. 311, p. 127608, 2021.

[53] A. Sayyah, M. N. Horenstein, and M. K. Mazumder, "Energy yield loss caused by dust deposition on photovoltaic panels," *Sol. Energy*, vol. 107, pp. 576–604, 2014.

[54] Z. A. Darwish, H. A. Kazem, K. Sopian, M. A. Al-Goul, and H. Alawadhi, "Effect of dust pollutant type on photovoltaic performance," *Renew. Sustain. Energy Rev.*, vol. 41, pp. 735–744, 2015.

[55] S. C. S. Costa, A. S. A. C. Diniz, and L. L. Kazmerski, "Dust and soiling issues and impacts relating to solar energy systems: literature review update for 2012–2015," *Renew. Sustain. Energy Rev.*, vol. 63, pp. 33–61, 2016.

[56] M. R. Maghami, H. Hizam, C. Gomes, M. A. Radzi, M. I. Rezadad, and S. Hajighorbani, "Power loss due to soiling on solar panel: a review," *Renew. Sustain. Energy Rev.*, vol. 59, pp. 1307–1316, 2016.

[57] F. M. Zaihidee, S. Mekhilef, M. Seyedmahmoudian, and B. Horan, "Dust as an unalterable deteriorative factor affecting PV panel's efficiency: why and how," *Renew. Sustain. Energy Rev.*, vol. 65, pp. 1267–1278, 2016.

[58] B. Figgis, A. Ennaoui, S. Ahzi, and Y. Remond, "Review of PV soiling measurement methods," in *Proc. 2016 Int. Renew. Sustain. Energy Conf. IRSEC 2016*, pp. 176–180, 2017.

[59] A. Shaju and R. Chacko, "Soiling of photovoltaic modules – review," *IOP Conf. Ser. Mater. Sci. Eng.*, vol. 396, no. 1, 2018.

[60] S. Mondal, A. K. Mondal , A. Sharma, *et al.*, "An overview of cleaning and prevention processes for enhancing efficiency of solar photovoltaic panels," *Curr. Sci.*, vol. 115, no. 6, pp. 1065–1077, 2018.

[61] S. Ghosh, V. K. Yadav, and V. Mukherjee, "Impact of environmental factors on photovoltaic performance and their mitigation strategies – a holistic review," *Renew. Energy Focus*, vol. 28, no. 00, pp. 153–172, 2019.

[62] R. Chawla, P. Singhal, and A. K. Garg, "Impact of dust for solar PV in Indian scenario: experimental analysis," in *Driving the Development, Management, and Sustainability of Cognitive Cities, no.* January, 2019, pp. 111–138.

[63] H. A. Kazem, M. T. Chaichan, A. H. A. Al-Waeli, and K. Sopian, "A review of dust accumulation and cleaning methods for solar photovoltaic systems," *J. Clean. Prod.*, vol. 276, p. 123187, 2020.

[64] M. Jaszczur, Q. Hassan, K. Styszko, *et al.*, "The field experiments and model of the natural dust deposition effects on photovoltaic module efficiency," *Environ. Sci. Pollut. Res.*, vol. 23, pp. S1199–S1210, 2018.

[65] Y. Charabi and A. Gastli, "Spatio-temporal assessment of dust risk maps for solar energy systems using proxy data," *Renew. Energy*, vol. 44, pp. 23–31, 2012.

[66] M. Hamidi, M. R. Kavianpour, and Y. Shao, "Numerical simulation of dust events in the Middle East," *Aeolian Res.*, vol. 13, pp. 59–70, 2014.

[67] J. Herrmann, K. Slamova, R. Glaser, and M. Köhl, "Modeling the soiling of glazing materials in arid regions with geographic information systems (GIS)," in *Energy Procedia*, vol. 48, pp. 715–720, 2014.

[68]   J. Herrmann, E. Klimm, M. Koehl, *et al.*, "Dust mitigation on PV modules in western Saudi Arabia," in *2015 IEEE 42nd Photovolt. Spec. Conf. PVSC 2015*, pp. 2–6, 2015.

[69]   S. Bouaddi, A. Ihlal, and A. Fernández-García, "Soiled CSP solar reflectors modeling using dynamic linear models," *Sol. Energy*, vol. 122, pp. 847–863, 2015.

[70]   M. García, L. Marroyo, E. Lorenzo, and M. Pérez, "Soiling and other optical losses in solar-tracking PV plants in navarra," *Prog. Photovoltaics Res. Appl.*, vol. 19, no. 2, pp. 211–217, 2010.

[71]   S. Bouaddi and A. Ihlal, "A study of the soiling patterns of CSP mirrors," in *2015 3rd International Renewable and Sustainable Energy Conference (IRSEC)*, 2015, pp. 1–4.

[72]   S. Bouaddi, A. Fernández-García, A. Ihlal, R. Ait El Cadi, and L. Álvarez-Rodrigo, "Modeling and simulation of the soiling dynamics of frequently cleaned reflectors in CSP plants," *Sol. Energy*, vol. 166, pp. 422–431, 2018.

[73]   S. Bouaddi, A. Ihlal, and A. Fernández-García, "Comparative analysis of soiling of CSP mirror materials in arid zones," *Renew. Energy*, vol. 101, pp. 437–449, 2017.

[74]   G. P. Smestad, T. A. Germer, H. Alrashidi, *et al.*, "Modelling photovoltaic soiling losses through optical characterization," *Sci. Rep.*, vol. 10, pp. 1–13, 2020.

[75]   N. Barth, B. Figgis, A. A. Abdallah, S. P. Aly, and S. Ahzi, "Modeling of the influence of dust soiling on photovoltaic panels for desert applications the example of the solar test facility at Doha, Qatar," in *Proc. 2017 Int. Renew. Sustain. Energy Conf. IRSEC 2017*, no. 1, pp. 1–6, 2018.

[76]   L. Zhou, D. B. Schwede, K. Wyat Appe, *et al.*, "The impact of air pollutant deposition on solar energy system efficiency: an approach to estimate PV soiling effects with the Community Multiscale Air Quality (CMAQ) model," *Sci. Total Environ.*, vol. 651, pp. 456–465, 2019.

[77]   C. Fountoukis, B. Figgis, L. Ackermann, and M. A. Ayoub, "Effects of atmospheric dust deposition on solar PV energy production in a desert environment," *Sol. Energy*, vol. 164, pp. 94–100, 2018.

[78]   M. Coello and L. Boyle, "Simple model for predicting time series soiling of photovoltaic panels," *IEEE J. Photovolt.*, vol. pp, pp. 1–6, 2019.

[79]   A. Y. Al-hasan and A. A. Ghoneim, "A new correlation between photovoltaic panel's efficiency and amount of sand dust accumulated on their surface," *Int. J. Sustain. Energy*, vol. 24, no. 4, pp. 187–197, 2005.

[80]   J. Zang and Y. Wang, "Analysis of computation model of particle deposition on transmittance for photovoltaic panels," *Energy Procedia*, vol. 12, pp. 554–559, 2011.

[81]   J. K. Kaldellis and M. Kapsali, "Simulating the dust effect on the energy performance of photovoltaic generators based on experimental measurements," *Energy*, vol. 36, no. 8, pp. 5154–5161, 2011.

[82]   H. Qasem, T. R. Betts, H. Müllejans, H. Albusairi, and R. Gottschalg, "Dust-induced shading on photovoltaic modules," *Prog. Photovolt. Res. Appl.*, vol. 22, pp. 218–226, 2012.

[83]   N. S. Beattie, R. S. Moir, C. Chacko, G. Buffoni, S. H. Roberts, and N. M. Pearsall, "Understanding the effects of sand and dust accumulation on photovoltaic modules," *Renew. Energy*, vol. 48, pp. 448–452, 2012.

[84]   M. Mesrouk and A. H. Arab, "Effet de la poussière sur le rendement des modules photovoltaïques," in *3rd Int. Semin. New Renew. Energies*, pp. 1–6, 2014.

[85]   L. Boyle, H. Flinchpaugh, and M. P. Hannigan, "Natural soiling of photovoltaic cover plates and the impact on transmission," *Renew. Energy*, vol. 77, no. 1, pp. 166–173, 2015.

[86]   N. Ketjoy and M. Konyu, "Study of dust effect on photovoltaic module for photovoltaic power plant," *Energy Procedia*, vol. 52, pp. 431–437, 2014.

[87]   T. Zarei and M. Abdolzadeh, "Optical and thermal modeling of a tilted photovoltaic module with sand particles settled on its front surface," *Energy*, vol. 95, pp. 51–66, 2016.

[88]   S. Pulipaka and R. Kumar, "Power prediction of soiled PV module with neural networks using hybrid data clustering and division techniques," *Sol. Energy*, vol. 133, pp. 485–500, 2016.

[89]   S. Pulipaka and R. Kumar, "Analysis of irradiance losses on a soiled photovoltaic panel using contours," *Energy Convers. Manag.*, vol. 115, pp. 327–336, 2016.

[90]   F. Mani, S. Pulipaka, and R. Kumar, "Characterization of power losses of a soiled PV panel in Shekhawati region of India," *Sol. Energy*, vol. 131, pp. 96–106, 2016.

[91]   A. G. Haddad, S. Member, and A. P. V. C. Model, "Modeling and analysis of PV soiling and its effect on the transmittance of solar radiation," in *2018 Adv. Sci. Eng. Technol. Int. Conf.*, pp. 1–5, 2018.

[92]   P. G. Piedra, L. R. Llanza, and H. Moosmüller, "Optical losses of photovoltaic modules due to mineral dust deposition: experimental measurements and theoretical modeling," *Sol. Energy*, vol. 164, pp. 160–173, 2018.

[93]   S. You, Y. J. Lim, Y. Dai, and C. H. Wang, "On the temporal modelling of solar photovoltaic soiling: energy and economic impacts in seven cities," *Appl. Energy*, vol. 228, no. June, pp. 1136–1146, 2018.

[94]   A. A. Hegazy, "Effect of dust accumulation on solar transmittance through glass covers of plate-type collectors," *Renew. Energy*, vol. 22, no. 4, pp. 525–540, 2001.

[95]   A. Heimsath and P. Nitz, "The effect of soiling on the reflectance of solar reflector materials – model for prediction of incidence angle dependent reflectance and attenuation due to dust deposition," *Sol. Energy Mater. Sol. Cells*, vol. 195, pp. 258–268, 2019.

[96]   P. Bellmann, F. Wolfertstetter, R. Conceição, and H. G. Silva, "Comparative modeling of optical soiling losses for CSP and PV energy systems," *Sol. Energy*, vol. 197, pp. 229–237, 2020.

[97]   D. D. P. A. Massi Pavan and A. Mellit, "The effect of soiling on energy production for large-scale photovoltaic plants," *Sol. Energy*, vol. 85, no. 5, pp. 1128–1136, 2011.

[98]   A. Massi Pavan, A. Mellit, D. De Pieri, and S. A. Kalogirou, "A comparison between BNN and regression polynomial methods for the evaluation of the

effect of soiling in large scale photovoltaic plants," *Appl. Energy*, vol. 108, pp. 392–401, 2013.

[99]   L. Cristaldi, M. Faifer, M. Rossi, *et al.*, "Simplified method for evaluating the effects of dust and aging on photovoltaic panels," *Meas. J. Int. Meas. Confed.*, vol. 54, pp. 207–214, 2014.

[100] P. D. Burton, B. H. King, and D. Riley, "Predicting the spectral effects of soils on high concentrating photovoltaic systems," *Sol. Energy*, vol. 112, pp. 469–474, 2015.

[101] B. Guo, W. Javed, B. W. Figgis, and T. Mirza, "Effect of dust and weather conditions on photovoltaic performance in Doha, Qatar," in *1st Workshop on Smart Grid and Renewable Energy, SGRE*, 2015.

[102] M. G. Deceglie, M. Muller, S. Kurtz, and Z. Defreitas, "A scalable method for extracting soiling rates from PV production data preprint," *Presented at the 43rd IEEE Photovoltaic Specialists Conference*, Portland, OR, June 5–10, 2016.

[103] W. Javed, B. Guo, and B. Figgis, "Modeling of photovoltaic soiling loss as a function of environmental variables," *Sol. Energy,* vol. 157, pp. 397–407, 2017.

[104] B. Laarabi, O. M. Tzuc, D. Dahlioui, *et al.*, "New correlation of PV modules soiling and outdoor conditions using artificial neural networks," in *2017 International Renewable and Sustainable Energy Conference (IRSEC)*, 2017, pp. 1–5.

[105] L. Micheli and M. Muller, "An investigation of the key parameters for predicting PV soiling losses," *Prog. Photovoltaics Res. Appl.*, vol. 25, no. 4, pp. 291–307, 2017.

[106] B. Laarabi, O. May Tzuc, D. Dahlioui, A. Bassam, M. Flota-Bañuelos, and A. Barhdadi, "Artificial neural network modeling and sensitivity analysis for soiling effects on photovoltaic panels in Morocco," *Superlattices Microstruct.*, vol. 127, no. 2019, pp. 139–150, 2017.

[107] M. H. Bergin, C. Ghoroi, D. Dixit, J. J. Schauer, and D. T. Shindell, "Large reductions in solar energy production due to dust and particulate air pollution," *Environ. Sci. Technol. Lett.*, vol. 4, no. 8, pp. 339–344, 2017.

[108] M. G. Deceglie, L. Micheli, and M. Muller, "Quantifying soiling loss directly from PV yield," *IEEE J. Photovoltaics*, vol. 8, no. 2, pp. 547–551, 2018.

[109] L. Micheli, M. G. Deceglie, and M. Muller, "Predicting photovoltaic soiling losses using environmental parameters: an update," *Prog. Photovoltaics Res. Appl.*, August, pp. 1–10, 2018.

[110] R. Conceição, H. G. Silva, and M. Collares-Pereira, "CSP mirror soiling characterization and modeling," *Sol. Energy Mater. Sol. Cells*, vol. 185, pp. 233–239, 2018.

[111] M. G. Deceglie, L. Micheli, and M. Muller, "Quantifying year-to-year variations in solar panel soiling from PV energy-production data," in *2017 IEEE 44th Photovolt. Spec. Conf. PVSC 2017*, pp. 1–4, 2017.

[112]   S. Bouaddi, I. Ahmed, and O. A. Mensour, "Modeling and prediction of reflectance loss in CSP plants using a non linear autoregressive model with exogenous inputs (NARX)," in *2016 International Renewable and Sustainable Energy Conference (IRSEC)*, 2016, pp. 706–709.

[113]   K. Chiteka, R. Arora, and S. N. Sridhara, "A method to predict solar photovoltaic soiling using models," *Energy Syst.*, vol. 52, p. 102126, 2019.

[114]   S. Shapsough, R. Dhaouadi, I. Zualkernan, S. Shapsough, R. Dhaouadi, and I. Zualkernan, "Using linear regression and back propagation neural networks to using linear regression and back propagation neural networks to predict performance of soiled PV modules predict performance of soiled PV modules," *Procedia Comput. Sci.*, vol. 155, no. 2018, pp. 463–470, 2019.

[115]   L. Micheli, E. F. Fernández, M. Muller, and F. Almonacid, "Extracting and generating PV soiling profiles for analysis, forecasting, and cleaning optimization," *IEEE J. Photovoltaics*, vol. PP, pp. 1–9, 2019.

[116]   W. Al-Kouz, S. Al-Dahidi, B. Hammad, and M. Al-Abed, "Modeling and analysis framework for investigating the impact of dust and temperature on PV systems' performance and optimum cleaning frequency," *Appl. Sci.*, vol. 9, no. 7, p. 1397, 2019.

[117]   J. Y. Kang, S. C. Yoon, Y. Shao, and S. W. Kim, "Comparison of vertical dust flux by implementing three dust emission schemes in WRF/Chem," *J. Geophys. Res. Atmos.*, vol. 116, no. 9, pp. 1–18, 2011.

[118]   N. Martín and J. M. Ruiz, "A new model for PV modules angular losses under field conditions," *Int. J. Sol. Energy*, vol. 22, no. 1, pp. 19–31, 2002.

[119]   P. Campagnoli, S. Petrone, and G. Petris, *Dynamic Linear Models with R*. New York, *NY*: Springer, 2009.

[120]   J. Honaker, G. King, and M. Blackwell, "AMELIA II: a program for missing data," *J. Stat. Softw.*, vol. 45, no. 7, pp. 1–54, 2011.

[121]   C. M. Jarque and A. K. Beda, "Efficient tests for normality, homoscedasticity and serial independence of regression residuals," vol. 6, pp. 255–259, 1980.

[122]   G. M. Ljung and G. E. P. Box, "On a measure of Lack of fit in time series models," *Biometrika*, vol. 65, pp. 297–303, 1978.

[123]   A. A. Hegazy, "Effect of dust accumulation on solar transmittance through glass covers of plate-type collectors," *Renew. Energy*, vol. 22, no. 4, pp. 525–540, 2001.

[124]   L. Micheli, D. Ruth, M. G. Deceglie, M. Muller, D. Ruth, and M. G. Deceglie, Time Series Analysis of Photovoltaic Soiling Station Data: Time Series Analysis of Photovoltaic Soiling Station Data: Leonardo Micheli, September, 2017.

[125]   US EPA Office of Research and Development, "CMAQ," 30-Jun-2017. Available: https://zenodo.org/record/1167892. [Accessed: 13-Mar-2019].

[126]   I. Bey, D. J. Jacob, R. M. Yantosca, *et al.*, "Global modeling of tropospheric chemistry with assimilated meteorology: model description and

evaluation," *J. Geophys. Res. Atmos.*, vol. 106, no. D19, pp. 23073–23095, 2001.

[127]   W. C. Skamarock, J. B. Klemp, J. Dudhia, *et al.*, "A description of the advanced research WRF version 3," *NCAR Tech. NOTE*, vol. NCAR/TN-47, p. 125, 2008.

[128]   L. Boyle, H. Flinchpaugh, and M. P. Hannigan, "Natural soiling of photovoltaic cover plates and the impact on transmission," *Renew. Energy*, vol. 77, no. 1, pp. 166–173, 2015.

[129]   L. Boyle, H. Flinchpaugh, and M. Hannigan, "Ambient airborne particle concentration and soiling of PV cover plates," in *2014 IEEE 40th Photovolt. Spec. Conf. PVSC 2014*, pp. 3171–3173, 2014.

[130]   C. A. Gueymard, "SMARTS2: a simple model of the atmospheric radiative transfer of sunshine: algorithms and performance assessment," *Rep. No. FSEC-PF-270-95*, pp. 1–84, 1995.

[131]   U. Jahn, D. Mayer, M. Heidenreich, *et al.*, "International Energy Agency Pvps Task 2: analysis of the Operational Performance of the Iea Database PV Systems," in *16th Eur. Photovolt. Sol. Energy Conf.* Glasgow, pp. 2673–2677, 2000.

[132]   A. Benatiallah, A. M. Ali, F. Abidi, D. Benatiallah, A. Harrouz, and I. Mansouri, "Experimental study of dust effect in multi-crystal PV solar module," *Int. J. Multidiscip. Sci. Eng.*, vol. 3, no. 3, pp. 3–6, 2012.

[133]   C. F. Bohren, "Multiple scattering of light and some of its observable consequences," *Am. J. Phys.*, vol. 55, no. 6, pp. 524–533, 1987.

[134]   S. Prahl and M. Keijzer, "A Monte Carlo model of light propagation in tissue," *Dosim. Laser Radiat. Med. Biol.*, vol. I, no.1989, pp. 102–111, 1989.

[135]   A. Heimsath, P. Lindner, E. Klimm, *et al.*, "Specular reflectance of soiled glass mirrors – study on the impact of incidence angles," *AIP Conf. Proc.*, vol. 1734, p. 130009, 2016.

[136]   A. Luque and S. Hegedus, *Handbook of Photovoltaic Science and Engineering*, New York, NY: Wiley, 2011.

[137]   G. Garson, "Interpreting neural-network connection weights," *Artif. Intell. Expert*, vol. 6, no. 7, pp. 47–51, 1991.

[138]   F. Pianosi and T. Wagener, "A simple and efficient method for global sensitivity analysis based on cumulative distribution functions," *Environ. Model. Softw.*, vol. 67, pp. 1–11, 2015.

[139]   Environment Protection Agency (EPA), "Air Data: Air Quality Data Collected at Outdoor Monitors Across the US."

[140]   Environment Protection Agency (EPA), "RSIG-Related Downloadable Data Files." Available: https://epa.gov/hesc/rsig-related-downloadable-data-files#faqsd. [Accessed: 20-Mar-2019].

[141]   A. Van Donkelaar, R. Martin, M. Brauer, *et al.*, "Global estimates of fine particulate matter using a combined geophysical-statistical method with information from satellites, models, and monitors," *Environ. Sci. Technol.*, vol. 50, no. 7, pp. 3762–3772, 2016.

[142]   "PRISM Climate Group, Oregon State University. PRISM gridded climate data." Available: http://prism.oregonstate.edu/. [Accessed: 20-Mar-2019].

[143]   G. A. Schmidt, M. Kelley., L. Nazarenko, *et al.*, "Configuration and assessment of the GISS ModelE2 contributions to the CMIP5 archive," *J. Adv. Model. Earth Syst.*, vol. 6, pp. 141–185, 2014.

[144]   M. H. Bergin, R. Greenwald, J. Xu, Y. Berta, and W. L. Chameides, "Influence of aerosol dry deposition on photosynthetically active radiation available to plants: a case study in the Yangtze delta region of China," *Geophys. Res. Lett.*, vol. 28, no. 18, pp. 3605–3608, 2001.

[145]   M. B. Kursa, A. Jankowski, and W. R. Rudnicki, "Boruta – a system for feature selection," *Fundam. Informat.*, vol. 101, pp. 271–285, 2010.

[146]   G. Sai Krishna and T. Moger, "Reconfiguration strategies for reducing partial shading effects in photovoltaic arrays: State of the art," *Sol. Energy*, vol. 182, pp. 429–452, 2019.

[147]   T. Fawcett, "An introduction to ROC analysis," *Pattern Recognit. Lett.*, vol. 27, no. 8, pp. 861–874, 2006.

Chapter 10

Dust soiling concentration measurement system based on image processing techniques

Hicham Tribak[1], Mehdi Gaou[1], Salma Gaou[1], Youssef Zaz[2] and Saad Motahhir[3]

Solar plants are often established in desert and arid areas characterized by harsh environmental conditions (e.g. dust scattering, rainfall scarcity). Dust soiling accumulation on photovoltaic (PV) panels adversely influences their efficiency in such a way, the PV power decreases considerably. Once dust soiling concentration exceeds a certain level, the cleaning process becomes a necessity. In this chapter, we propose a novel image processing-based system that allows achieving two-fold measurements, starting with, the quantification of dust-soiling concentration on solar panels' surface, then, estimating the corresponding produced power decrease rate of the targeted solar panels. The proposed system compares the energy production of two solar panels, the first one is kept clean, whereas the second one undergoes an under-control dust accumulation (DA) process. This system acquires two images of basic telltales placed near the panels, in such a way that they receive the same DA as the two solar panels. By observing that the contrast variation of the captured telltale images follows up the variation of energy production, we propose a process, based on image analysis, that ensures the prediction of the lost production proportion caused by DA. The proposed system turned out more beneficial compared to the existing systems, which require complex and expensive equipment. The performed experiments have been undertaken in the field under real-like DA conditions (simulated thanks to an own-made dust particles blower) that commonly affect solar plant installation.

10.1 Introduction

Nowadays, producing energy based on solar sources becomes very competitive compared to fossil fuels. Countries with wide non-arable lands and high solar

[1]Polydisciplinary Faculty of Ouarzazate, Ibn Zohr University, Morocco
[2]Faculty of Science, Abdelmalek Essaadi University, Morocco
[3]ENSA, SMBA University, Morocco

irradiance opt for PV or CSP Technologies by dint of their high production potential. However, the productivity of solar panels and the reflectivity of mirrors decrease significantly with the accumulation of dust and soiling.

Solar panel production efficiency mainly depends on several factors: cells technology (monocrystalline, polycrystalline), weather conditions (Clear sky, Clouds, Fog, etc.), Panels fasten method (fixed, mono-axial, or biaxial solar tracking), period, ambient temperature, season, geographical position, equipment ageing, and so on. In addition to those mentioned above, another factor must be dealt with, especially in the arid area, i.e. dust soiling.

Dust soiling highly impacts the optics of solar panel devices, causing a great degradation of the conversion efficiency of solar radiance to electrical power [1,2]. Indeed, once the dust-soiling rate attains a certain value, the panels cleaning process must immediately be triggered to maintain a profitable production level. Based on the existent financial studies, most of the proposed solar panel state evaluation techniques turned out to be quite costly, taking into consideration that several approaches used infrared imaging analysis to assess the outer surface of solar panels [3]. However, besides their high operating cost, the evoked approaches have not proved a high precision in terms of dust soiling rate quantification. On the other hand, other approaches have been proposed, Ref. [4] proposed a mathematical formula that allows finding a linear relation (correlation) between the efficiency degradation proportion of the PV panel and the corresponding dust concentration rate (g/m2). Refs. [5,6] performed a set of measurements to find the relationship between the inclination angle of the solar panels and the corresponding dust accumulation (DA) rate. Ref. [7] developed an outdoor soiling microscope that captured images on a 10-min scale-time basis. The microscope measures deposited dust particles (whose size is greater than 10 μm^2) in outdoor conditions. Ref. [8] based on the instantaneous electrical characteristics (current and voltage) related to the solar panels to estimate their production decrease rate caused either by solar cell degradation or by cloud flow. Ref. [9] used two power indicators (thermal and miscellaneous capture losses) to identify the PV solar cells' faults. Ref. [10] proposed an intelligent system based on a Takagi-Sugeno-Kahn Fuzzy Rule-Based System (TSK-FRBS) which allows deducing PV defects, by comparing the instant power production of the PV panel in normal functioning with the real generated power. On the one hand, other approaches were further interested in proposing an efficient dirt cleaning process rather than dust soiling estimation among others, Refs. [11,12] used electrodynamic screens technique (made of transparent plastic sheets based on electrodes) in which a self-panels cleaning process is performed. Furthermore, Ref. [13] proposed a dry cleaning device that browses vertically each upper solar panel surface and removes dust particles. Ref. [14] based on infrared image analysis through which IR images of solar panels are captured and analyzed. This allows to localize the defective cells and critical dirt presence. In [15], solar panel cracks and structural defects are detected through successive image processing treatments i.e. Gauss–Laplacian transform and contour detection. However, this approach did not propose a DA measurement process, making it less profitable. Certainly providing a huge solar plant with such dust quantification systems requires a colossal investment, hence, devising an alternative cheaper, and more efficient technique seemed to be worthwhile.

In this chapter, an experimental and a theoretical study has been carried out and crowned with an integrated system, which allows quantifying the DA rate on solar panels' surface, using an image processing-based system, implemented on an embedded system platform. Indeed, the proposed system deduces the DA rate through two telltales (i.e. a telltale is a rectangular piece of chapter divided into two equal areas half black and half white) which are placed near the solar panels, at the same tilt angle and undergone the same DA soiling concentration. Based on the difference between the mean intensities of the black and white areas (expressed as image contrast) of the said telltales, the corresponding DA rate is deduced as well as the corresponding panels' power decrease rate caused by the DA phenomenon. An overall diagram of the proposed approach is presented in Figure 10.1. The remainder of this chapter is organized as follows: the proposed system is presented

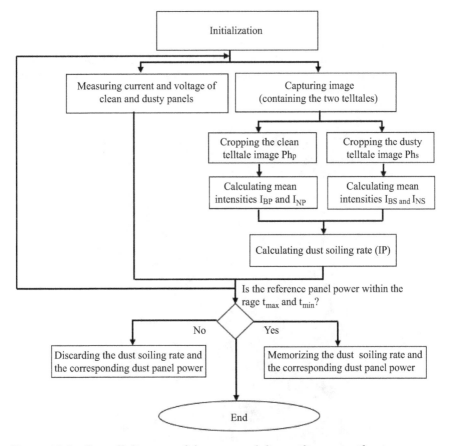

Figure 10.1 Overall diagram of the proposed dust soiling quantification algorithm

in Section 10.2. The obtained results are reported in Section 10.3. We conclude this chapter and giving an outlook of future improvement in Section 10.4.

10.2 Proposed approach

To concretize our approach and find a precise correlation between dust soiling rate and the corresponding power decrease rate of PV modules, we developed a smart system, devised according to solar plants monitoring needs i.e. simulating as closely as possible the real dust effect on PV modules. The proposed system components are depicted in detail in Figure 10.2. The evoked component utilities along with the corresponding technical specifications are described in Section 10.2.1.

10.2.1 Detailed system components description

The proposed system is compounded by the following elements (depicted in Figure 10.2):

- In Figure 10.2(j), a processing unit based on an embedded system platform (Raspberry Pi 3 and Arduino), through which the main processing of the proposed system is ensured. This unit is composed of a microcontroller (Arduino) provided with two current sensors, two voltage sensors, and a temperature

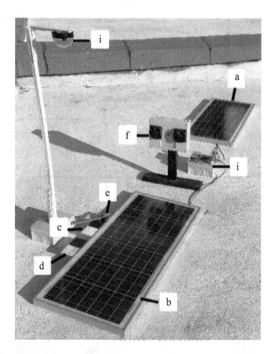

Figure 10.2   Schematic set up of the proposed system

sensor. The evoked unit is connected to both testing and reference panels to get synchronized electrical measurements.

- In Figure 10.2(b), a testing panel that continuously (on a 30-s basis) undergoes a dust-blowing process, this indeed allows a progressive increase of DA density on the panel outer surface.
- In Figure 10.2(a), a reference panel has identic technical characteristics as the testing one. The reference panel is placed near the testing panel, however, it remains clean during all the performed tests (since it is continuously wiped), namely, both testing and reference panels produce the same power at time t_0 (i.e. the initial moment wherein the two panels are clean). It should be noted that the reference panel is exclusively used to select the set of testing panel measurements (voltage, current, power) to be taken into account (i.e. we retain only testing panel electrical measurements in which the corresponding reference panel output power approximates its optimal value). This allows making sure that the testing panel power decrease is mainly due to the dust soiling accumulation. We note that the reference panel is only used during the testing phase to find representative curves relating dust soiling rates and the corresponding panel power decrease rates. Once the curves are established, the reference panel is retired.
- Two telltales (i.e. reference telltale in Figure 10.2(c) and testing telltale in Figure 10.2(d)) were installed on the right edge of the testing panel and at the same tilt angle, in such a way that both telltales receive the same DA concentration as the testing panel. However, the reference telltale is regularly wiped using an automatic cleaning system (Figure 10.2(e)). The evoked telltales are simply made of rectangular plasticized chapter characterized by a specific texture as shown in Figure 10.4, i.e. two equal areas 50% black and 50% white. We recall that in image processing standards, the black color is defined by an intensity that is equal to 0, whereas the white color is defined as 255. Based on this, a relationship between DA and image contrast has been found i.e. when the telltale is clean, the difference of intensities between the two areas is equal to 255 (i.e. DA rate is equal to 0%), whereas when the telltale is wholly dusty, the said difference is equal to 0 (i.e. DA rate of 100%).
- A dust-blowing system in Figure 10.2(f) allows a homogeneous blow of dust particles toward the testing panel as well as the two telltales. The dust particles are stored in a small reservoir. The latter is integrated into the said dust-blowing system.
- A video capturing system (based on Raspberry Pi 3) is provided with an HD camera, through which the images of evoked telltales are captured (Table 10.1).

In Figure 10.3, we display the internal composition of the used processing unit in Figure 10.2(j). As shown in Figure 10.3, this unit is compounded of a Raspberry Pi 3 (Model B), an HD camera, two current sensors, two voltage sensors, and a temperature/ humidity sensor.

Table 10.1   Technical specifications of the used PV panels

Parameter	Specification
Maximum power (P_{max})	50 W
Open-circuit voltage (V_{oc})	24.8 V
Short-circuit current (I_{sc})	2.70 A
Maximum power voltage (V_{mp})	21.0 V
Maximum power current (I_{mp})	2.39 A
Operating temperature	−40 °C to 85 °C
Maximum system voltage	1,000 V DC
Tolerance	+3/−3%
Maximum series fuse rating	10 A

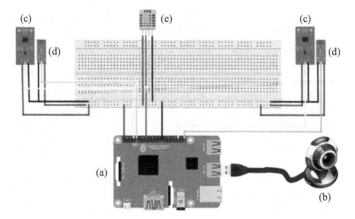

Figure 10.3   Detailed description of the used processing unit: (a) Raspberry Pi 3, (b) HD camera, (c) current sensor, (d) voltage sensor, and (e) temperature sensor

10.2.2   Handling constraints are taken into consideration

To get precise and reliable results, many constraints have been taken into consideration, namely:

- Using two polycrystalline PV panels (testing and reference panels) with the same technical specifications, to be sure that they produce the same electrical power when they are operated under identic operating conditions.
- Mounting the two PV panels at the same tilt and orientation angles.
- Taking the electrical measurements related to the testing panel on an accelerated basis to avoid power fluctuation caused by sun position changing and temperature fluctuation.
- Synchronizing the output powers (related to the testing panel) with the corresponding captured telltale images. Taking into consideration that the telltales

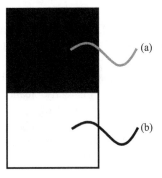

Figure 10.4 Telltale structure. (a) Telltale black area and (b) Telltale white area

images capturing is shifted to some extent compared to the corresponding measured output panel powers. This shift is due to the processing frequency difference between the used processing unit and the video-capturing system.

• Testing sensors (voltage, current, and temperature) reliability.
• Making sure that the power decrease of the testing panel is mainly owing to DA phenomenon. This verification is conducted by exploiting the produced powers of the reference panel as a precise indicator.

The electrical measurements related to the two panels have been taken in a short period and in the same daytime under optimal weather conditions, namely, clear sky, moderated temperature and humidity. As aforementioned, these factors allow for avoiding power fluctuation of the used panels, as well as reaching their optimal produced power (Figure 10.4).

10.2.3 A detailed description of the proposed approach

In real environmental and operating conditions, dust particles are to some extent deposited in a homogenous manner on solar panels' surfaces. To simulate as closely as possible the DA effect on solar panel performance, we used our own-made dust blower which allows us to blow dust particles (stored in a small tank) toward the desired area. As previously mentioned, the said dust propellant blows progressively and homogenously dust particles towards the testing panel as well as the two telltales. In the first step, the proposed system performs two synchronized tasks, namely, taking a set of electrical measurements (voltage, current, power) of the testing and reference panels along with an image covering the testing panel and its two associated telltales (testing and reference telltales), afterward, each telltale image is cropped separately to be analyzed in the next stage.

To further facilitate the proposed process explanation, the two telltale images are labeled, namely, the mean intensities of the black and white areas of the testing telltale are labeled I_{NS} and I_{BS}, respectively. The mean intensities of the black and white areas of the reference telltale are labeled I_{NP} and $I_{BP,}$ respectively. In the initial state, when the two telltales are clean, the mean intensities of their white areas (I_{BS} and I_{BP}) are equal to 255, whereas the mean intensities of their black areas (I_{NS} and I_{NP}) are equal to 0. Based on the performed experiment results, we found that the more dust particles fall on the testing telltale, the more progressively the mean intensities values of its black and white areas change i.e. the mean intensity of its black area increases, whereas, related to its white area decreases. As regards the reference telltale, it is permanently cleaned using the aforementioned cleaning system, hence, the mean intensities of its white and black areas remain unchangeable. Based on the experiment results analysis we observed that there was a strong relationship between DA concentration and telltale image contrast change.

From this observation, the DA rate can be calculated from the difference (contrast) between the mean intensities of the white and black areas of the testing telltale as defined by (10.1). In Figure 10.5, we presented the testing telltale with different DA rates. For instance, in Figure 10.5(d) the mean intensity of the white area is equal to 255, whereas that of the black area is worth 0, hence the difference (i.e. contrast) is equal to 255. Since the contrast is maximum, the corresponding DA rate is equal to 0%. In Figure 10.5(a), the contrast is equal to 0, seeing that dust particles totally cover the testing telltale surface, consequently, the corresponding DA rate is equal to 100%. The use of the formula (10.1) becomes inaccurate in case of luminosity variation, hence, other parameters have been integrated to make it stable enough towards such factors. In this regard, we integrated a stabilizing factor, which is defined as the division of 255 by the contrast related to the reference telltale. By doing so, the new formula becomes (10.2). Now, to convert the obtained contrast value to DA rate (i.e. find a correlation between contrast and dust concentration proportion), we used the formula (10.3). For further simplification, (10.3) is in turn simplified. By doing so, the DA rate (denoted DR) is now calculated directly via the formula (10.4).

$$Dif_1 = I_{BS} - I_{NS} \tag{10.1}$$

$$Dif_2 = \frac{255\,(Dif_1)}{I_{BP} - I_{NP}} \tag{10.2}$$

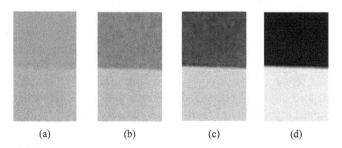

    (a)   (b)   (c)   (d)

Figure 10.5   Testing telltale representation with different dust soiling rates.
(a) Wholly dusty telltale, i.e. 100% of DA rate. (b) 70% of DA rate.
(c) 30% of DA rate. (d) Totally clean telltale, i.e. 0% of DA rate

$$Dif_3 = \left(1 - \frac{Dif_2}{255}\right) \times 100 \tag{10.3}$$

$$\text{Dust Rate (DR)} = \frac{(I_{BP} - I_{BS}) + (I_{NS} - I_{NP})}{I_{BP} - I_{NP}} \times 100 \tag{10.4}$$

At the end of each experiment, we obtain a couple of measurements i.e. DA rate along with the corresponding testing panel output power (which has been previously measured). These measurements are verified before being stored in such a way if the corresponding output power of the reference panel is bounded between two pre-defined thresholds, the evoked measurements are retained, otherwise discarded. Once done, the dust blower system is activated to increase dust soiling. The experiments have been reiterated several times. By the end, representative correlative curves between DA rates and output panel powers are established in Figure 10.6.

10.3 Experiments and results

From a total of more than 100 measurements, we present here a sample of 15 experiment results. The latter are represented in Figure 10.6, in which DA rates (right Y-axis) are expressed in terms of the corresponding produced panel powers (left Y-axis). The X-axis stands for the indices of the performed experiments. By analyzing Figure 10.6, we observe that the more the DA rate increase, the more output panel power diminishes, namely when the DA rate reaches 60%, the panel power completely decreases, i.e. panel power drops to 0 W.

The obtained experimental results can be represented differently since in some cases, solar plant supervisors are rather interested in the power decrease rate. For this reason, we presented in Figure 10.7, a new curve representation, expressed as

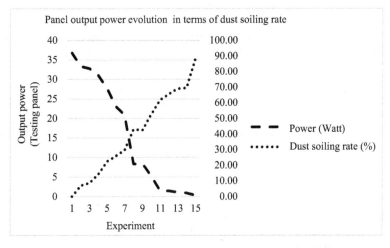

Figure 10.6 Panel output power evolution expressed in terms of dust soiling rate

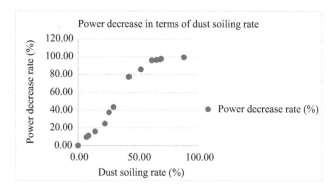

Figure 10.7   Power decrease proportion expressed in terms of dust soiling rate

power decrease rates in terms of DA rates. Based on this curve, one can observe that when the DA rate is ranged between 0% and 60%, the power decrease rate diminishes smoothly, however, once the DA rate exceeds 60% (critical DA rate), the power decrease rate rises at once to 100%, consequently, a cleaning process must be performed once DA rate nears 60%.

10.4   Conclusion

In this chapter, we presented an image processing-based system, for quantifying DA rate along with the corresponding power decrease rate of PV devices. The experiments have been conducted using two solar panels; a reference panel (which remains clean during experiments) and a testing panel. The testing panel was associated with two telltales, through which DA rates were quantified.

To study exclusively dust soiling impact on PV modules and ignore other phenomena, several precautions have been taken into account; especially the use of output powers related to the reference panel as an indicator, namely if the said powers fluctuate, we deduce that other phenomena occur (other than dust soiling) and consequently the corresponding experiment measurements are discarded.

DA rate was quantified using the aforementioned telltales, based on the correlation between telltale image contrast and dust soiling concentration. Indeed the evoked correlation is found via our proposed formula. The latter has been adjusted to be invariant towards illumination variation.

At the end of the experiments, we established a curve of correspondence between DA rates and their related power decrease rates.

References

[1]   T. Sarver, A. Al-Qaraghuli, and L. L. Kazmerski, "A comprehensive review of the impact of dust on the use of solar energy: History, investigations, results, literature, and mitigation approaches," *Renewable Sustainable Energy Reviews*, vol. 22, pp. 698–733, 2013.

[2]   M. Mani and R. Pillai, "Impact of dust on solar photovoltaic (PV) perfor-
      mance: Research status, challenges and recommendations," *Renewable
      Sustainable Energy Reviews*, vol. 14(9), pp. 3124–3131, 2010.

[3]   G. Acciani, G. B. Simione, and S. Vergura, "Thermographic analysis of
      photovoltaic panels", In *International Conference on Renewable Energies
      and Power Quality*, 2010, www.icrepq.com/icrepq'10/634-Acciani.pdf

[4]   A. Y. Al-Hasan and A. A. Ghoneim, "A new correlation between photo-
      voltaic panel's efficiency and amount of sand dust accumulated on their
      surface," *International Journal of Sustainable Energy*, vol. 24(5), pp. 187–
      197, 2005. doi: 10.1080/14786450500291834

[5]   H. Hottel and B. Woertz, "The performance of flat plate solar heat collec-
      tors," *Transactions of ASME*, vol. 12, pp. 91–104, 1942.

[6]   H. P. Garg, "Effect of dust on transparent covers in flat plate solar energy
      collectors," *Solar Energy*, vol. 15, pp. 299–302, 1974.

[7]   B. Figgis, A. Ennaoui, B. Guo, W. Javed, and E. Chen, "Outdoor soiling
      microscope for measuring particle deposition and resuspension," *Journal of
      Solar Energy*, vol. 137, pp. 158–164, 2016.

[8]   D. Nilsson, "Fault detection in photovoltaic systems," *KTH Computer
      Science and Communication*, 2014.

[9]   A. Chouder and S. Silvestre, "Automatic supervision and fault detection of
      PV systems based on power losses analysis", *International Journal of
      Energy Conversion and Management*, vol. 51, no. 10, pp. 1929–1937, 2010.

[10] P. Ducange, M. Fazzolari, B. Lazzerini, and F. Marcelloni, "An intelligent
      system for detecting faults in photovoltaic fields," In *2011 11th International
      Conference on Intelligent Systems Design and Applications*, Cordoba, 2011,
      pp. 1341–1346.

[11] M. K. Mazumder, R. Sharma, A. S. Biris, J. Zhang, C. Calle, and M. Zahn,
      "Self-cleaning transparent dust shields for protecting solar panels and other
      devices," *International Journal of Particulate Science and Technology*,
      vol. 25, no. 1, pp. 5–20, 2007.

[12] M. Mazumder, M. Horenstein, J. Stark, *et al.*, "Characterization of electro-
      dynamic screen performance for dust removal from solar panels and solar
      hydrogen generators," In *IEEE Transactions on Industry Applications*,
      vol. 49, no. 4, pp. 1793–1800, 2013.

[13] M. Meller and E. Meller, "Solar panel cleaning system and method," Patent
      number US 8500918 B1, 2013.

[14] E. Kaplani, "Detection of degradation effects in field-aged c-Si solar cells
      through IR thermography and digital image processing," *International
      Journal of Photoenergy*, vol. 2012, Article ID 396792, 11 pages, 2012.

[15] Z. Fu, Y. Zhao, Y. Liu, *et al.*, "Solar cell crack inspection by image pro-
      cessing," In *Proceedings of 2004 International Conference on the Business
      of Electronic Product Reliability and Liability* (IEEE Cat. No. 04EX809),
      2004, pp. 77–80.

Chapter 11

Anatomization of dry and wet cleaning methods for general to rural and remote installed of solar photovoltaic modules

M. Palpandian[1], D. Prince Winston[2], B. Praveen Kumar[3] and T. Sudhakar Babu[4]

Dust accumulation on the PV module restricts solar radiation and reduces the efficiency of the PV module. To improve the efficiency, it is essential to remove the dust deposited on the PV module thereby improving the performance. The main concern of the cleaning system is the additional cost of the equipment. This chapter proposes a low-cost drying and cleaning method using a vacuum cleaner and pressurized water pump to clean the dust accumulated on the module. In this work, the considered experimental setup consists of an 80 W polycrystalline solar PV system with three sets of the array (sets A, B, & C) installed at Kamaraj College of Engineering and Technology, Tamil Nadu, India. The experimental set-up is exposed to solar radiation for a week with set A left uncleaned and sets B and C cleaned by dry and wet cleaning methods. It was found that for the case I wet cleaning has improved the efficiency by 1.21% and for case II by 0.92% compared to dry cleaning. The wet clean inning method reduces the operating temperature of the module compared to the dry cleaning method. The proposed work is a low-cost method that involves a low cost, less manpower, and is more conventional for a roof-top installed in a remote location, where the readiness of skilled labor is not accessible.

11.1 Introduction

The growth of PV systems attained enormous growth over the last decade because of government scheme formation, subsidies, and development in PV technology [1–4]. The performance of the PV module is affected by environmental factors such

[1]Department of EEE, SCAD College of Engineering and Technology, India
[2]Department of EEE, Kamaraj College of Engineering and Technology, India
[3]Department of EEE, Vardhaman College of Engineering, India
[4]Department of EEE, Chaitanya Bharathi Institute of Technology, India

as ambient temperature, incident irradiation, wind, humidity, tilt angle and orientation, partial shading, dust, dirt, snow, and mounting techniques. These factors often overlap with each other and lead to major performance degradation. Dust, soiling, and pollutants are site dependent which significantly influence the performance of the module [5].

The scattering of a dust particle in the atmosphere reduces the intensity of solar radiation reaching the surface of the module resulting in the deterioration in the conversion efficiency of the module. Further, the reduction in the efficiency depends on the size and composition of the suspended particle. Meanwhile, the density of dust deposition depends on the time of exposure. The composition of the dust particles may differ depending on the region. In dry climate areas, the main source of dust is the soil, whereas in urban areas depend on the pollutants emitted from vehicles and industries. The dust accumulated on the PV module installed in an industrial area has reduced the daily efficiency [6,7].

The size of the dust particle plays a major role in the absorption, scattering, and reflectance of the incident solar radiation on the module. The fine particles are uniformly distributed and can occupy a larger area than the coarse particles thereby significantly shielding the incident solar radiation. El-Shobokshy and Hussein [8] studied the impacts of the physical and chemical composition of the dust particle and found that smaller dust particles degrade the performance more than larger ones.

The removal of dust from the surface of the module depends on the wind velocity and tilt angle. In the horizontally installed modules, wind speed greater than 50 m/s cannot completely remove the dust particles. The cleaning of fine dust particles (50 μm) by the wind is ineffective due to the adhesion force that exists on the module surface. Biryukov [9] examined the samples of the natural dust particles and observed that the larger size particles (20–40 μm) covered 50% of the module surface. Therefore, the size of the dust accumulated on the surface increases with the time of exposure.

Rainfall is considered the natural cleaning agent for eradicating dust accumulation on the PV module. Javed *et al.* [10] observed that the 4 mm rain is found to be a threshold to eradicate the dust deposition effectively and anything beyond have not been effective. Similarly, in [10], 5 mm of rain is enough to clean and restore the performance of the module. However, the cleaning of dust is effective with an increase in inclination angle [11]. A reduction in dust deposition from 15.84 g/m^2 to 4.48 g/m^2 was recorded with an increase in inclination angle from 0° to 90° [12]. From the literature, it is clearly understood that periodical cleaning should be done on the PV system to restore its performance. The frequency of cleaning depends on the location of the site, local environmental conditions, wind patterns, atmospheric dust concentrations, and human activities.

Several methods available in practice to clean the PV modules are mechanical cleaning, robot cleaning, electro-dynamic screen, and anti-dust coating. In the mechanical cleaning method, the dust is removed by using a brush, blower, and water spraying. The frequent cleaning by brushes can cause scratches on the surface of the module. In the blowing method, the air is blown using a compressor can able to remove the dust accumulation on the module. The blowing technique is more efficient than the brushing method. In the water spraying method, water is sprayed on the surface of the module and cleaned by using the wiper. This method can reduce the operating temperature of the module and improve its performance. This mechanical method requires labor to clean the modules and the efficiency of cleaning is not up to the mark [13].

Robot cleaning can be employed in areas, which lack sufficient rainfall, and daily power losses exceed 20%. The module is cleaned by three movements namely zig-zag, spiral, and anti-fall movements. This method can able detect dust by using the sensors and can clean the dust, dirt, and mud efficiently with less amount of water and without scratches. The robotics systems consist of bulky and heavy structures and require more energy to clean the modules. Hence, it increases the complexity of large-scale systems and also increases the cost of the small-scale system [14].

In the electro-dynamic method, the screen is placed on the module surface and the high voltage is phase shifted to create a traveling wave that is applied between the electrodes for generating the electric field to repel the dust deposition to the edge of the modules. Afterward, the screen is moved to the adjacent module. The advantage of this method is it does not require water and is more suited for dry areas. In addition, it requires high voltage, which is dangerous for the cleaning staff [15].

In the coating method, the surface of the module is coated with a transparent anti-dust coating such as hydrophobic and hydrophilic. The hydrophobic coating has to be frequently cleaned with water to remove the dust sticking to the coating. Otherwise, the coating should be reapplied; this may increase the cost and affects the efficiency. The hydrophilic coating can be achieved by the nano-film of titanium oxide, chemical coating, and nano-patterned glass surface. The degradation of the coating results in more accumulation of dust on the surface. Hence, it needs regular washing during the rain [16].

Based on the literature review, this chapter experimentally investigates the influences of dry and wet cleaning methods on the mono-crystalline module. Table 11.1 represents the main drawback of the existing cleaning methods as complexity, cost, maintenance, and labor requirements require high cost, and manpower, which increased the cost of the PV system. To overcome the above shortcomings, this chapter proposes a low-cost dry and wet cleaning system

Table 11.1   Comparison of cleaning methods

Parameter	Mechanical cleaning	Robot cleaning	Electro-dynamic screen	Anti-dust coating
Cost	Low	High	High	High
Maintenance	Less	High	High	High
Complexity	Less	High	High	High
Requirement of labor	High	High	High	High

employing a vacuum cleaner and pressurized water to remove the dust accumulated on the PV module. The dry cleaning method employs a vacuum cleaner to clean the surface of the module. In wet cleaning, the pressurized water utilized for cleaning also acts as a coolant for reducing the operating temperature by 8 °C and increases the module efficiency by 3% [10]. The cleaning methods reported in the literature review are complex, and the high cost requires high manpower. The proposed work is a low-cost method that involves a low cost, less manpower, and is more convenient for a roof-top installed in a remote location, where the availability of skilled labor is not available. The main advantage of the proposed cleaning techniques, it eliminates the need for moving parts, anti-coating, and guard rails. In addition, a comprehensive analysis is made of the cleaning methods in comparison with the uncleaned PV array.

This chapter divides into four sections: Section 11.1 elaborates on the influence of the performance degradation of dust deposition on PV modules and its cleaning methods for restoring its performance. Section 11.2 elaborates on the experimental setup. In Section 11.3, experimental procedures for the dry and wet cleaning methods are clearly explained. In Section 11.4, the results and discussion is presented including the PV performance and hotspot formation during wet cleaning. Section 11.5 summarizes the main findings and conclusions.

11.2   Experimental setup

The main objective of this chapter is to eradicate the dust deposited on the PV module thereby restoring its performance. Most of the cleaning methods proposed by the previous researchers have involved equipment that is highly costs. This chapter proposes cost-effective dry and wet cleaning methods for a PV array to restore performance. The schematic diagram of the solar PV setup is given in Figure 11.1.

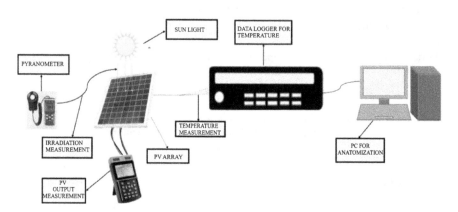

Figure 11.1   Schematic diagram of the PV setup

An experimental setup has been conducted to investigate the impact of dust accumulation on the performance of mono-crystalline PV arrays. The experimental setup consists of three sets of PV arrays (A, B, and C) installed at Kamaraj College of Engineering and Technology (9.6728°N, 77.9659°E), Virudhunagar, Tamil Nadu, India. Each set consists of eight 10 W mono-crystalline PV modules connected in series employed at a tilt angle of 45° (Figure 11.2). The peak power of the PV array is 80 W. The electrical characteristics of the PV modules are represented in Table 11.2.

In the proposed method, the cleaning system consists of a dry and wet cleaning method to clean the dust accumulated on the PV array. In the dry cleaning method, a vacuum cleaner (300 W, 230 V AC) is used to suck the dust deposited on the surface of the module. The wet cleaning method consists of an aluminum water tank of 0.15 m^3 capacity, and a water pump (100 W, 12 V DC).

The water pump is used to suck the water from the water tank through the filter to avoid the suction of the dust and sprayed over the surface of the module via a nozzle. The pressure of the water is adjusted using a nozzle maintained at 3 bar. The range of spray angle of the nozzle varies between 15° and 145°. Afterward, the cleaning water is collected at the bottom of the module and returned to the storage tank.

Figure 11.2   Experimental setup of the PV system used

Table 11.2   Electrical characteristics of the PV module

Parameter	Ratings
Maximum power (P_{max})	10 W
Open circuit voltage (V_{oc})	21.96 V
Short circuit current (I_{sc})	0.59 A
The maximum voltage at P_{max} (V_m)	18.25 V
Maximum current at P_{max} (I_m)	0.55 A

11.3   Experimental procedure

The experimental setup consists of three sets (A, B, and C) of PV array to validate the effectiveness of the cleaning methods. In set A, the PV array is left uncleaned, while in set B the array is cleaned by dry cleaning with a vacuum cleaner, and set C is cleaned by using wet cleaning with pressurized water to clean the dust accumulated on the surface of the modules. To investigate the performance of dust accumulation on the PV array, experimentation has been conducted for two days.

Case I: On 10 March 2021, three sets of PV arrays (A, B, and C) are cleaned and exposed to solar radiation for a week. On 17 March 2021, set A is left uncleaned, set B is cleaned by using the dry cleaning method and whereas, and set C is cleaned by the wet cleaning method. Case II: Similarly, on 18 March 2021, three sets of PV arrays is cleaned and exposed to solar radiation for a week. On 25 March 2021, set A is left uncleaned, set B is cleaned by using the dry cleaning method and whereas set C is cleaned by the wet cleaning method. After the completion of each cleaning process, the PV array is exposed to solar radiation from 7.00 AM to 6.00 PM, and their respective array voltage, current, and output power are recorded. The incident irradiation on the PV module is measured by the solar meter. The infrared thermometer is used to measure the operating temperature of the PV module at a specific point throughout the experiment. From the measurements, percentage power loss (11.1), fill factor (11.2), efficiency (11.3), and efficiency loss (11.4) of the PV array is calculated as shown below:

$$\% \text{ Powerloss} = \frac{P_{uc} - P_{dc,wc}}{P_{uc}} \times 100 \tag{11.1}$$

$$\text{Fillfactor} = \frac{V_{mp}I_{mp}}{V_{oc}I_{sc}} \tag{11.2}$$

$$\text{Efficiency}(\eta) = \frac{\text{Maximum output power (W)}}{\text{Incident solar radiation} \left(\frac{W}{m2}\right) \times \text{module surface area}(m^2)} \tag{11.3}$$

$$\% \text{ Efficiency loss}(\eta_{loss}) = 1 - \frac{\text{Maximum power of uncleaned PV array}}{\text{Maximum power of cleaned PV array}} \times 100 \tag{11.4}$$

11.4   Result and discussion

11.4.1   Case: I

Figure 11.3(a) shows the output power of the PV array (Case: I) for the uncleaned, dry and wet cleaning methods. At 03.00 PM, the peak power has recorded as 48.30 W, 53.80 W, and 60.50 W for uncleaned, dry, and wet cleaning methods at irradiation of 812 W/m^2. The efficiency of an uncleaned, dry and wet cleaned array is 7.65%,

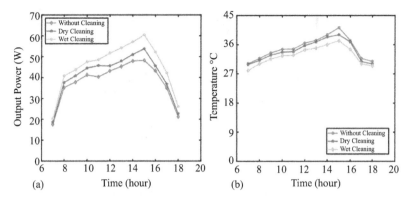

Figure 11.3 (a) Output power vs. time and (b) temperature (°C) vs. time for the case I

8.52%, and 9.59% (Table 11.3), respectively. The dry and wet cleaning method has reduced the efficiency loss to 32.75% and 24.38% compared to the uncleaned PV array of 39.63%. Figure 11.3(b) shows the variation of module temperature with respect to time. The average module temperature of the uncleaned, wet, and dry cleaned array is 34.94 °C, 34.12 °C, and 32.68 °C.

Figure 11.4(a) and (b) shows the efficiency and % efficiency loss for case I. The uncleaned, dry and wet cleaned PV array has produced average efficiency of 8.18%, 8.79%, and 9.75%, respectively. In the uncleaned array, the accumulated dust restricts the incident irradiation from reaching the PV cell and has obtained an average efficiency loss of 52.41%. In dry cleaning, the vacuum cleaner removes the accumulated dust and reduces the efficiency to 48.79%, whereas the wet cleaning method has reduced the operating temperature of the module, thereby reducing the average efficiency loss to 43.15% respectively.

For the case I, when compared to the uncleaned PV array, the dry cleaning method has improved the maximum power, % power loss, efficiency and reduction in a % efficiency loss of 3.61%, 6.91%, 0.77%, and 4.52% whereas for wet cleaning is 9.26%, 17.67%, 1.97%, and 11.58%, respectively.

11.4.2 Case: II

Figure 11.5(a) shows the output power of the PV array for the uncleaned, dry and wet cleaning process. At 01.00 PM, the peak power is recorded as 51.48 W, 54.45 W, and 57.42 W for uncleaned, dry, and wet cleaning methods at an irradiation of 827 W/m^2. Figure 11.5(b) shows the variation of module temperature with respect to time. The average operating temperature (Table 11.4) of the uncleaned, wet, and dry cleaned array is 34.55 °C, 33.89 °C, and 33.00 °C.

The efficiency of the uncleaned, dry, and wet cleaned array is 7.65%, 8.47%, and 8.93%, respectively. The dry and wet cleaning method has reduced the efficiency loss to 31.94% and 28.23% compared to the uncleaned PV array of 35.65%.

Table 11.3 Cleaning process for case 1

Time (Hours)	Irradiation (W/m²)	Uncleaned module								Dry cleaning								Wet cleaning							
		Module Temperature (°C)	Voltage (V)	Current (A)	Power (W)	% Power loss	Fill factor	Efficiency (η)	% Efficiency loss (% η loss)	Module Temperature (°C)	Voltage (V)	Current (A)	Power (W)	% Power loss	Fill factor	Efficiency (η)	% Efficiency loss (% η loss)	Module Temperature (°C)	Voltage (V)	Current (A)	Power (W)	% Power loss	Fill factor	Efficiency (η)	% Efficiency loss (% η loss)
7:00	270	30.2	35.10	0.50	17.50	78.13	0.17	8.34	78.13	29.9	35.00	0.54	18.80	76.50	0.18	8.96	76.50	28.1	34.90	0.59	20.40	74.50	0.20	9.72	74.50
8:00	545	31.8	34.10	1.04	35.50	55.88	0.34	8.33	55.88	31.2	34.00	1.11	37.70	52.88	0.36	8.90	52.88	30.1	33.80	1.21	40.90	48.88	0.39	9.66	48.88
9:00	600	33.5	34.00	1.12	37.90	52.63	0.37	8.13	52.63	32.8	33.80	1.21	40.90	48.88	0.39	8.77	48.88	31.6	33.60	1.30	43.80	45.25	0.42	9.39	45.25
10:00	648	34.6	33.80	1.22	41.30	48.38	0.40	8.20	48.38	33.7	33.60	1.33	44.60	44.25	0.43	8.86	44.25	32.6	33.40	1.42	47.60	40.50	0.46	9.45	40.50
11:00	650	34.7	33.80	1.19	40.40	49.50	0.39	8.00	49.50	33.9	33.50	1.37	45.80	42.75	0.44	9.07	42.75	32.8	33.40	1.46	48.60	39.25	0.47	9.62	39.25
12:00	690	36.6	33.70	1.28	43.20	46.00	0.42	8.06	46.00	35.8	33.50	1.36	45.60	43.00	0.44	8.50	43.00	34.4	33.20	1.56	51.80	35.25	0.50	9.66	35.25
13:00	712	37.4	33.50	1.35	45.40	43.25	0.44	8.20	43.25	36.9	33.40	1.43	47.90	40.13	0.46	8.66	40.13	35.0	33.00	1.64	54.20	32.25	0.52	9.79	32.25
14:00	744	39.1	33.40	1.44	48.00	40.00	0.46	8.30	40.00	38.4	33.20	1.54	51.10	36.13	0.49	8.84	36.13	36.1	32.80	1.74	57.20	28.50	0.55	9.89	28.50
15:00	812	41.3	33.40	1.45	48.30	39.63	0.47	7.65	39.63	39.1	33.00	1.63	53.80	32.75	0.52	8.52	32.75	37.3	32.60	1.85	60.50	24.38	0.58	9.59	24.38
16:00	714	37.4	33.70	1.29	43.30	45.88	0.42	8.00	45.88	36.9	33.50	1.36	45.70	42.88	0.44	8.24	42.88	34.6	33.10	1.58	52.30	34.63	0.50	9.42	34.63
17:00	534	31.8	34.10	1.02	35.00	56.25	0.34	8.43	56.25	30.8	34.00	1.08	36.90	53.88	0.36	8.89	53.88	30.1	33.70	1.26	42.30	47.13	0.41	10.19	47.13
18:00	316	30.9	34.90	0.61	21.30	73.38	0.21	8.67	73.38	30.1	34.80	0.65	22.80	71.50	0.22	9.28	71.50	29.5	34.60	0.76	26.20	67.25	0.25	10.67	67.25
AVG	693	34.94	33.96	1.13	38.08	52.41	0.37	8.18	52.41	34.12	33.75	1.22	40.97	45.79	0.40	8.79	45.79	32.68	33.51	1.36	45.48	43.15	0.44	9.75	43.15

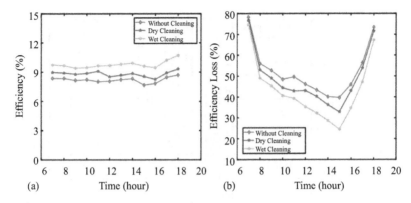

Figure 11.4   (a) Efficiency vs. time and (b) efficiency loss vs. time for the case I

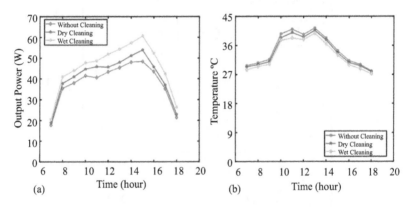

Figure 11.5   (a) Output power vs. time and (b) temperature (°C) vs. time for case II

The uncleaned, dry, and wet cleaned PV array has produced average efficiency (Figure 11.6(a)) of 7.99%, 8.88%, and 9.54, respectively. The uncleaned PV array obtained an average efficiency loss (Figure 11.6(b)) of 56.31% whereas the dry and wet cleaned arrays obtained 52.30% and 48.46%, respectively. For case II, when compared to the uncleaned PV array, the dry cleaning method has improved the maximum power, % power loss, efficiency, and reduction in % efficiency loss of 4.01%, 7.12%, 1.01%, and 5.01% whereas for wet cleaning is 7.86%, 13.95%, 1.93%, and 9.82%, respectively.

The wet cleaning method reduces the operating temperature of the module compared to the dry cleaning method. Therefore, the wet cleaning method is found to be superior to restore the performance of the PV array. The difference in maximum power, efficiency, and efficiency loss of case I and case II reveals that the variation in performance of the PV array depends on the local environmental conditions, dust size, composition, and distribution.

Table 11.4 Cleaning process for case II

Time (Hours)	Irradiation (W/m²)	Uncleaned module								Dry cleaning								Wet cleaning							
		Module Temperature (°C)	Voltage (V)	Current (A)	Power (W)	% Power loss	Fill factor	Efficiency (η)	% Efficiency loss (% η loss)	Module Temperature (°C)	Voltage (V)	Current (A)	Power (W)	% Power loss	Fill factor	Efficiency (η)	% Efficiency loss (% η loss)	Module Temperature (°C)	Voltage (V)	Current (A)	Power (W)	% Power loss	Fill factor	Efficiency (η)	% Efficiency loss (% η loss)
7:00	240	29.7	35.20	0.44	15.49	80.64	0.15	8.30	80.64	29.3	35.10	0.47	16.50	79.38	0.16	8.84	79.38	28.4	35.10	0.51	17.90	77.62	0.17	9.60	77.62
8:00	404	30.5	34.70	0.74	25.68	67.90	0.25	8.18	67.90	30.0	34.50	0.80	27.60	65.50	0.27	8.79	65.50	29.4	34.40	0.86	29.58	63.02	0.29	9.42	63.02
9:00	570	31.6	34.20	1.02	34.88	56.40	0.34	7.87	56.40	30.9	33.90	1.14	38.65	51.69	0.37	8.72	51.69	30.1	33.80	1.23	41.57	48.03	0.40	9.38	48.03
10:00	778	39.5	33.40	1.42	47.43	40.72	0.46	7.84	40.72	38.3	33.40	1.54	51.44	35.71	0.50	8.51	35.71	37.5	33.20	1.64	54.45	31.94	0.53	9.00	31.94
11:00	808	40.9	33.32	1.50	49.98	37.53	0.48	7.96	37.53	39.9	33.30	1.61	53.61	32.98	0.52	8.54	32.98	38.2	33.23	1.69	56.16	29.80	0.54	8.94	29.80
12:00	773	39.3	33.30	1.46	48.62	39.23	0.47	8.09	39.23	38.5	33.32	1.54	51.31	35.86	0.50	8.54	35.86	37.7	33.21	1.66	55.13	31.09	0.53	9.18	31.09
13:00	827	41.3	33.43	1.54	51.48	35.65	0.50	8.01	35.65	40.6	33.20	1.64	54.45	31.94	0.53	8.47	31.94	39.7	33.00	1.74	57.42	28.23	0.55	8.93	28.23
14:00	740	39.2	33.60	1.30	43.68	45.40	0.42	7.59	45.40	37.6	33.30	1.48	49.28	38.40	0.48	8.57	38.40	36.4	33.00	1.68	55.44	30.70	0.53	9.64	30.70
15:00	635	34.3	33.90	1.17	39.66	50.42	0.38	8.04	50.42	33.6	33.70	1.29	43.47	45.66	0.42	8.81	45.66	32.9	33.30	1.47	48.95	38.81	0.47	9.92	38.81
16:00	502	31.2	34.30	0.92	31.56	60.56	0.30	8.09	60.56	30.6	34.10	1.03	35.12	56.10	0.34	9.00	56.10	29.9	34.00	1.13	38.42	51.98	0.37	9.85	51.98
17:00	304	30.0	35.00	0.54	18.90	76.38	0.18	8.00	76.38	29.6	34.90	0.63	21.99	72.52	0.21	9.31	72.52	28.7	34.70	0.69	23.94	70.07	0.23	10.13	70.07
18:00	195	28.1	35.40	0.34	12.04	84.96	0.12	7.94	84.96	27.8	35.30	0.41	14.47	81.91	0.14	9.55	81.91	27.1	35.20	0.45	15.84	80.20	0.15	10.45	80.20
AVG	565	34.35	34.15	1.03	34.95	56.31	0.34	7.99	56.31	33.89	34.00	1.13	38.16	52.30	0.37	8.80	52.30	33.00	33.85	1.23	41.23	48.46	0.40	9.54	48.46

The energy consumption of the dry and wet cleaning is represented in Table 11.5. The dry and wet cleaning method consumes 0.01 kWh per cycle and 0.0037 kWh per cycle. The thermal image of the wet cleaning process (Figure 11.7) reveals that there is a possibility of the occurrence of the hot spot due to the improper cleaning and quality of the demineralized water used in the wet cleaning method. However, in this experiment, the hotspot is not found in the dry cleaning method. Due to improper cleaning of the wet cleaning method, adhesion of the dust deposition increases which may not be blown away by the wind. Hence, this increases the operating temperature of the module and leads to

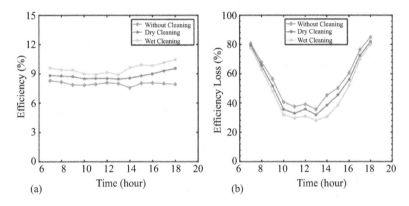

Figure 11.6   (a) Efficiency vs. time and (b) efficiency loss vs. time for case II

Table 11.5   Energy consumed for the cleaning methods for both the cases

Equipment's	Power rating (W)	Operating period (min)	Energy consumed (kWh)
Vacuum cleaner	300	2	0.01
Water pump	100	2	0.0033
Solenoid value	12	2	0.0004

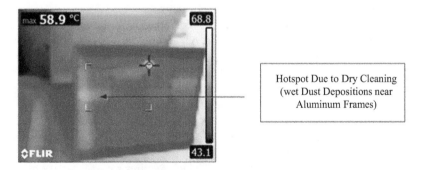

Hotspot Due to Dry Cleaning (wet Dust Depositions near Aluminum Frames)

Figure 11.7   Thermal image of wet cleaned panel

the hot-spot formation that may leads to early degradation and permanent damage to the module.

11.4.3   Observation and suggestions

From the experimental results, it is observed that dry cleaning is found to be simple and has less maintenance. The dry cleaning methods are idle for the hot areas where there is a scarcity of water and the adhesion between the dust and the surface of the module is low. The main drawback of the dry cleaning method is a thin layer of dust can exists on the surface of the module even after cleaning. This may decrease the frequency of cleaning and the impact on the output power is minor. However, the cost of the cleaning system depends on the frequency of cleaning.

In the wet cleaning methods, cleaning the modules with demineralized water can reduce the operating temperature of the module and increase its efficiency. The efficiency of the cleaning methods depends on the quality of water, pressure, and temperature utilized for cleaning. The efficiency of cleaning mainly depends on the quality of the water, sufficient pressure should be maintained to eradicate the dust, the temperature of the water used should be almost the same as the module temperature, and otherwise, it may lead to cracks on the glass surface.

11.5   Conclusion

The performance degradation of dust deposition is studied on three sets of PV arrays. In this study, experimental measurements were conducted to analyze the performance of the impact of dry and wet cleaning techniques on mono-crystalline modules. It is observed that the dust deposition has reduced the intensity of solar irradiance absorbed by the PV module. From the experimental results, the following is concluded:

- For the case I, the wet cleaning method has improved the maximum power, efficiency, and reduction in an efficiency loss of 5.65%, 1.21%, and 7.06% compared to dry cleaning.
- Similar for case II, the wet cleaning method has improved the maximum power, efficiency, and reduction in an efficiency loss of 3.85%, 0.92%, and 4.81%, respectively, compared to dry cleaning.
- The wet cleaning method reduces the operating temperature of the module compared to the dry cleaning method. Therefore, the wet cleaning method is found to be superior.
- The thermography image reveals that the improper cleaning of the module by wet cleaning results in the formation of the hotspot.
- The difference in the enhancement of maximum power, efficiency, and reduction of efficiency loss for case I and case II reveals that the reduction in performance of the PV array depends on the local environmental conditions, dust size, composition, and distribution.
- This system is designed for PV farms, which requires a cost-effective cleaning system for restoring performance.

- Further, the present study will help the policy and decision-makers to understand the effects of dust accumulation and mitigation techniques.
- In the future, the modules can be coated with an anti-dust coating like super-hydrophilic and super-hydrophobic materials to make the glass surface non-adhesive and can operate in both dry and wet dusty conditions.

References

[1]   Rivotti P, Karatayev M, Mourão ZS, Shah N, Clarke ML, and Konadu DD (2019) *Energy Strategy Review.* 24: 261–267.

[2]   Narasipuram RP (2021) Analysis, identification and design of robust control techniques for ultra-lift Luo DC–DC converter powered by fuel cell. *International Journal of Computer Aided Engineering and Technology* 14(1): 102–129.

[3]   Narasipuram RP (2018) Optimal design and analysis of hybrid photovoltaic-fuel cell power generation system for an advanced converter technologies. *International Journal of Mathematical Modelling and Numerical Optimisation* 8(3): 245–276.

[4]   Rajanand Patnaik N and Ravindranath Tagore Y (2016) Design and evaluation of PUC (packed U cell) topology at different levels and loads in terms of THD. *European Journal of Advances in Engineering and Technology* 3(9): 33–43.

[5]   Chaichan MT and Kazem HA (2017) *Generating Electricity Using Photovoltaic Solar Plants in Iraq.* Berlin: Springer.

[6]   Hammad, BK, Rababeh SM, Al-Abed MA, and Al-Ghandoor AM (2013) Performance study of on-grid thin-film photovoltaic solar station as a pilot project for architectural use. *Jordan Journal of Mechanical and Industrial Engineering* 7(1): 1–9.

[7]   Hottel H and Woertz B (1942) Performance of flat-plate solar-heat collectors. *Transactions of the American Society of Mechanical Engineers* 64: 64–91.

[8]   El-Shobokshy MS and Hussein FM (1993) Effect of dust with different physical properties on the performance of photovoltaic cells. *Solar Energy* 51: 505–511. https://doi.org/10.1016/0038-092X(93)90135-B.

[9]   Biryukov SA (1996) Degradation of optical properties of solar collectors due to the ambient dust deposition as a function of particle size. *Journal of Aerosol Science,* 27: S37–S38. https://doi.org/10.1016/0021-8502(96)00091-2.

[10] Javed W, Guo B, Wubulikasimu Y, and Figgis B (2016) Photovoltaic performance degradation due to soiling and characterization of the accumulated dust. In *Proceedings of the IEEE International Conference on Power and Renewable Energy,* vol. 1, pp. 5–9.

[11] Kimber A, Mitchell L, Nogradi S, and Wenger H (2006) The effect of soiling on large grid-connected photovoltaic systems in California and the Southwest Region of the United States. In *Proceedings of the 4th IEEE World Conference on Photovoltaic Energy Conference,* vol. 2, pp. 2391–2395.

[12] Elminir HK, Ghitas AE, Hamid RH, El-Hussainy F, Beheary MM, and Khaled MA-M (2006) Effect of dust on the transparent cover of solar collectors. *Energy Conversation Management* 47(18): 3192–3203. https://doi.org/10.1016/j.enconman.2006.02.014.

[13] Syafiqa A, Pandeya AK, Adzmana NN, and Nasrudin AR (2018) Advances in approaches and methods for self-cleaning of solar photovoltaic panels. *Solar Energy* 162: 597–619. https://doi.org/10.1016/j.solener.2017.12.023.

[14] Michele GA, Pierluigi BZ, Andrea DM, and Elia P (2020) Autonomous robot for cleaning photovoltaic panels in desert zones. *Mechatronics* 68: 102372. https://doi.org/10.1016/j.mechatronics.2020.102372.

[15] Hiroyuki K (2019) Electrostatic cleaning equipment for dust removal from soiled solar panels. *Journal of Electrostatics* 98: 11–16. https://doi.org/10.1016/j.elstat.2019.02.002.

[16] Bernard AR, Eriksen R, and Horenstein MN (2018) Dust settles, we don't: the electrodynamic screen—a self-cleaning technology for concentrated solar power mirrors and photovoltaic panels. *MRS Energy & Sustainability* 5: 11. https:// doi:10.1557/mre.2018.12

Chapter 12

Suryashtmikaran – an Internet of Things-based photovoltaic module cooling and cleaning device

Harshil Sathwara[1], Meet Suraiya[1] and Smita Joshi[2]

Solar rooftop program has become very popular in the domestic, industrial, and agricultural sector. The high conversion efficiency of the solar panel is obtained if they are operated at Standard Test Condition (STC). As a result of shading due to dust particles and temperature rise above the STC of the solar panel, there is a significant reduction in power generation, conversion efficiency, and life of the cell. A controlled water spraying system using a NodeMCU board was found to be a good and economic solution to the current problem. The Suryashtmikaran system is built on the Adafruit IO cloud service, through which users may navigate real-time PV array temperature and water level of tanks. The purpose of this study is to cool and clean the PV panels with the least quantity of water and energy. The result shows the efficiency and power output reaching a maximum of 17.47% and 113.75 W, respectively, when the panel was cleaned and cooled from 9.81% and 64.60 W before cooling and cleaning. Suryashtmikaran with these user-friendly features might be a boon for investors and startups to design and develop various solar-based devices.

12.1 Introduction

There is an increased energy requirement for industrial, commercial, and home purposes due to economic expansion, the rise of facilities, and the development of infrastructure. The impact of global warming and greenhouse gas emissions have roughly doubled over the past several decades due to the industrial revolution, technological breakthroughs, higher rates of energy consumption worldwide, and a rise in the population [1]. Climate change is a major source of worry throughout the world, prompting a push for renewable energy. Among all energy systems, solar PV systems are the most tempting solution for generating electricity due to their low maintenance costs, simplicity of installation, and high efficiency. Solar panel efficiencies are measured at STC, which include solar irradiation of 1,000 W/m^2

[1]Deparment of Electrical Engineering, G H Patel College of Engineering and Technology, India
[2]Deparment of Applied Science and Humanities, G H Patel College of Engineering and Technology, India

and the cell temperature at 25 °C. Unfortunately, in the field, STC is rarely fulfilled, and most solar photovoltaic installations do not operate at 25 °C constantly. More crucially, when the temperature rises, the efficiency of the majority of solar panels decreases [2]. Solar energy is absorbed by these PV panels, with a small fraction (15–20%) being converted to electrical energy. The kind of PV cell and the surrounding environment has a big impact on conversion efficiency. PV system performance is significantly influenced by both internal and external elements, including structural qualities, age, radiation, shadow, temperature, wind, pollution, and cleanliness [3]. Any type of climatic change results in variations in solar radiation and ambient temperature, which affect solar PV output performance. The solar PV panel's surface temperature is substantially greater than the surrounding air, reaching a temperature of 40 °C. When the surface temperature increases, the PV panel's electrical efficiency may decline. PV panel production drops by 0.2–0.5% for every 1 °C increase in panel surface temperature [4]. To maximise the electrical output of PV panels, a cooling system is essential. In addition to the heating, dust buildup on the PV panel has also been a problem, reducing PV panel efficiency. One study found that over the course of a year, the average daily energy loss brought on by dust accumulating on PV module surfaces is about 4.4%. Long stretches of dry weather may cause daily energy losses of more than 20%. The chemical and physical properties of dust particles vary depending on a variety of environmental factors. Isolated dust is affected by air, humidity, temperature, wind velocity, and how the dust settles on PV cells [5]. Hence, it becomes equally important to clean the PV panel at regular intervals. In this work, the researchers developed an inexpensive microcontroller-based water-spraying solar system. The installation site and major expense involved in PV module cooling systems have both been cited as major reasons for the poor adoption of solar cooling devices. As a result, increasing solar PV array efficiency at a comparatively affordable cost was one of the main objectives. The thermal control water spraying system uses a little quantity of water to cool the glass while cleaning it when it notices an increased temperature on the solar panel. The method enables the solar PV module to produce more energy by preventing it from rising over a temperature threshold that restricts efficiency.

In this chapter, the authors have developed an automated system to cool and clean the solar panel in order to improve the efficiency of the solar panel. A controlled water spraying system using a NodeMCU board was found to be a good and economic solution to the current problem. The purpose of this study is to cool and clean the PV panels with the least quantity of water and energy. A significant drop in temperature of the solar panel, improvement in efficiency and thus, the improvement in power output have been noted. All these have been explained via various I–V and P–V characteristics as well as through various plots of efficiency, power output, temperature drop and wind effect in both the conditions, i.e. before and after implementing the system.

12.2   Literature review

Numerous research projects and tests have been carried out to improve the effectiveness of solar systems. Solar panels are an issue because ambient temperature variation affects the surface temperature of the solar panel. By adding more cells, tracking the sun's position, and cooling the solar panels, among other methods, many solar-powered networks only improve their efficiency.

12.2.1 Dust cleaning methods

An important study topic for investigating more advanced cleaning methods and systems is dust cleaning on PV panels. The solar cell was protected while the efficiency of the PV module was boosted. The self-cleaning Nano-film, electro-static dust removal, mechanical dust removal, and natural dust removal were all examined by the authors [6]. For the best power production in an industrial plant, a solar panel cleaning method based on a linear piezoelectric actuator is utilised. To effectively remove the dust layer from the solar module surface, a wiper is connected to the actuator for a linear motion. This cleaning technique is lightweight and compact [7]. However, the solution is limited to industrial owners and might not be economical for many residential and commercial solar panel owners. Reduced solar panel cleaning costs represent a significant research problem for solar plant profitability. The authors' main focus was on enhancing solar plant cleaning procedures in semi-arid climate conditions. Various cleaning techniques are used, and the findings show that water and brush washing is the most efficient technique. The research data, which revealed an average efficacy of 98.8% during rainy seasons and 97.2% during dry seasons, validated the theory [8]. It was thoroughly investigated how air dust particles affected the effectiveness of the PV model. Using a scanning electron microscope, the collected dust samples were examined, and the images obtained were processed to identify the nature and topography of the dust sample particles [9]. A study was done to create an active self-cleaning surface technology that combines mechanical vibration with nano characteristics. A hydrophobic background is covered by hydrophilic curved rungs in an anisotropic ratchet conveyor (ARC) [10]. Land-based PV panels are the focus of the majority of research. But from both technical and financial angles, an evaluation of the cleaning of floating PV systems has been conducted [11].

12.2.2 Temperature cooling methods

In the study, a variety of cooling techniques have been explored to increase the efficiency of solar panels [12]. In this chapter, one such cooling technique is employed. In one of the investigations, a cooling rate model was developed to determine how long it would take to use water spraying to cool the PV panels to their operating temperature. A mathematical model was used to calculate the heating rate of the PV panels and determine when to start cooling [13]. This mentioned cooling technique investigated requires more automation for commercial application. Due to its high latent heat of fusion and shape stability, poly-ethylene glycol/expanded graphite form-stable phase change material (FSPCM) was created and used for cooling solar PV panels in another experiment [14]. In one of the investigations, a computational fluid dynamics (CFD) simulation model was built to test a passive cooling method that uses fins attached to the back of the PV module as well as a method to improve airflow at the back of the PV module by cutting holes in the frame [15]. The research does not help with the dust cleaning aspect. To develop an alternative cooling liquid, however, that would optimise the cooling process while maintaining the fluid properties for a long recycling life, there is now a lot of research being done in this area. The quantity of wind blowing to cool the solar will not be sufficient. In this article, water is employed as the cooling medium for PVs. One study found that one of the best strategies to lessen the temperature rise of the solar cells is to submerge the PV module by few millimetres [16]. Although this submerging system has the advantage of being versatile, it is not appropriate for power generation because hot water is not needed. In one study, researchers increased system efficiency by 8–15% by substituting the top glass surface of the solar panel with water that covered the top surface of the PV panel by 1 mm [17]. The research differs from the present work considering the automation used with the help of a microcontroller by the current chapter authors. One of the research mentions an automated water-spraying technique that was employed in the study to lower the high temperature created on the glass of solar panels. The minimum and maximum working efficiency temperature of cells can be easily known by looking at the datasheet provided by respective PV module manufacturers. Under conditions of high ambient temperature, the temperature of solar panels may be controlled using recognised cooling technologies [18]. However, the research has not taken any steps to harvest the water.

Hence after a thorough literature review, the following research gaps are founded by the authors:

1. Various mechanical methods such as the implementation of wipes and linear piezoelectric actuators are implemented by various researchers for cleaning solar panels, but these methods are restricted only till cleaning and cannot be used for cooling purposes. Not only that, but due to the incorporation of mechanical components into the system, more maintenance is required which adds up to the recurring costs, making the system costly.
2. Water and brush-based methods for panel cleaning are available, but there is no control over the usage of water as well as no proper water harvesting is implemented leading to wastage of water.
3. Floating-based cleaning methods and submerging panels for cooling purposes are both limited by their applications (no widespread usage found for commercial and residential users).
4. Fins-based methods of cooling are restricted by their purpose of only bringing down the temperature of the PV module, and they do not help with cleaning purposes.
5. The various solutions available are cleaning the panel or cooling it down, solving only one issue at a time. Simultaneous solutions lack features of water harvesting and proper automation of the system which includes timer-based control spraying for the least utilisation of energy and water and panel condition monitoring.

All these research gaps have been resolved all at once in this chapter which can be understood well in the upcoming sections of the chapter.

12.3 Methodology

The test was conducted in the month of April (2022) at G. H. Patel College of Engineering and Technology (22.56° N, 72.92° E). The sample which was collected from the date 4 April to 7 April 2022 was analysed and the results are shown in the next section. The readings were taken at the interval of 30 min.

12.3.1 Block diagram

Figure 12.1 shows the circuit connection of the Suryashtmikaran. The Suryashtmikaran uses a NodeMCU microcontroller board for the control operation. MLX90614 sensor is used to measure solar panel temperature whereas the water level of the harvesting tank is measured by HCSR-04 (ultrasonic sensor). HCSR-04 sensor gives the reading of the height of the water tank. The height is used to calculate the volume of a cylinder. All the sensors readings are then sent to NodeMCU, which then controls both the pumps, i.e. main pump and water harvesting pump as well as sends the measured parameters to the Adafruit IO cloud. Since the system is based on a microcontroller, it becomes easy to install and provides a comparatively cheaper solution as compared to those available in the market. Also, there are no mechanical or moving parts, thus minimum or no further maintenance costs are added to the system.

12.3.2 Working procedure

Figure 12.2 shows the flowchart of the Suryashtmikaran, explaining its working. The working of Suryashtmikaran can be well understood by the following steps:

Step 1: Provide a 5V power supply to the NodeMCU (ESP8266) microcontroller board, all the sensors and relays.

Step 2: Try connecting NodeMCU to Wi-Fi.

Step 3: Once the connection is established, the MQTT client class is set up.

Step 4: MLX90614 IR temperature sensor is then initialised.

Step 5: Connect NodeMCU to Adafruit IO cloud service.

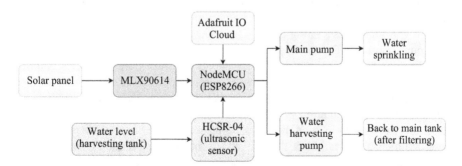

Figure 12.1   Block diagram representing circuit connection of Suryashtmikaran

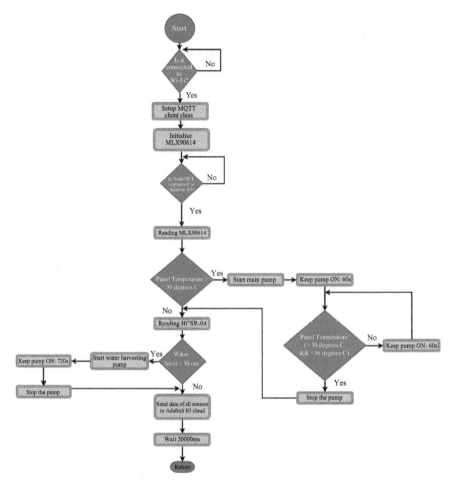

Figure 12.2   Working of Suryashtmikaran

Step 6: MLX90614 IR temperature sensor measures the temperature of the solar panel. If the panel temperature is more than 30 °C, then it starts the main pump. This pump has a rating of 20 W, connected to AC mains through the relay. The pump is turned ON for 60 s.

Step 7: Again after that, panel temperature is measured by an MLX90614 temperature sensor, and if the temperature is found in the range of 30–36 °C, then the pump is turned OFF. Or else the pump is kept further ON for 60 s and step 7 is repeated until the condition is not satisfied. This threshold range of temperature was selected keeping in mind the water temperature available on various household and commercial bases. Also, the aim was to use the least amount of water and energy usage, because the continuous operation of the water pump may lead to more energy consumption.

Step 8: If the condition in step 6 is satisfied, i.e. the temperature is below 30 °C, then the volume of the water harvesting tank is measured using HCSR-04.

Step 9: A threshold of 30 cm height is kept for the HCSR-04 sensor. So if the measured value from HCSR-04 is less than 30 cm, then it means that the water harvesting tank is being filled up by the water coming from the solar panel after the sprinkling process. Hence the main tank is getting emptied and requires to be filled up.

Step 10: The water harvesting pump is then turned ON for 720 s. After that, it is turned OFF.

Step 11: If the threshold of HCSR-04 in step 10 is not satisfied, then the NodeMCU sends all the measured readings, i.e. solar panel temperature and volume of water harvesting tank are sent to Adafruit IO cloud service.

Step 12: Wait for 20 s and repeat the procedure from step 1.

12.3.3 Experimental setup

A solar panel testing kit was developed by the researchers for simplifying the procedure of making the connections while taking readings for the I–V and P–V curves (Table 12.1). This kit not only helps make the connection faster and easier but also ensures that no loose connections are left. Researchers simply have to place probes of both the DMM in the connectors and connect the ends of the rheostat as per the circuit diagram drawn on the kit. Then researchers have to take readings of solar panel voltage and current by varying the resistance through rheostat as per the standard procedure [19].

Figure 12.3 Experimental setup

Table 12.1   List of instruments/components used in the experimental setup as shown in Figure 12.3

Sr. no.	Instruments/ component	Use
1	Digital Multimeter as voltmeter	To measure the voltage of the panel
2	Digital Multimeter as ammeter	To measure the current of the panel
3	Rheostat	For varying the load
4	Pyranometer	To measure solar irradiation
5	Pyranometer display	To show pyranometer reading digitally
6	Solar panel testing kit	For experimenting
7	Main water tank 3 layer	200 L capacity, to store cool water
8	Water tank 3 layer	To store water that came after cleaning and cooling the solar panel, thus ensuring the harvesting of water
9	Water sprinkling system	Made by drilling holes at equal intervals, in a UPVC pipe. The holes drilled in the pipe are made in such a way that the complete area of the solar panel is covered with proper pressure at a faster rate. This makes sure that minimum water is utilised
10	Water harvesting system	To ensure no water is wasted while cooling and cleaning the panel. The water sprinkled on the solar panel gets collected through this system and stored in a water harvesting tank
11	Pt-100	For measurement of solar panel temperature, the temperature of water in the main water tank and the temperature of the water in the water harvesting tank
12	Pt-100 channel display	To show Pt-100 reading digitally
13	Suryashtmikaran	For controlling the systems
14	Power supply	Power supply to provide AC mains power supply to the pumps placed in both the water tanks as well as to the Suryashtmikaran
15	Solar panel 160 W	For the experiment purpose

12.4   Results and discussion

The performances of the solar panel were tested with different parameters affecting the efficiency of the panel and analysis was done by plotting graphs of input power vs. output power, current vs. voltage, temperature vs. power, power vs. voltage and efficiency with and without cooling and cooling the panels as well as the effect of wind velocity.

12.4.1   Effect of cleaning and cooling on power output of solar panel

Figure 12.4 shows the graphical relation of power output with dust and without cooling, power output without dust and with cooling, along with the variation of solar radiation (measured using a pyranometer in terms of W/m^2) with respect to time. The maximum difference in power output was found to be 74.29 W at 13:00 h and the minimum difference was 8.99 W at 14:30 h. During the testing (of 160 W panel), the maximum power output was 64.60 W when the panel was covered with dust and without any kind of cooling, while the maximum power output was 113.75 W at 13:00 h when the panel was cleaned and cooled. This indicates an increase in power output due to cleaning as well as cooling down of the panel.

12.4.2 Temperature effect on power output of solar panel

Figure 12.5 shows the graphical relation of power output and temperature concerning time. The maximum temperature drop was found to be 27 °C at 12:00 h and

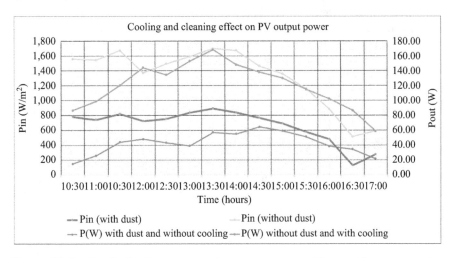

Figure 12.4   Graph showing variation in power output and input with respect to time

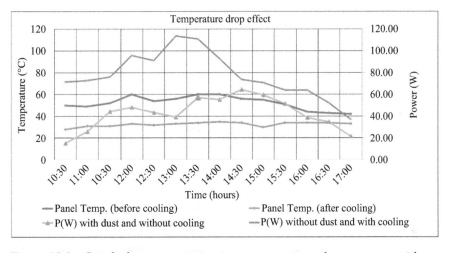

Figure 12.5   Graph showing variation in power output and temperature with respect to time

the minimum drop was 9 °C at 16:30 h. The temperature of the panel was measured using a Pt-100 sensor, which was placed on the glass of the panel. The maximum temperature reached was 60 °C at 12:00 h, 13:30 h and 14:00 h, when the panel was not cooled. Whereas, the maximum temperature was 35 °C at 14:00 h when the panel was cooled down. The temperature coefficient is the temperature effect on PV performance. As a result, as the temperature rises, power output decreases. The percentage of temperature coefficient, which measures changes in production when temperatures deviate from typical ranges of 25 °C, does so. Near the equator, the panel temperature reaches 70 °C on bright days. Poor currents and a local hotspot are the results of extreme temperatures. As a result, panel temperatures exceeding 45 °C shorten their lifespan and affect their electrical performance. Therefore, it is essential to remove heat from the PV using suitable cooling systems to reduce the negative impacts of cell temperature and maintain the operating temperature within the range recommended by the manufacturer.

12.4.3   Efficiency improvement

Figure 12.6 shows the efficiency of the PV module calculated at the point of maximum power in both cases. When the panel was covered with dust and no cooling system was implemented, then the maximum efficiency was found to be 9.81%. Whereas, after cleaning and cooling the panel, the maximum efficiency attained was 17.47%. The difference in efficiencies was found to be in the range of 11.47 (max) and 2.18 (min). This indicates the improvement in efficiency due to the incorporation of Suryashtmikaran.

12.4.4   I–V and P–V characteristics

Figures 12.7–12.10 show the I–V and P–V characteristics of the solar PV panel plotted from the readings taken from the date 4 April to 7 April, from 10:30 h to

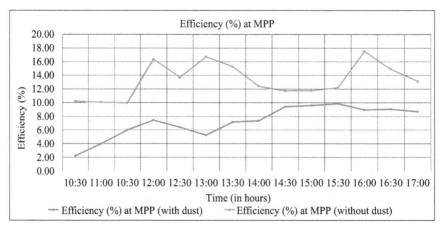

Figure 12.6   Graph showing percentage efficiency at maximum power point (MPP)

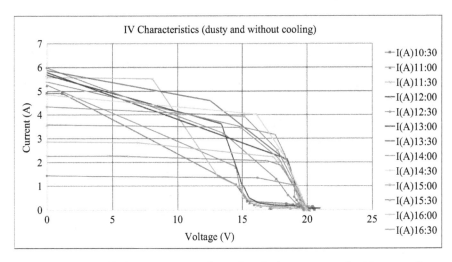

Figure 12.7   I–V characteristics of panel with dust cover and without cooling

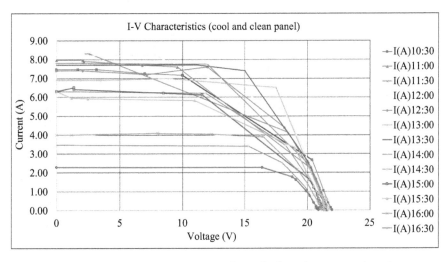

Figure 12.8   I–V characteristics of panel after cleaning and cooling

17:00 h, after 30 min of interval. The number of photons absorbed by the semiconductor material affects the short circuit current (I_{SC}), which is proportional to light intensity [20]. As a result, the conversion efficiency is largely constant, and the power output is typically linked with the irradiance, however, it drops as the cell temperature rises. The open-circuit voltage (V_{OC}) varies only slightly with light intensity [21]. The V_{OC} reduces dramatically as the panel temperature rises above 25 °C, while the short-circuit current, I_{SC}, increases only minimally.

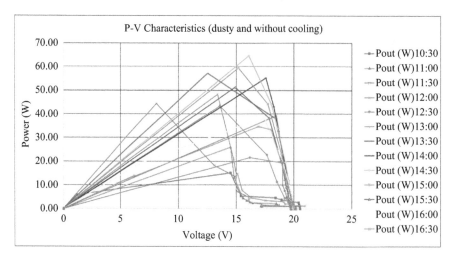

Figure 12.9   P–V characteristics of panel with dust cover and without cooling

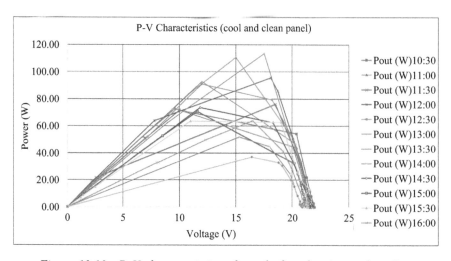

Figure 12.10   P–V characteristics of panel after cleaning and cooling

12.4.5   Effect of wind

Figure 12.11 gives the readings of wind velocity measured using an anemometer. The notion that wind velocity has a direct influence on solar PV efficiency may be incorrect. It does, however, play a significant role in PV generation. The temperature of the solar cell drops while the wind blows [22]. The wind cools the solar panels, generating less oscillation of the electrons and allowing the electrons to transmit more energy while ascending. A detailed study to determine the

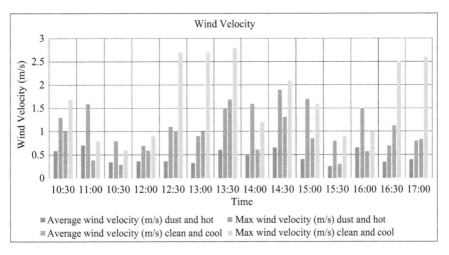

Figure 12.11   Wind velocity data

temperature of PV modules as a function of solar irradiation, ambient temperature, and wind velocity is presented in one such study [23].

12.5   Conclusion

The purpose of this study is to cool and clean the PV panels with the least utilisation of water and energy, at an affordable cost, which is an advancement fulfilling all the research gaps mentioned in the literature review section. As per the algorithm, a non-pressurised cooling system based on spraying the PV panels with water occasionally has been developed. According to the results of the water sprayed solar algorithm, there is a significant improvement in the power generated by the solar panel, with the efficiency reaching a maximum of 17.47% when cleaned and cooled from 9.81% when the panel was heated and shaded with dust. During the testing (of 160 W panel), the maximum power output was 64.60 W and the maximum temperature was 60 °C when the panel was covered with dust and without any kind of cooling, while the maximum power output was 113.75 W and the maximum temperature was 35 °C when the panel was cleaned and cooled. For the solar panel to work efficiently, it is important that their operational temperature is maintained at a lesser magnitude (near to STC) as well as cleaned at regular intervals. When opposed to adding more solar panels to boost energy output, this design offers substantial space savings. Regarding the future scope, the researchers are planning to add real time clock (RTC) and automation systems into the system. RTC can be set as per the requirement to cool and clean the panel at regular intervals whereas, through an automation system, one can control the pumps as per their wish through their smartphone. This design may be put into action by improving the prototype into a commercially viable product.

Acknowledgement

The authors would like to express their gratitude to Dr H. B. Soni, Principal of GCET, and Dr Ritesh Patel, HOD of the Electrical Department. The authors wish to acknowledge Mr Dhaval Kalathiya for setting up the test facility. The authors would also like to express their gratitude to IIT Bombay and MNRE for supplying the solar panel kit. The Student Start-Up Innovation Program (SSIP) Cell at the G.H. Patel College of Engineering and Technology is appreciative of a grant of INR 35,000/- (ID-GCETSSIP2021-22EE02) to set up the Suryashtmikaran and execute the experiment.

References

[1]   V. Karthikeyan, C. Sirisamphanwong, S. Sukchai, S. K. Sahoo, and T. Wongwuttanasatian, "Reducing PV module temperature with radiation based PV module incorporating composite phase change material," *J. Energy Storage*, vol. 29, no. November 2019, p. 101346, 2020, doi: 10.1016/j.est.2020.101346.

[2]   O. Dupré, R. Vaillon, and M. A. Green, *Thermal Behavior of Photovoltaic Devices: Physics and Engineering.* New York, NY. Springer, 2017.

[3]   R. J. Mustafa, M. R. Gomaa, and M. Al-dhaifallah, "Environmental impacts on the performance of solar photovoltaic systems," *Sustain.*, vol. 12, no. 2, pp. 1–17, 2020, https://doi.org/10.3390/su12020608.

[4]   C.-Y. Huang, H.-C. Sung, and K.-L. Yen, "Experimental study of photovoltaic/thermal (PV/T) hybrid system," *Int. J. Smart Grid Clean Energy*, vol. 2, no. 2, pp. 148–151, 2013, doi: 10.12720/sgce.2.2.148-151.

[5]   Z. Darwish, H. Kazem, K. Sopian, M. Alghoul, and M. Chaichan, "Impact of some environmental variables with dust on solar photovoltaic (PV) performance: review and research status," *Int. J. Energy Environ.*, vol. 7, no. 4, pp. 152–159, 2013.

[6]   G. He, C. Zhou, and Z. Li, "Review of self-cleaning method for solar cell array," *Procedia Eng.*, vol. 16, pp. 640–645, 2011, doi: 10.1016/j. proeng.2011.08.1135.

[7]   X. Lu, Q. Zhang, and J. Hu, "A linear piezoelectric actuator based solar panel cleaning system," *Energy*, vol. 60, no. January 2019, pp. 401–406, 2013, doi: 10.1016/j.energy.2013.07.058.

[8]   A. Raza, A. R. Higgo, A. Alobaidli, and T. Zhang, "Water recovery in a concentrated solar power plant," in *International Conference on Concentrating Solar Power and Chemical Energy Systems*, 2016, vol. 1734, no. May 2016, doi: 10.1063/1.4949255.

[9]   A. Hussain, A. Batra, and R. Pachauri, "An experimental study on effect of dust on power loss in solar photovoltaic module," *Renewables Wind. Water, Sol.*, vol. 4, no. 1, 2017, doi: 10.1186/s40807-017-0043-y.

[10] D. Sun and K. F. Böhringer, "An active self-cleaning surface system for photovoltaic modules using anisotropic ratchet conveyors and mechanical

vibration," *Microsystems Nanoeng.*, vol. 6, no. 1, p. 87, 2020, doi: 10.1038/s41378-020-00197-z.

[11] R. Zahedi, P. Ranjbaran, G. B. Gharehpetian, F. Mohammadi, and R. Ahmadiahangar, "Cleaning of floating photovoltaic systems: a critical review on approaches from technical and economic perspectives," *Energies*, vol. 14, no. 7, p. 2018, 2021, doi: 10.3390/en14072018.

[12] P. Dwivedi, K. Sudhakar, A. Soni, E. Solomin, and I. Kirpichnikova, "Advanced cooling techniques of P.V. modules: a state of art," *Case Stud. Therm. Eng.*, vol. 21, no. June, p. 100674, 2020, doi: 10.1016/j.csite.2020.100674.

[13] K. A. Moharram, "Enhancing the performance of photovoltaic panels by water cooling," *Ain Shams Eng. J.*, vol. 4, no. 4, pp. 869–877, 2013, doi: 10.1016/j.asej.2013.03.005.

[14] S. K. Marudaipillai, B. K. Ramaraj, R. K. Kottala, and M. Lakshmanan, "Experimental study on thermal management and performance improvement of solar PV panel cooling using form stable phase change material," *Energy Sources, Part A Recover. Util. Environ. Eff.*, vol. 00, no. 00, pp. 1–18, 2020, doi: 10.1080/15567036.2020.1806409.

[15] J. Kim and Y. Nam, "Study on the cooling effect of attached fins on PV using CFD simulation," *Energies*, vol. 12, no. 4, p. 758, 2019, doi: 10.3390/en12040758.

[16] G. M. Tina, M. Rosa-clot, P. Rosa-clot, and P. F. Scandura, "Optical and thermal behavior of submerged photovoltaic solar panel: SP2," *Energy*, vol. 39, no. 1, pp. 17–26, 2012, doi: 10.1016/j.energy.2011.08.053.

[17] S. Krauter, "Increased electrical yield via water flow over the front of photovoltaic panels," *Sol. Energy Mater. Sol. Cells*, vol. 82, pp. 131–137, 2004, doi: 10.1016/j.solmat.2004.01.011.

[18] D. Ramere, "Efficiency improvement in polycrystalline solar panel using thermal control water spraying cooling," *Procedia Comput. Sci.*, vol. 180, pp. 239–248, 2021, doi: 10.1016/j.procs.2021.01.161.

[19] D. Dirnberger, *Photovoltaic Module Measurement and Characterization in the Laboratory*. New York, NY: Elsevier Ltd., 2017.

[20] W. Luo, Y. S. Khoo, P. Hacke, *et al.*, "Potential-induced degradation in photovoltaic modules: a critical review," *Energy Environ. Sci.*, vol. 10, no. 1, pp. 43–68, 2017, doi: 10.1039/c6ee02271e.

[21] B. V Chikate and Y. Sadawarte, "The factors affecting the performance of solar cell," *Int. J. Comput. Appl. Sci. Technol.*, vol. 4, pp. 975–8887, 2015.

[22] J. K. Kaldellis, M. Kapsali, and K. A. Kavadias, "Temperature and wind speed impact on the efficiency of PV installations. Experience obtained from outdoor measurements in Greece," *Renew. Energy*, vol. 66, pp. 612–624, 2014, doi: 10.1016/j.renene.2013.12.041.

[23] C. Schwingshackl, M. Petitta, J.E.Wagner, *et al.*, "Wind effect on PV module temperature: Analysis of different techniques for an accurate estimation," *Energy Procedia*, vol. 40, pp. 77–86, 2013, doi: 10.1016/j.egypro.2013.08.010.

Chapter 13

Robust control for energy storage system dedicated to solar-powered electric vehicle

Marwa Ben Saïd-Romdhane[1,2], Sondes Skander-Mustapha[1,3]
and Ilhem Slama-Belkhodja[1]

Nowadays, among the objectives of solar-powered electric vehicles are to promote and support electric mobility growth in rural areas. This is because charging stations in these areas are few and far between. In addition, the decrease of city obstacles like buildings and tunnels allows this kind of vehicle to work with high performance which makes it ideal for these areas. In this chapter, the control and energy management of a solar-powered electric vehicle energy storage system is investigated. The proposed system is composed of a photovoltaic system as a renewable energy source, batteries, and supercapacitors as storage systems. The role of the photovoltaic system is to charge the battery or supply the auxiliary loads when the battery reaches its fully charged state. Supercapacitors act in repetitive charge and discharge. Their role is to supply fast power demand. They are charged through the direct current bus essentially by photovoltaic energy. But if shading prevents photovoltaic production, the battery takes over. In this case, some loads can be shed. This chapter proposes a global solution to control this system. First, an optimal frequency separation energy management strategy is adopted to ensure a good power distribution between each component of the energy storage system. Internet of Things-based wireless battery management system is adopted to define the battery and the supercapacitor state of charge. Second, an H-infinity-based controller is proposed for the energy storage system power converters to enhance stability in solar-powered electric vehicles. The suggested controller offers robust stability by ensuring the perfect rejection of disturbances that come from direct current bus fluctuations and parameters variations. Simulations achieved under MATLAB® software are presented and discussed to validate the effectiveness and high performance of both the proposed H-infinity control and energy management strategy.

[1]Université de Tunis El Manar, Ecole Nationale d'Ingénieurs de Tunis, LR11ES15 Laboratoire de Systèmes Electriques, Tunisia
[2]Université de Gabès, Institut Supérieur des Sciences Appliquées et de Technologie de Gabès, Tunisia
[3]Université de Carthage, Ecole Nationale d'Architecture et d'Urbanisme, Tunisia

13.1 Introduction

Motivated by the rapid evolution towards a sustainable energy alternative especially for transport sectors and the global pressure to reduce carbon emissions and fossil fuel consumption, the expansion and the use of electric vehicles (EVs) are on the rise [1]. With their acceptable energy efficiency, reduced working noise and minimal CO_2 emissions, this kind of vehicle will be one of the most promising energy alternatives for the future in the field of transport [2].

Moreover, thanks to the significant progress of photovoltaic technology and to exploit inexhaustible and clean energy sources, EVs have integrated photovoltaic (PV) modules generally placed on the EV roof. These solar-powered electric vehicles (SEVs) allow for decreasing transport costs as well as saving the environment by making it possible to achieve "zero pollution". However, one of the most significant drawbacks of this vehicle is the shading phenomenon, especially in urban areas [3,4]. This shading mainly affects the SEV energy storage system (ESS) [5]. It should be noted here that the shading phenomenon is reduced in rural areas, which promotes the use of solar-powered electric vehicles.

Many authors investigate the electrical vehicle ESS [6–8]. The option with fuel cells is widely studied but the most important technical barrier is the heavyweight caused by fuel cell devices and hydrogen tank which compromise electric vehicle energy efficiency. The alternative with battery and supercapacitor is more realistic for the case of SEVs, but even with this choice, several difficulties must be overcome.

To overcome these problems, both energy management and control of the SEV ESS should be carefully selected. The frequency separation method is usually employed to manage the ESS behaviors in EVs. This is due to its flexibility in various driving conditions as well as its adaptative online response [9,10]. On the other hand, many control strategies were proposed in the literature to ensure robust control of the SEV ESS [11,12].

In this chapter, the frequency separation approach and the H-infinity (H_∞) control are proposed to ensure the optimal operating mode of the SEV ESS. In fact, the frequency separation strategy offers complementarity between the different storage units. Moreover, the H_∞ control applied to the ESS direct current to direct current (DC/DC) converters ensures robust stability by being robust even for load variation, parameters change, and DC bus fluctuations [13,14].

This chapter is organized as follows. First, in Section 13.2, the ESS in a solar-powered electric vehicle is described. Then, in Section 13.3, the H_∞ control for ESS is presented with details on the weighting functions tuning. After that, the energy storage devices and the energy management strategy are presented in Sections 13.4 and 13.5, respectively. Simulation results achieved under the MATLAB®-Simulink® software tool are given in Section 13.6. The obtained simulation results prove the performance, efficiency, and robustness of both the proposed energy management strategy and H_∞ control of the energy storage system in case of the solar-powered electric vehicles.

13.2 ESS description

Figure 13.1(a) presents the main electrical parts of the solar-powered electric vehicle which are the rooftop PV array, the ESS, the DC/DC power converters, the DC bus, the three-phase inverter, the electric motor, and the auxiliary systems. For the rooftop PV array, the most used types are polycrystalline silicon, monocrystalline silicon, and thin film solar panels that include copper indium gallium selenide, cadmium telluride and amorphous silicon [15]. Each type of solar cells has different characteristics. In general, monocrystalline silicon solar panels are chosen for SEVs, thanks to their long lifetime and highest efficiency [16]. The ESS is composed of a battery and a super-capacitor. The supercapacitor is used as a second energy storage device to take advantage of the fast charging and discharging [17]. There are four battery choices which are nickel metal hydride, lithium polymer, lead acid, and lithium–ion [18]. For SEVs, it is preferable to use a lithium–ion battery due to its size and weight advantages, fast and efficient charging in addition to its high energy density [19]. The most used electric motor in SEVs is Brushless DC Motor (BLDCM) and Permanent Magnet Synchronous Motor (PMSM) [20]. For the auxiliary systems, they can be divided into

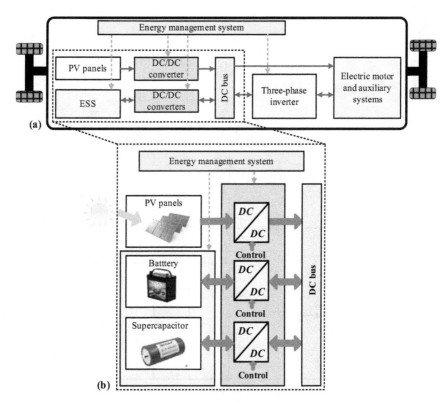

*Figure 13.1   The electrical system of the solar-powered electric vehicle.
(a) General configuration and (b) studied system*

two categories: necessary and luxury auxiliary systems. The necessary ones include heating, horn, and lights. The luxury ones include satellite navigation, brake booster, sound system, air conditioner, power steering, etc.

This chapter will focus only on the first four parts as depicted in Figure 13.1(b). The ESS is composed of a lithium battery and a supercapacitor (SC). The PV array is linked to a maximum power point tracking (MPPT) unidirectional boost converter to ensure maximum output power. The goal of this converter is to provide the MPPT for solar panels while transforming the low PV array voltage to a high-voltage necessary for the DC bus. The battery stores energy when the vehicle decelerates or stops while PV panels provide excessive solar energy. In addition, the battery provides energy when the load demands more power than the PV power delivery capacity. It should be noted here that the load is the electric motor and the auxiliary systems as shown in Figure 13.1(a).

The controlled DC/DC boost converter of the battery and the supercapacitor (SC) are depicted in Figure 13.2(a) and (b), respectively. These converters are composed of two insulated-gate bipolar transistor (IGBT) (S_1 and S_2), an inductor (L_0), a capacitor (C_0), and a load resistor (R_0) as depicted in Figure 13.2. Following [21], the boost converter transfer function is expressed as follows.

$$G(s) = \frac{V_{dc}(s)}{d(s)} = \frac{V_{Bat}}{(1-d)^2 R_0 C_0} \frac{s - \frac{(1-d)^2 R_0}{L_0}}{s^2 + \frac{s}{R_0 C_0} + \frac{(1-d)^2}{L_0 C_0}} \tag{13.1}$$

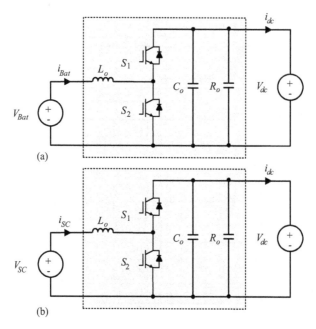

(a)

(b)

Figure 13.2   DC/DC boost converter. (a) Battery and (b) SC

where *d* is the duty ratio which varies between 0 and 1. V_{Bat} and V_{SC} are the input voltage of the DC/DC boost converter of the battery and the SC, respectively. V_{dc} is the DC bus voltage and *s* denotes the *Laplace* variable.

13.3 H∞ control for ESS

In the present work, the PV panels DC/DC converter is regulated via an MPPT algorithm as shown in Figure 13.3. For the ESS DC/DC boost converters, only the battery current i_{Bat} and the SC current i_{SC} are regulated as depicted in Figure 13.3. These currents are controlled based on H∞ control algorithm. Figure 13.4 presents the regulation of the current i_{Bat} and the same principle is valid for i_{SC}. It should be noted that the H∞ controller ensures robust stability and performance, thanks to its high rejection capacity of disturbances. In fact, the main objective of this controller is to regulate with high robustness the i_{Bat} and the i_{SC} currents even in presence of an external nose as well as different parameters variations.

Figure 13.5 shows the H∞ battery current control block. Figure 13.6 presents the closed-loop system composed of the H∞ controller *K* and the system transfer function *G*. *y* and *z* denote the measured and the regulated outputs, respectively. *u* and *w* denote the regulated and the exogenous inputs, respectively. The weighting functions that characterize the H∞ control are denoted $W_{i(i=1,2,3)}(s)$. $W_1(s)$, $W_2(s)$, and $W_3(s)$ are the weighting functions for the error tracking, the system transfer function and the system robust performance, respectively. The objective of the H∞

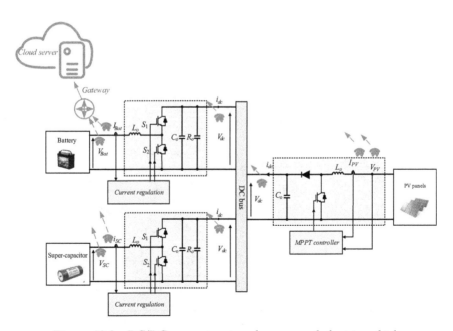

Figure 13.3 DC/DC converters in solar-powered electric vehicle

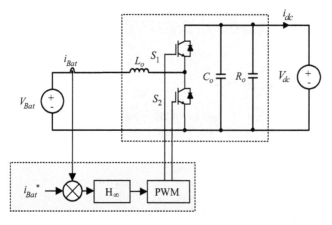

Figure 13.4 H_∞ control for the battery current i_{Bat}

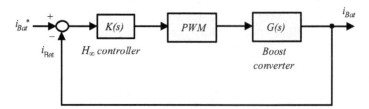

Figure 13.5 H_∞ battery current control block

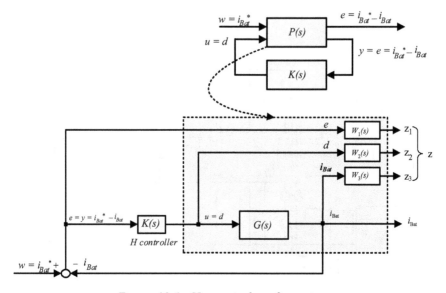

Figure 13.6 H_∞ control configuration

control is to synthesize a controller $K(s)$ so that the H_∞ gain from w to z is less than γ as depicted on (13.2). γ presents the optimality level and it is between zero and one. As shown in (13.2), three conditions should be ensured. In this equation, $S(s)$, $R(s)$, and $T(s)$ are the sensitivity transfer function, the control effort transfer function, and the complementary sensitivity transfer function, respectively. These transfer functions are given by (13.3), (13.4), and (13.5), respectively. The relation between $S(s)$ and $T(s)$ is given by (13.6). In the following, the weighting functions are tuned considering the transfer functions $T(s)$, $S(s)$, and $R(s)$ and (13.2):

$$\|T_{wz}\|_\infty < \gamma \iff \left\| \begin{matrix} W_1 S(s) \\ W_2 R(s) \\ W_3 T(s) \end{matrix} \right\|_\infty < \gamma \tag{13.2}$$

$$S(s) = \frac{1}{1 + G(s)K(s)} \tag{13.3}$$

$$R(s) = S(s)K(s) \tag{13.4}$$

$$T(s) = \frac{K(s)G(s)}{1 + K(s)G(s)} \tag{13.5}$$

$$T(s) + S(s) = 1 \tag{13.6}$$

The limitations and the specifications required by the designer are determined by the weighting functions. General guidelines for weighting functions tuning are detailed in [21,22]. In this chapter W_1, W_2, and W_3 are selected as:

$$W_1(s) = \frac{1}{K_1} \frac{s + \omega_c K_1}{s + \omega_c k_1} \tag{13.7}$$

$$W_2(s) = \frac{1}{k_2} \frac{s + \frac{\omega_c}{K_2}}{s + \frac{\omega_c}{k_2}} \tag{13.8}$$

$$W_3(s) = K \tag{13.9}$$

where K_1, K_2, k_1, k_2, and K are given by the following equations:

$$K_1 = K_2 = e^{-\frac{\log(10)G_m}{20}} \tag{13.10}$$

$$k_1 = k_2 = \varepsilon \tag{13.11}$$

$$K \ll \varepsilon \tag{13.12}$$

In (13.10), ω_c, G_m, and ε denote the desired cutoff frequency, the gain margin and the steady-state error, respectively. Once, the weighting functions are tuned, an H_∞ controller $K(s)$ can be defined thanks to the mixsyn MATLAB software function and based on the obtained weighting functions $W_1(s)$, $W_2(s)$, $W_3(s)$, and the system transfer function $G(s)$.

13.4   Energy storage devices

To decide the appropriate energy storage devices for electrical vehicles, several criteria should be considered such as lightweight, affordable, and long-life [23,24]. In the case of using several storage components, the complementarity between them must also be respected for solar-powered electric vehicle applications to simultaneously provide high power and high energy storage. Indeed, as the manufacturing technology varies depending on the type of storage elements, their behavior is also different. For example, capacitors that store energy using electrostatic charge at their electrode–electrolyte terminals depending on the affected voltage. This approach provides energy for a short time which is beneficial in case of vehicle acceleration or braking. Batteries technology is different. Their storage process is done using electrochemical reactions in the whole substance of the electrode. This leads to a big quantity of storage energy [23]. The Ragone plot is a practical tool to guide users in hybridizing several storage elements (Figure 13.7). It is a system composed of two orthogonal axes presenting the power density in (W/kg) and the energy density in (Wh/kg). Storage devices are placed in this reference system according to their physical characteristics.

In this study, lithium–ion batteries and supercapacitors are used, since they are placed on either side in the Ragone plot, which ensures their complementarities. The hybridization of supercapacitors and lithium–ion batteries is a good alternative for electrical vehicles, especially in the case of urban traffic. Compared to Ni–Cd and lead acid batteries, lithium–ion ones are characterized by higher specific energy (100–265 Wh/kg) [25], thanks to their great energy density electrode materials [26]. Supercapacitors participate by their higher power density and reliability in addition to their long life characteristic [27]. This option induces the reduction of system peak power.

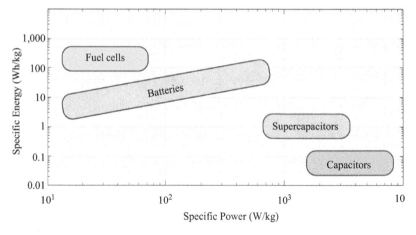

Figure 13.7   Ragone plot

13.5   Energy management strategy

To benefit from the hybridization of storage elements, it is mandatory to ensure an overall energy management strategy that respects the physical characteristics of each device. The frequency separation (FS) approach is a good candidate for that. The principle of the frequency separation strategy is to assign the more adapted energy frequency bandwidth to each storage component. Low frequencies are assigned to batteries with respect to their physical characteristics marked by elevated energy density and low power density. While high frequencies are assigned to supercapacitors, which are specified by low energy density and an elevated power density. To accomplish the frequency separation, it is necessary to carefully choose the employed filter cut-off frequency. As defended, the battery should work under low-frequency bandwidth as the SC operates with high-frequency one. The FS working assumption is described in Figure 13.8. The photovoltaic system is the principal energy source. The missing power is provided by the storage systems, with respect to the physical characteristic of each device as presented in (13.13). The main benefit of the frequency separation scheme is to improve battery and supercapacitor lifetimes:

$$P_{Load} - P_{PV} = P_{SC} + P_{Bat} \tag{13.13}$$

The low pass filter transfer function is presented by (13.14). Where T_s, τ, and k are the sample time, the time constant, and the filter gain, respectively:

$$F(z) = k \frac{\frac{T_s}{\tau} z^{-1}}{1 + \left(\frac{T_s}{\tau} - 1\right) z^{-1}} \tag{13.14}$$

The filter Bode plot presented in Figure 13.9 shows the filter frequency response. It is practically horizontal for low frequencies (less than 1 Hz for the first case and less than 10 Hz for the second case). For this operating range, all power signal is delivered to the output, with a gain of near unity. When the signal reaches the cut-off frequency value ($F_{cut-off}$), it decreases to zero according to the fixed slope.

Internet of Things (IoT)-based wireless battery management system (WBMS) is adopted to establish the battery and the SC State of Charge (SoC$_{Bat}$ and SoC$_{SC}$).

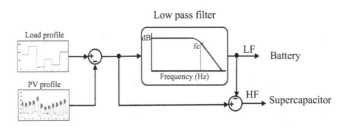

Figure 13.8   Frequency separation strategy principle

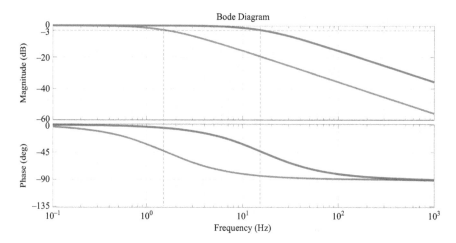

Figure 13.9   Low pass filter Bode plot for two values of cut-off frequency (red curve corresponds to 10 Hz and blue curve corresponds to 1 Hz)

In fact, IoT-WBMS allows for reducing the wire-harness concerns as well as interferences and communication power losses.

A leader node (selected by the Leader Election Algorithm) employs the Message Queuing Telemetry Transport (MQTT), to collect measured values (the current, the voltage, and the temperature) [28–30]. Then it sends SoC_{Bat} and SoC_{SC} values to an intermediate module, which decides whether to allow or not the frequency separation module to continue discharging or charging the battery.

13.6   Simulation results

In this section, simulation results performed under MATLAB/Simulink environment are presented to validate the performances of both the proposed H_∞ controller and the proposed energy management strategy for an energy storage system dedicated to a solar-powered electric vehicle. The specification of the employed energy storage system supercapacitor and the battery is given in Tables 13.1 and 13.2, respectively.

The main goal of the H_∞ controller is to make the output current track its reference with the rapid transient response and to ensure the disturbance rejection. The current tracking performance is verified by varying the battery reference current i_{Bta}* as shown in Figure 13.10. The reference current value varies from 15 A to 30 A and then to 10 A. For this test, the load resistance R_0 is equal to 10 Ω. As it can be seen in Figure 13.10, the output current follows perfectly its reference with a fast settling time and without any unacceptable peak or overshoot. Consequently, the obtained H_∞ controller ensures good tracking performance with the rapid transient response for energy storage systems dedicated to a solar-powered electric vehicle.

Table 13.1   Specification of the supercapacitor TC2700 of
Maxwell Technologies

Description	Value	Unit
Capacitance	2,700	F
Voltage		
Continues	2.5	V
Peak	2.7	V
Stored energy	2.3	Wh
Series resistance	0.7	m Ω
Weight	600	g

Table 13.2   Specification of the power brick lithium–ion
battery pack

Description	Value	Unit
Capacity	70	Ah
Voltage	12.8	V
Stored energy	896	Wh
Internal resistance	40	m Ω
Weight	9.8	kg

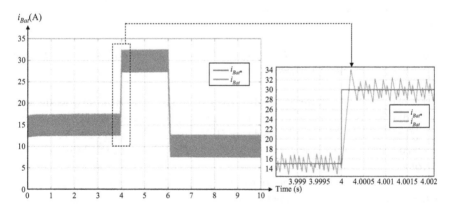

Figure 13.10   Transient performances of the battery current i_{Bat} *for different*
reference current

On the other hand, to verify the robustness of the proposed controller, tests with various load resistance R_0 and various values of the input voltage V_{Bat} are presented. Figure 13.11(a) shows the battery current i_{Bat} for different values of load resistance R_0. The taken values of this resistance are 10 Ω, 100 Ω, and 150 Ω. As presented in this figure, the battery current i_{Bat} is well regulated with good transient

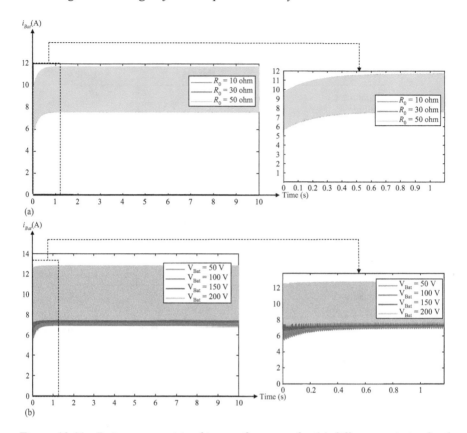

Figure 13.11   Battery current tracking performance for (a) different resistive load R_0 and (b) different voltage input V_{Bat}

performance even for different load resistances. Figure 13.11(b) presents the evolution of the current battery i_{Bat} for different values of the input voltage V_{Bat} shown
in Figure 13.11(b), the input voltages are 50 V, 100 V, 150 V, and 200 V. It should
be noted based on this figure, that the H_∞ controller offers good results even in case
of voltage input change.

The last test that shows the robust performance of the designed H_∞ controller
is the robustness against parameter uncertainty. Figure 13.12 presents the battery
current i_{Bat} response when the inductance L_0 varies from its nominal value to $\pm50\%$
of its nominal value. It is clear from Figure 13.12, that the output current i_{Bat}
perfectly follows its reference, without any noticeable ringing or overshoot.
Figure 13.13 shows the evolution of the battery current i_{Bat} when C_0 varies from its
nominal value to $\pm50\%$ of its nominal value. As presented in this figure, the current
i_{Bat} is well regulated even in the case of capacitor variation. These tests prove the
robustness of the H_∞ controller in presence of parameter uncertainty.

A variable load is considered to validate the operation of the frequency
separation EMS. As presented in Figure 13.14, fast current demands are supplied

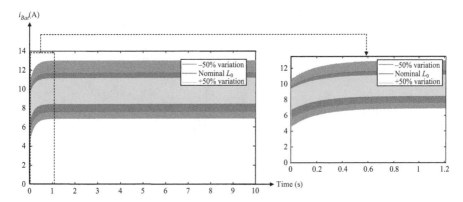

Figure 13.12   Battery current tracking performance when L_0 varies from 2 mH to 3 mH (2 mH ± 50%)

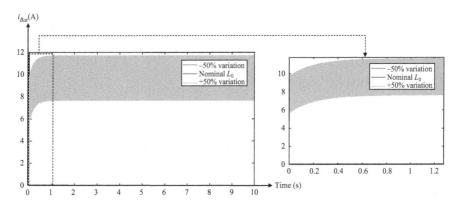

Figure 13.13   Battery current tracking performance when C_0 varies from 800 μF to 2,400 μF (1,600 μF ± 50%)

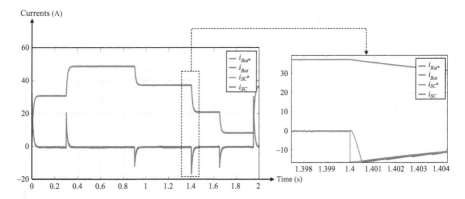

Figure 13.14   Battery and SC current profile with frequential separation approach

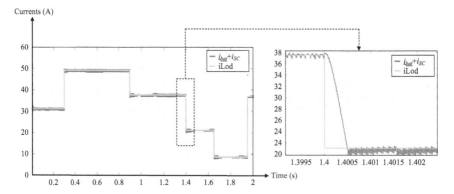

Figure 13.15   Superposition of the load current and the sum of the currents of the storage devices (battery and SC)

by the SC and the rest is supplied by the batteries. Figure 13.15 presents the load current and the sum of the battery and SC current to validate the efficiency of the applied EMS.

13.7   Conclusion

Solar-powered electric vehicles are considered as the vehicle of the future. This is due to its ability to decrease carbon emissions, low cost and dependency on fossil fuel. Moreover, they are considered as typical electric mobility in rural areas due to the lack of charging stations in these areas in addition to the decrease of city obstacles which are considered as the main drawback of this kind of vehicle. However, several challenges remain to be solved. Among these challenges, the impact of shading in rural areas and stability issues of the energy storage system. In this chapter, both H_∞ and frequency separation energy management strategies are proposed for the solar-powered electric vehicle energy storage system. The obtained simulation results show the optimal operation of the energy management system. In fact, the obtained H_∞ controller offers a fast transient response in addition to robust stability by ensuring the perfect rejection of disturbances that come from DC bus fluctuations and parameters variations. Moreover, it shows that the proposed frequency separation energy management strategy ensures an optimal distribution between the energy storage system components which are the battery and the supercapacitor. In addition, to prevent the wire-harness problems and communication power losses, SoCBat and SoCSC computing are performed via IoT-based WBMS.

For future works, an experimental set-up will be developed to test both the proposed H_∞ control and the energy management strategy for an energy storage system dedicated to a solar-powered electric vehicle. The experimental setup will be composed of a PV panel emulator, battery emulator, load emulator, and OPAL-RT platform.

Acknowledgment

This work was supported by the Tunisian Ministry of Higher Education and Scientific Research under Grant LSE- ENIT-LR 11ES15 and funded in part by PAQ-Collabora (PAR & I-Tk) program.

References

[1] Husain I., Ozpineci B., Islam S., Gurpinar E., and Su G. J. 'Electric drive technology trends, challenges, and opportunities for future electric vehicles'. *Proceedings of the IEEE*. 2021; 109(6): 1039–59.

[2] Ben Said-Romdhane M., Skander-Mustapha S., and Slama-Belkhodja I. 'Analysis study of city obstacles shading impact on solar PV vehicle'. In *2021 4th International Symposium on Advanced Electrical and Communication Technologies (ISAECT)*; 2021. pp. 01–06.

[3] Arun, P. and Mohanrajan, S. 'Effect of partial shading on vehicle integrated PV system'. In *2019 3rd International Conference on Electronics, Communication and Aerospace Technology (ICECA)*; 2019.

[4] Li H., Yang D., Su W., Lü J., and Yu, X. 'An overall distribution particle swarm optimization MPPT algorithm for photovoltaic system under partial shading'. *IEEE Transactions on Industrial Electronics*. 2019; 66(1): 265–75.

[5] Satpathy P. R., Mahmoud A., Panigarhi S. K., and Sharma R. 'Partial shading effect on the performance of electric vehicle-integrated solar PV system'. In Mahapatra S., Shahbaz M., Vaccaro A., and Emilia Balas V. (eds.), *Advances in Energy Technology. Advances in Sustainability Science and Technology*, Springer, Singapore, 2021.

[6] Ben Said-Romdhane M. and Skander-Mustapha S. 'A review on vehicle-integrated photovoltaic panels'. In Motahhir S. and Eltamaly A. M. (eds.), *Advanced Technologies for Solar Photovoltaics Energy Systems. Green Energy and Technology*, *Springer*, Cham, 2021.

[7] Soares dos Santos G., José Grandinetti F. 'Augusto Rocha Alves R. and de Queiróz L. W. 'Design and simulation of an energy storage system with batteries lead acid and lithium-ion for an electric vehicle: battery vs. conduction cycle efficiency analysis', *IEEE Latin America Transactions*. 2020; 18(8): 1345–52.

[8] Sangswang A. and Konghirun, M. 'Optimal strategies in home energy management system integrating solar power, energy storage, and vehicle-to-grid for grid support and energy efficiency'. *IEEE Transactions on Industry Applications*. 2020; 56(5): 5716–28.

[9] Snoussi J., Elghali S. B., Benbouzid M., and Mimouni, M. F. 'Optimal sizing of energy storage systems using frequency-separation-based energy management for fuel cell hybrid electric vehicles.' *IEEE Transactions on Vehicular Technology*. 2018; 67(10): 9337–46.

[10]   Slouma S., Skander-Mustapha S., Slama-Belkhodja I., and Machmoum M. 'Frequency separation model based on infinite-impulse response filter applied to hybrid power generation intended for residential sector'. *International Journal of Renewable Energy Research (IJRER)*. 2019; 9(1): 118–27.

[11]   Ding H., Hu Z., Song Y., Hu X., and Liu Y. 'Coordinated control strategy of energy storage system with electric vehicle charging station'. In *2014 IEEE Conference and Expo Transportation Electrification Asia-Pacific (ITEC Asia-Pacific)*; 2014, pp. 1–5.

[12]   Podder A. K., Chakraborty O., Islam S., Manoj Kumar N., and Alhelou H. H. 'Control strategies of different hybrid energy storage systems for electric vehicles applications'. *IEEE Access*. 2021; 9: 51865–95.

[13]   Ben Said-Romdhane M., Skander-Mustapha S., and Slama-Belkhodja, I. 'PV system inverter control design based on H∞ control'. In *2020 11th International Renewable Energy Congress (IREC)*; 2020, pp. 1–6.

[14]   Bai Z., Yan Z., Wu X., Xu J., and Cao B. 'H∞ control for battery/super-capacitor hybrid energy storage system used in electric vehicles'. *International Journal of Automotive Technology*. 2019; 20: 1287–96.

[15]   Reichmuth S. K., Siefer G., Schachtner M., Mühleis M., Hohl-Ebinger J., and Glunz, S. W. 'Measurement uncertainties in I–V calibration of multi-junction solar cells for different solar simulators and reference devices'. *IEEE Journal of Photovoltaics*. 2020; 10(4): 1076–83.

[16]   Maehlum M. A. 'Which solar panel type is best? Mono- vs. polycrystalline vs. thin film'. *Energy Informative*. 2021: 29. http://energyinformative.org/best-solar-panelmonocrystalline-polycrystalline-thin-film/.

[17]   Şahin M. E., Blaabjerg F., and Sangwongwanich A. 'A comprehensive review on supercapacitor applications and developments'. *Energies*. 2022; 15: 674.

[18]   Pattnaik M., Badoni M., Yogesh T., and Singh H. P. 'Analysis of electric vehicle battery system'. In *2021 4th International Conference on Recent Developments in Control, Automation & Power Engineering (RDCAPE)*, 2021, pp. 540–43.

[19]   ElMenshawy M., ElMenshawy M., Massoud A., and Gastli, A. 'Solar car efficient power converters' design'. In *2016 IEEE Symposium on Computer Applications & Industrial Electronics (ISCAIE)*, 2016, pp. 177–82.

[20]   Ben Said-Romdhane M. and Skander-Mustapha S. 'A review on vehicle-integrated photovoltaic panels'. In Motahhir S. and Eltamaly A.M. (eds.), *Advanced Technologies for Solar Photovoltaics Energy Systems. Green Energy and Technology*. Springer, Cham, 2021.

[21]   Boukerdja M., Chouder A., Hassaine L., Bouamama B. O., Issa W., and Louassaa K. 'H∞ based control of a DC/DC buck converter feeding a constant power load in uncertain DC microgrid system'. *ISA Transactions*. 2020;105 :278–95.

[22]   Mishra D. and Mandal S. 'Voltage regulation of DC–DC boost converter using H-infinity controller'. In *2020 International Symposium on Devices, Circuits and Systems (ISDCS)*; 2020. pp. 1–5.

[23]   Lee S. C. and Jung W. Y. 'Analogical of the Ragone plot and a new categorization of energy devices'. *Energy Procedia.* 2016; 88: 526–30.

[24]   McCloskey B. D. 'Expanding the Ragone plot: pushing the limits of energy storage'. *The Journal of Physical Chemistry Letters.* 2015; 6(18): 3592–93.

[25]   Manthiram A. 'A reflection on lithium-ion battery cathode chemistry'. *Nature Communications.* 2020; 11(1) : 1–9.

[26]   Mesbahi T., Bartholomeüs P., Rizoug N., Sadoun R., Khenfri F., and Le Moigne, P. 'Advanced model of hybrid energy storage system integrating lithium-ion battery and supercapacitor for electric vehicle applications'. *IEEE Transactions on Industrial Electronics.* 2020; 68(5): 3962–72.

[27]   Yu S., Lin D., Sun Z., and He D. ' Efficient model predictive control for real-time energy optimization of battery-supercapacitors in electric vehicles'. *International Journal of Energy Research.* 2020; 44(9): 7495–506.

[28]   Samanta A. and Williamson S. S. 'A survey of wireless battery management system: topology, emerging trends, and challenges'. *Electronics.* 2021; 10 (18), 2193.

[29]   Mohammadi F. and Rashidzadeh R. 'An overview of IoT-enabled monitoring and control systems for electric vehicles'. *IEEE Instrumentation & Measurement Magazine.* 2021; 24(3), 91–97.

[30]   Faika T., Kim T., and Khan M. 'An Internet of Things (IoT)-based network for dispersed and decentralized wireless battery management systems'. In *IEEE Transportation Electrification Conference and Expo (ITEC)*; 2018. pp. 1060–64.

Chapter 14

Influence of energy management in solar photovoltaic system by block chain technologies for rural and remote areas

S. Raj Anand[1], A.D. Dhass[2] and Ram Krishna[3]

In the recent trend, energy management is one of the criteria in various sectors (industrial, agricultural, residential, and commercial accessibilities). The energy conservation is a key role for consuming and utilization of power in all the sectors but in the case of conserve energy, but still, it is an uncontrolled consumption process in various sectors. The existing systems of farmers are utilizing the energy in the agricultural activities, but the consumption of energy is not measured by the farmers and most of them can be an unutilized source of power or loss of energy through various electrical and thermal processes. The unutilized usage of energy will be used for other agricultural activity for pumping of water (irrigation activity). Aforementioned problems can be minimized by block chain technology (digital technology) in rural and remote areas. It is a decentralized mechanism of sharing the energy in the present situation of agriculture and it is vital role for sharing the energy in all the way by decentralized access. In this system, energy will be shared to all the farmers and consumption of energy is calculated based on their utilization in the farmland. If consumption power is more utilize by the farmers, automatically the information would be transferred to their mobile based on the consumption from the slab. This system is very helpful to the government for analyzing and tracking the consumption of power utilization not only by farmers but also control the unwanted energy level in rural area (home and other sectors). The result of the system would be shown that quality of the service will be provided to the farmers in addition to that maximize the throughput for saving and control the utilization of energy. This chapter summarizes the recent trends in solar photovoltaic system in agricultural applications and discussed the techniques and algorithm for energy conservation by implementation of block chain technology.

[1]Department of Computer Science and Engineering, Vemu Institute of Technology, India
[2]Department of Mechanical Engineering, IITE, Indus University, India
[3]Department of Metallurgical and Materials Engineering, National Institute of Technology, India

14.1   Introduction

Sustainable energy has been a hot topic in the twenty-first century. Wind and solar electricity are getting more and scarcer, while contaminating our climate. As the world's population continues to grow, so does our need for energy. A significant increase in this type of exploitation has occurred since the industrial revolution. Future generations must have access to an adequate amount of energy. Many developing countries still rely on coal and petroleum as their primary sources of power. The availability of these materials will be depleted in a short period of time. In order to address this issue, we must turn to environmentally friendly forms of energy production like solar and wind power. Renewable energy sources must be utilized to meet the rising need for energy. Wind, solar, and geothermal energy harvesting has already given them an advantage over the competition. However, the most important problem with the current model is that it is solely utilized as a backup. A few examples include solar-powered street lighting and ATMs that are still using antiquated technology. A network must be established for each and every renewable energy generator in order to distribute this resource. Examples of solar-powered homes can be seen across the slums and villages of India. Current systems force homes to use all the power they generate regardless of how much energy they use. This limits the quantity of electricity that can be generated by solar panels.

Decentralization is made possible through the deployment of blockchain technology, which connects all network members without requiring any prior knowledge of one another or the presence of an intermediate. Consensus-based algorithm, which relies on the confidence of all nodes, may work without a middleman because of decentralization. Several methods exist, but the most common are Proof of Work (PoW) and Proof of Stake (PoS) [1].

Problems with the Power Farmers program's expected benefits necessitate attention. Instability in the grid and unequal access are two major problems. Power technical challenges continued to be a problem for rural areas of farmers with small-scale solar systems, especially in remote places. Grid voltage fluctuations, regular outages, fuses tripping on installed inverter systems, and even burnouts due to lightning storms and power surges are all examples of these problems. Any inverter systems connected to the grid and activated by local voltages that exceed their acceptable operating limitations are promptly shut off.

It was more likely that families (rural and remote area) with a wide variety of socioeconomic resources would be eligible and hence have a better chance of benefiting from the program. Two examples of this kind of resource are social connections and education. Education and social ties can both help with the timing and method of submitting an application. Thus, the Power Farmers project has boosted the incomes of financially well-off people as much as it has provided a crucial source of income for those who are less fortunate (which would help secure a livelihood in remote rural locations). The Power Farmers initiative has not enabled food farmers to become energy suppliers, but it has increased the overall number of new energy producers [2].

Rural populations in need of affordable electricity can witness a substantial improvement in education, economic activity such as cottage industries and commerce, communications, access to national and worldwide media and information, safety, and enjoyment. In today's hyper competitive and linked culture, an affordable supply of electricity is crucial. The use of blockchain technology can benefit rural impoverished communities by reducing transaction costs, increasing transparency and traceability of consumer payment data, and building trust between parties. A peer-to-peer (P2P) business model can also open the door to new companies and services that can disrupt all aspects of the energy market. In underdeveloped nations, a P2P infrastructure and the utilization of innovative, low-cost renewable energy is one of the most intriguing use cases for decentralized energy [3].

In many ways, SolarCoin resembles other cryptocurrencies like Bitcoin because of the blockchain's decentralization. Unlike other cryptocurrencies, SolarCoin is based on the production of vertically produced solar energy, which is both economically and environmentally advantageous.

SolarCoin is a novel phenomenon because of the unique qualities of blockchain technology:

1. a further incentive for solar power companies to produce;
2. natural resources are safeguarded by digital money;
3. it is the first non-governmental, decentralized, and international solar energy stimulus initiative.

It is one of the best ways to ensure agricultural productivity while also benefiting the community by reducing electricity prices to some extent or in other cases entirely [4].

Another energy-related issue where blockchain can make a significant impact is the trading of carbon emissions and green certificates. Using blockchain technology, a system has been put in place to ensure the integrity of monitoring, verification, and reporting procedures in carbon trading. Research demonstrates that central registration is the biggest stumbling point when it comes to developing clean development methods [5].

There is no need for third parties on a blockchain because transactions are recorded in a digital ledger. Data can be exchanged, processed, stored, and displayed in a way that can be understood by humans through the use of certain software systems. There are timestamps and transaction data included in each block's header in the original bitcoin design [6]. Logistics systems based on Logistics system based on Information and Communication Technology (ICT) and BT keep track of every item. ICT focuses on market efficiency when it comes to consumption, with earlier articles explaining ways to optimize the balance between supply and demand to reduce price dispersion. While Blockchain Technology (BT) is used in consumer research to demonstrate the product characteristics that buyers use to make judgments, there is no discussion over which elements effect purchase decisions, and this is an opportunity for future research [7].

Blockchain is not simply a single technology, but rather an integrated system encompassing several research results and technologies. Consensus, encryption, smart contracts, and distributed data storage are all highlighted in Figure 14.1.

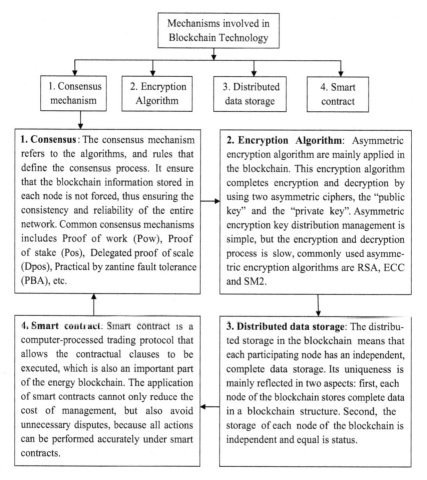

Figure 14.1   Process of mechanism in blockchain technology [8–12]

P2P services, smart meters, and solar power are all part of the energy block-chain platform in Figure 14.2. For example, a microgrid is a smart system that uses solar energy to power other homes in the neighborhood. However, a blockchain verifies the storage and transit of intermediary data. There is a visual representation of this concept in Figure 14.2 [13].

Legal entities like associations have the power to develop their own photo-voltaic parks. The group needs to be able to sell the energy it generates to both its members and its neighbors in order to do so. The solar park's foundation will be completed with the money earned from this project. Registering in the SOLARCOIN database is all it takes for farmers and the organizations that represent them to be entered to win SolarCoin acquiring and reinvesting the revenues of the purchase of energy and other resources. The SolarCoin Foundation will require information on solar panel acquisitions depending on current operational conditions.

Additional energy generated will be sold using the new blockchain process. Developing a software application that tracks the performance of each user is a primary objective of the organization. The association owns the equipment that will be given to members of the association as commodities on the ground where they are located. To maximize land use and to keep water clean, the panels should be placed above the canal. Members whose energy requirements cannot be met by their own production, or the national grid will be able to buy some of the association's excess energy (Figure 14.3) [14–16]. In this chapter, a systemic study of blockchain technology is employed in the enhancement of efficiency in solar photovoltaic system for agricultural and rural and remote area categories. Different modes have to be considered for analysis that provides the importance of electrical energy consumption incorporated blockchain techniques (digital technologies) that are implemented to enhance performance.

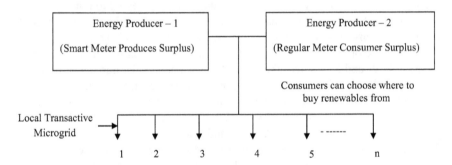

Figure 14.2   Smart renewable energy and P2P blockchain service [13]

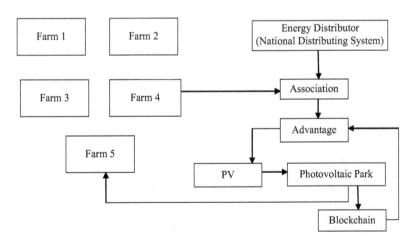

Figure 14.3   Proposed diagram for the software platform used in activity profitable [14–16]

14.2   Challenges in blockchain technology

Blockchain technology is one of the challenge areas in power sector by using of decentralization in the electrical power grids. The decentralization access of energy sharing methodology is very useful to many applications especially in the rural and remote areas and agricultural field to distribute the energy to the various locations with decentralized access. Due to environmental and reliability of the distribution, the energy level has been measured wherever appropriate location of the people can be utilized. It avoids the electricity demand, because P2P energy markets develop the renewable generation [17].

In electricity power generation, the PoW is a technique which has been used to verify whether it is authorized users that are sharing the energy level among the users who are already authorized. When a new block of user is connected to previous block users for accessing the energy level, the PoW has been confirmed the authentication without any spam data. If each block is accepted to transfer the data from one user to other, the PoS has been activated to confirm the transaction. In this connection, consensus mechanism is integrated all the centralized and cloud trans-action for developing the solar system in the particular location or in many locations by using PoW and PoS. Algorithm 1 has used to connect the number of users with decentralized access for making its confirmation for every transaction with PoS [18].

This mechanism is mainly focused on protecting the data from malware. In Figure 14.4, five entities have been used to make the transaction such as Data Owner, Cloud Storage Service Provider (CSP), Third Party Auditor (TPA), PoW, and PoS. The accumulation of energy has been transferred securely through the cloud database to different systems which defined in Figure 14.4. Data owner of the system has created a block and consecutively more blocks are created based on the user to create

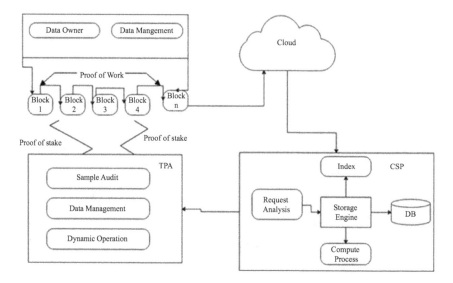

Figure 14.4   Working models of consensus mechanism [19]

the transaction to different systems. Each block is verified by the id of users with hash value and N number of blocks is connected with the help of a consensus algorithm named as PoW. If the owner of the data has ready for transferring the data to the destination, the cloud service provider (CSP) stores all the data based on the index to the storage engine. Based on the request, each information will be verified by TPA mechanism to provide to the particular destination. In this scenario, PoS is used to verify the transaction to each block. Each and every block knows the data which is transferred from one block to another block, but authenticated user can access between source and destination.

14.3 PoW

PoW has described about the consensus mechanism which is used to create the block in the transaction to make a chain for accessing the energy. Every transaction has made a chain with every block to authenticate the users. In this scenario, the PoW is used to determine the transactions to be confirmed and produce new blocks to share the energy for other resources. Based on the procedure of (14.1), (14.2), and (14.3), Algorithm 1 has shown when the numbers of blocks are created for every transaction, all the resources are verified that whether blocks are confirmed or not.

If blocks are confirmed, then each block is broadcasted to other nodes for providing the authorized transaction. The senders of every node identify the destination by decentralized application provider (DAP) with hash id. It has been created at the time of transaction which is performed by an authorized node. The resources have been verified by the spam or denial of service (DOS) attacks by secure hash algorithm (SHA) in each block with hash id. From (14.4), (14.5), and (14.6), Y is the block which is maximum for comparison with the newly created block by hash key value k. With PoW, resource competes against each node to complete the transactions on the network. From the working model of Figure 14.5, we can see how energy level is consumed for various resources to trade the power between producer and consumer. Figure 14.5 shows the overall power trading with decentralized access between electricity producer and electricity consumer [20,21].

This solar system is applicable to all the industries, agricultural, and other rural sector and it can be can be implemented in remote areas. The consumption of energy will save the cost and it is shared by the energy level with high performance. Improving the quality of the services has led to the success of every location with cost effective system of solar power.

14.4 Secure hash consensus techniques

A hash is contained the four tuples such as X, Y, K, and H for which the following conditions are satisfied. Where

1. X is a number of messages which needs to be transferred
2. Y is a finite set of possible authentication tags.
3. K is the key space to generate the finite set of possible keys.

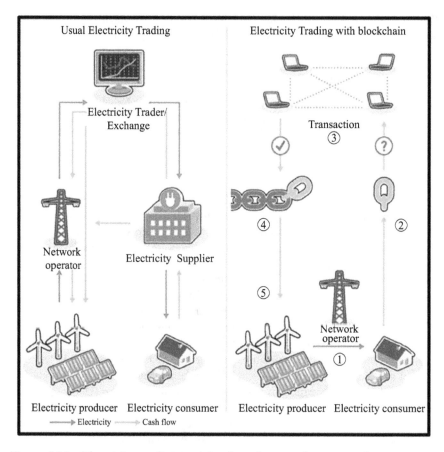

Figure 14.5   Electricity trading models of producer and consumer by consensus mechanism [22]

Suppose that hk: X→Y is an unkeyed hash function. Let x∈X, and define h = Y (x). If a hash function define to be secured, the following methodology needs to be solved.

Preimage:

Instance: A hash function h: X→Y and an Element y∈Y. Find: x∈X such that h (x) = y   (14.1)

Second Preimage:

Instance: A hash function h: X→Y and an element x∈X.

Find: x′∈X such that x ≠ x′ and h(x) = h(x′).   (14.2)

Collision:

Instance: A hash function h: X→Y

Find: x, x′∈X such that x ≠ x′ and h(x)= h(x′).   (14.3)

Algorithm 1. Secure hash transaction of blockchain (SHTB)

Step 1: Create blockchain with node URL for accessing the chain.
Step 2: Create New
Block(N)→ function(n, p, h);
  Where n→
  nonce. p→
  Previous Block.
  h→hash.
if (p∈N)
  index←chainlength
  +1; N←index;
    Create new transactions with amount, Sender and receiver id; TransactionId
      (T$_i$)←N.
  //create transaction in every block.
  Index(N)$^{\rightarrow}$T$_i$.
  else
  Transaction id has not created.

Step 3: Hash value h create secure value of Sha256 i.e. h(k).
Step 4: Begin //Proof of work
  if (*chainlength*> max*chainlength*) maxChainLength = *index*; newLongest
      Chain = *chainle.*
  newPendingTransactions = T$_i$
  if(!newLongestChain||(newLongestChain&&bitcoin.
chainIsValid(newLongestChain)))
    blockHash = h.correctBlock = getBlock(blockHash); block: correctBlock.
    "Current chain has not been replaced."
else
    bitcoin.chain = newLongestChain; bitcoin.pendingTransactions=newPending
    Transactions note: 'This chain has been replaced.'
    End

Cost function X is a message, Y is a block for storing the message which is digested
by hash and k is an Integer as a Key space:

$$F : C [0, Y_{max}][0, N] \rightarrow \{True, False\} \tag{14.4}$$

$$(X, Y, k) \rightarrow F(X, Y, k) \tag{14.5}$$

if Hash(X | Y | k) starts with D zero,

$$F(X, Y, k) = True \tag{14.6}$$

else false.

14.5   Results and discussion

In the real-time application, the decentralized mechanism of the transactions has been produced very efficiently without any collision. In view of client server communication, 50% of the data could be reliable when the transaction has been performed from source to destination. However, man in the middle attacker has monitored all the transactions and overloads the buffer size for making DOSs. Even though the data are stored in the cloud to access from the centralized server to various nodes. In this connection, 70% of the data has been accessed from the cloud. It has not been guaranteed to access the data with original indented access. The data will be captured before it can be accessed to the destination. Table 14.1 shows the mode of transaction to acquire the security levels for accessing the data. Figure 14.3 also depicts how the security level has been predicted from unauthorized access. In solar power system, the constant ratio of client server communication and centralized access control of data was described to be less than 70% only. The remaining percentage could be utilized for spam messages such as DOS and other unauthorized access. It is also shown that the decentralized data sharing communication has been established at 90% of secure transaction between N numbers of blocks. Remaining 10% of transaction perform the authentication based on the user interaction with each block of the solar power system.

The chain has performed in each block for sharing the energy with the secure way of transaction because each resource has to verify the data before the transaction would be created in the block. This kind of transaction has produced the reliability and scalable transaction between source and destination. The security of consensus mechanism has adopted with blockchain technology by decentralized applications, it creates a necessity for blockchain network to establish the communication with other networks group of transaction. In the scenario, the integration of producer and consumer, cloud and decentralized communication have been implemented in N number of nodes with the reliable transformation (Figure 14.6).

Table 14.2 shows the reliable transformation ratio of each node between centralized and decentralized data. Figure 14.7 describes that the comparison of the reliable transformation with 1,000 nodes for making 95% data has been secured and remaining 5% of the data has verified by various resources whether the block is created or not for transferring the data. PoS is used to verify the transactions successfully without the congestion.

In the case of producer, consumer, and centralized server, information is being secured in the data level but not in the communication, there is no PoS mechanism

Table 14.1   Mode of transaction between the senders and receiver [23]

Sl. no.	Accessing the data in various modes of transfer	Percentage of security level
1	Producer and consumer	50
2	Access the data through centralized power sector	70
3	Decentralized access of electricity power	99

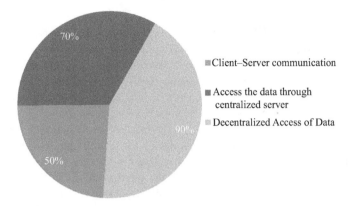

Figure 14.6 Security levels in various modes of transaction [24–26]

Table 14.2 Reliable transformation ratio uses number of users

S. no.	No of users	Reliable transformation ratio		
		Producer consumer	Centralized	Decentralized
1	100	10	10	100
2	250	20	30	100
3	500	30	40	100
3	750	40	40	95
4	1,000	50	50	95
5	1,250	50	60	90
6	1,300	50	70	90

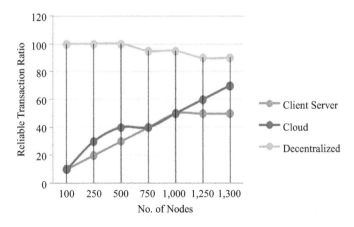

Figure 14.7 Reliable transaction ratio in centralized and decentralized node

for verifying the transaction. Both PoW and PoS have been successfully used in the consensus mechanism of decentralized transaction which is defined in Algorithm 1.

The unauthorized can reveal all the information for making the centralized data by their own. In order to view, the security level has been created inefficiently in all the nodes. Figure 14.8 describes the average error rate of reliable transaction. The storage of each data in the blockchain has to identify the error rate with hash Id. This data value is connected with every block until the genesis block as a first block could be reaches d. It shows that for up to 1000 nodes, the average error rate is 0.06 on average. The other mode of operation such as centralizing, and producer and consumer communication has been defined that the average value is 0.90. In this scenario, PoS is verified successfully with these error rates [27]. Finally, the PoW has been verified with 100% for testing of 1,400 nodes and has been communicated in various locations to share the energy level in the environment. Figure 14.9 shows

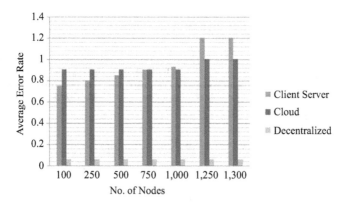

Figure 14.8   Average error rate of reliable transaction ratio of energy level

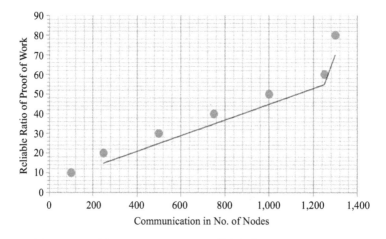

Figure 14.9   PoW between nodes as resources for solar power transaction

the reliable amount of data that is transferred through each block with the success rate of 95%. The hash value is verified in each block for accessing the data only between authorized users without any collision. In the decentralized consensus mechanism, the data has been transferred through multiple resources to produce the scalability and better throughput through Quality of Service (QoS).

14.6   Conclusion

In this chapter, the prediction of the data in decentralized mechanism of blockchain technology of 95% is reliable for making the transaction by using PoW and PoS. This approach has addressed some privacy issues of solving the untrusted data stored in cloud computing and producer and consumer communication. These two modes of operation have been implemented in decentralized access of blockchain technology to authenticate the data. The integration of these modes has produced the reliable data with large amounts of nodes. The remaining 5% has utilized for hash keys. These hash value was implemented in each block of secure transaction. The decentralized consensus mechanism has produced the less error rate (i.e.) 0.06 for number of nodes performed with multiple tasks. Thus, the blockchain technology has been implemented with PoW and PoS to produce the QOS for sharing the energy level in various applications and the secure way of transaction is scalable. Legal, regulatory, energy, or power loss during distribution and long-distance issues are among the many hurdles that blockchain technology faces in the energy industry to provide energy for rural and remote areas. Additional research and implementation will be solved and provides the energy with least amount to the people of rural areas by using digital technologies.

References

[1]   Jain, R. and Dogra, A. (2019, July). Solar energy distribution using blockchain and IoT integration. In *Proceedings of the 2019 International Electronics Communication Conference* (pp. 118–123).

[2]   Partridge, T. (2020). "Power farmers" in north India and new energy producers around the world: three critical fields for multiscalar research. *Energy Research & Social Science*, 69, 101575.

[3]   Thomason, J., Ahmad, M., Bronder, P., *et al.* (2018). Blockchain— Powering and empowering the poor in developing countries. In *Transforming Climate Finance and Green Investment with Blockchains* (pp. 137–152). London: Academic Press.

[4]   Enescu, F. M., Bizon, N., and Ionescu, V. M. (2019, June). Use of Blockchain Technology in Irrigation Systems of small farmers' association. In *2019 11th International Conference on Electronics, Computers and Artificial Intelligence (ECAI)* (pp. 1–6). New York, NY: IEEE.

[5]   Brilliantova, V. and Thurner, T. W. (2019). Blockchain and the future of energy. *Technology in Society*, 57, 38–45.

[6]   Kamilaris, A., Fonts, A., and Prenafeta-Bold, F. X. (2019). The rise of blockchain technology in agriculture and food supply chains. *Trends in Food Science & Technology*, 91, 640–652.

[7]   Liu, W., Shao, X. F., Wu, C. H., and Qiao, P. (2021). A systematic literature review on applications of information and communication technologies and blockchain technologies for precision agriculture development. *Journal of Cleaner Production*, 298, 126763.

[8]   Wu, J. and Tran, N. K. (2018). Application of blockchain technology in sustainable energy systems: an overview. *Sustainability*, 10(9), 3067.

[9]   Tencent Blockchain Solution White Paper. https://wenku.baidu.com/view/66dac81666ec102de2bd960590c69ec3d5bbdbfb.html (accessed on 23 October 2017).

[10]   Ding, W., Wang, G. C., Xu, A. D., Chen, H. J., and Hong, C. (2018). Research on key technologies and information security issues of energy blockchain. *Proceeidngs of the CSEE*, 38, 1026–1034, 1279.

[11]   Wu, G., Zeng, B., Li, R., and Zeng, B. (2017). Research on application mode of blockchain technology in response to resource transaction in integrated demand side. *Proceedings of the CSEE*, 37, 3717–3728.

[12]   O'Dwyer, K.J. and Malone, D. (2013). Bitcoin mining and its energy foot-print. In *Proceedings of the 25th IET Irish Signals & Systems Conference 2014 and 2014 China-Ireland International Conference on Information and Communications Technologies* (ISSC 2014/CIICT 2014), Limerick, Ireland, 26–27 June 2013; pp. 280–285.

[13]   Huh, J. H. and Kim, S. K. (2019). The blockchain consensus algorithm for viable management of new and renewable energies. *Sustainability*, 11(11), 3184.

[14]   Enescu, F. M., Bizon, N., Onu, A., *et al.* (2020). Implementing blockchain technology in irrigation systems that integrate photovoltaic energy generation systems. *Sustainability*, 12(4), 1540.

[15]   Cryptocurrencies, Blockchain and Pollution. http://ecoprofit.ro/criptomonedele-blockchainul-si-poluarea/ (accessed on 1 September 2019).

[16]   Could SolarCoin Help ACWA Boost Its Profits From Solar? https://www.greentechmedia.com/articles/read/could-solarcoin-help-acwa-boost-its-profits-from-solar (accessed on 1 September 2019).

[17]   Yoon, H.-J. (2018). A survey on consensus mechanism for blockchain. *International Journal of Advanced Research in Computer and Communication Engineering* 7, 1–6.

[18]   Ismail, L. and Materwala, H. (2019). A review of blockchain architecture and consensus protocols: use cases, challenges, and solutions. *Symmetry*, 11, 1–47.

[19]   Singh, S. (2017). Blockchain: future of financial and cyber security. In *2016 2nd International Conference on Contemporary Computing and Informatics* (IC3I), *IEEE Xplore* (pp. 14–17).

[20]   Abeyratne, S. A. and Radmehr, P. (2016). Blockchain ready manufacturing supply chain using distributed ledger. *International Journal of Research In Engineering and Technology*, 5, 1–11.

[21] Zhang, S. and Lee, J.-H. (2019). Analysis of the main consensus protocols of blockchain. *ScienceDirect*, 8, 1–5.

[22] Thwin, T. T. and Vasupongayya, S. (2019). Blockchain Based Access Control Model to Preserve Privacy for Personal Health Record Systems, Hindawi, Security and Communication Networks, 2019 (pp. 1–16).

[23] Wang, Q., Wang, C., Ren, K., Lou, W., and Li, J. (2013). Enabling public auditability and data dynamics for storage safety measures in cloud computing. *IEEE Transactions on Parallel with Distributed Systems*, 22, 847–859.

[24] Zhu, X. (2019). Research on blockchain consensus mechanism and implementation. *IOP Conf. Series: Materials Science and Engineering*, 569, 1–6.

[25] Cai, Y. and Zhu, D. (2020). Fraud detections for online businesses: a perspective from blockchain technology. *Springer Open, Financial Innovation*, 2, 1–10.

[26] Kaynak, B., Kaynak, S., and Uygun, O. (2020). Cloud manufacturing architecture based on public blockchain technology. *IEEE Access*, 8, 2163–2177.

[27] Mika, B. and Goudz, A. (2021). Blockchain-technology in the energy industry: blockchain as a driver of the energy revolution with focus on the situation in Germany. *Energy Systems*, 12, 285–355. Springer Link, http://dx.doi.org/10.1007/978-3-662-60568-4, 2020

Index

Printed in the USA
CPSIA information can be obtained
at www.ICGtesting.com
JSHW061418280823
47398JS00003B/42